13th Annual International Astrophysics Conference 2014

Voyager, IBEX and the Interstellar Medium

Journal of Physics: Conference Series Volume 577

Myrtle Beach, South Carolina, USA
10 March 2014

Editor:

G. P. Zank

ISBN: 978-1-63439-910-4
ISSN: 1742-6588

Printed from e-media with permission by:

Curran Associates, Inc.
57 Morehouse Lane
Red Hook, NY 12571

Printed by Curran Associates, Inc. (2015)

Institute of Physics
Dirac House, Temple Back
Bristol BS1 6BE UK

Phone: 44 1 17 929 7481
Fax: 44 1 17 920 0979

techtracking@iop.org

Additional copies of this publication are available from:

Curran Associates, Inc.
57 Morehouse Lane
Red Hook, NY 12571 USA
Phone: 845-758-0400
Fax: 845-758-2634
Email: curran@proceedings.com
Web: www.proceedings.com

TABLE OF CONTENTS

Journal of Physics: Conference Series **577** (2015) 011001 doi:10.1088/1742-6596/577/1/011001

Preface

G P Zank
Editor, *Proceedings of the 13th Annual International Astrophysics Conference*, Center for Space Plasma and Aeronomic Research (CSPAR) and Department of Space Physics, University of Alabama in Huntsville, Huntsville AL 35805, USA

E-mail: garyp.zank@gmail.com

The 13th Annual International Astrophysics Conference was held in scenic Myrtle Beach, South Carolina, USA, during the week of 10-14 March 2014. The meeting drew nearly 80 participants from all over the world, representing a wide range of interests and expertise in the interplanetary medium, the solar wind, observations, and theory. The theme of the meeting was *Voyager, IBEX, and the Interstellar Medium*. This decade may one day be viewed as the golden age in the exploration of the large-scale heliosphere and the local interstellar medium (LISM). Voyager 1 and 2 and IBEX are yielding remarkable new discoveries about the boundaries of the solar wind – LISM region and the interstellar medium. Hitherto, our basic understanding of the interstellar medium has been provided by telescope observations across multiple wavelengths that are typically integrated over many parsecs. For the first time, with these three spacecraft, we are making detailed measurements of the interstellar plasma, energetic particles (charged and neutral), magnetic field, and plasma waves in situ or with very short integration distances. IBEX provides insight into the global characteristics of the very local interstellar medium and Voyager 1 has just crossed the heliopause and is now in the interstellar medium. Remarkable results can be anticipated as discoveries over the next decade are made and the physics of the interstellar medium unfolds. As described in the papers in this volume, the new observations are already challenging theoretical models. The 13th Annual International Conference focused on the physics of the solar wind – LISM boundaries and the emerging physics of the local interstellar medium. To address this, astrophysicists and space physicists assembled to share their combined expertise to address in a highly interdisciplinary fashion the physics of the interaction between the solar wind and the interstellar medium.

We thank Adele Corona and ICNS for her continued excellent organization of the AIAC meetings and her help in providing the logistical support for this volume of papers. I should also like to thank Laxman Adhikari for his help with formatting a number of the submitted manuscripts.

Peer review statement

All papers published in this volume of *Journal of Physics: Conference Series* have been peer reviewed through processes administered by the proceedings Editors. Reviews were conducted by expert referees to the professional and scientific standards expected of a proceedings journal published by IOP Publishing.

13th Annual International Astrophysics Conference: Voyager, IBEX, and the Interstellar Medium IOP Publishing
Journal of Physics: Conference Series **577** (2015) 012001 doi:10.1088/1742-6596/577/1/012001

Turbulence transport within the Heliosphere

L Adhikari[1,2], G P Zank[1,2], R Bruno[3], D Telloni[3], P Hunana[2], and A Dosch[2]

[1]Department of Space Science, University of Alabama in Huntsville, Huntsville, AL 35899, USA
[2]Center for Space Plasma and Aeronomic Research (CSPAR), University of Alabama in Huntsville, Huntsville,
AL 35899, USA
[3] AA(INAF-IAPS Istituto di Astrofisica e Planetologia Spaziali, Via del Fosso del Cavaliere 100, 00133 Roma, Italy

E-mail: `la0004@uah.edu`

Abstract. This work continues the investigation of turbulence transport throughout the supersonic solar wind initiated in Zank et al 1996 [27] and Zank et al 2012 [20]. [20] developed a system of six coupled transport equations that describe the transport of energy corresponding to forward propagating (g) and backward propagating modes (f), the residual energy (E_D), and the correlation lengths corresponding to forward propagating modes (λ^-), backward propagating modes (λ^+), and the correlation length (λ_D) for residual energy. These models can be applied to both sub-Alfvénic (such as the lower corona) and super-Alfvénic (e.g., supersonic solar wind and inner heliosheath) flows. The correlation lengths calculated from our model are in good agreement with those observed. The evolution of related parameters is also calculated from 0.29 AU to 5 AU.

1. Introduction

Turbulence is ubiquitous in space and astrophysical plasmas such as, the solar corona, the solar wind, accretion disks, the interstellar medium and so forth. The scattering of solar energetic particles ([1, 2]), the generation of the fast solar wind from open field regions of the solar corona ([3, 4, 5]), the observed non-adiabatic radial solar wind temperature profile ([6 - 12]), are all thought to be due to the effects of turbulence. The solar wind is an almost collisionless magnetized plasma, and is regarded as a huge laboratory for studying turbulence ([13, 14]). The theoretical and observational study of solar wind turbulence, despite intensive attention over the past decades, remains at the forefront of solar physics research. Here, we focus on one aspect, which is the evolution of fundamental turbulence quantities as a function of heliocentric distance.

Velocity and magnetic field fluctuations in the solar wind near the Sun are observed ([15, 16]) to be highly correlated, and satisfy the relation $\delta\mathbf{v} = \pm\delta\mathbf{B}/\sqrt{4\pi\rho}$, where $\delta\mathbf{v}$ and $\delta\mathbf{B}$ are velocity and magnetic field fluctuations, respectively and ρ is the solar wind density. These correlations are stronger at lower than higher frequencies. The sign in front of $\delta\mathbf{B}$ depends on $sign[-\mathbf{k}\cdot\mathbf{B}]$, where $\mathbf{k}$ and $\mathbf{B}$ are the wave vector and background magnetic field, respectively ($\delta\mathbf{v}$ and $\delta\mathbf{B}$ are parallel when the average interplanetary magnetic field is directed inwards with respect to the Sun and they are antiparallel when it is directed outwards). There exist both types of forward and backward propagating modes below the Alfvén critical point (i.e., it is a point where the solar wind speed and Alfvén speed are equal); however the backward propagating modes in this region move faster than the solar wind speed, and hence always propagate towards the Sun. So, the only modes that escape from the sub-Alfvénic region are forward propagating modes. Thus, backward propagating modes beyond the Alfvén critical point are not of solar origin. Due to the presence of oppositely directed propagating modes, nonlinear interactions excite the cascade (i.e., the transfer of energy from larger scales to smaller scales) so that smaller scales energy eventually dissipate as heat energy. Usually the spectra of the fluctuations of the magnetic field in the solar wind exhibits a Kolmogorov-like scaling with a spectral slope of $-5/3$ in wavenumber space, and is due to the in situ generation of a turbulence cascade by dynamical processes ([14, 17]). An Iroshikov-Kraichnan-like scaling of $k^{-3/2}$, where k is the wave number ([18, 19]), is also observed sometimes in the solar wind.

The difference between the fluctuating kinetic and magnetic energies is known as the residual energy. The correlation between the square root of these two quantities is known as the cross helicity. The cross-helicity describes the equipartition of energy between forward and backward propagating modes. It decreases with increasing heliocentric distance, and is zero if the forward and backward propagating modes have the same energy. The heliocentric evolution of the energy in the forward and backward modes, and the corresponding residual energy and cross-helicity, is not particularly well understood observationally, and certainly not theoretically. Our purpose here is to solve the Zank et al 2012 [20] model that describes the transport of turbulence with heliocentric distance, and to compare the predicted correlation length with observations of the correlation length in the super-Alfvénic solar wind.

2. Background

The origin of the solar wind is the solar corona from where the solar wind expands supersonically into the heliosphere. The structure of the solar wind is very complex, and solar wind parameters, such as the solar wind velocity, the plasma density, the solar wind temperature, and the magnetic field fluctuate around their mean values. However, one can separate the mean field from the fluctuating fields, allowing us to express the magnetic and velocity fields as $\underline{\mathbf{B}} = \mathbf{B} + \mathbf{b}$ and $\underline{\mathbf{U}} = \mathbf{U} + \mathbf{u}$, where $\mathbf{B}$ and $\mathbf{U}$ are mean fields, and $\mathbf{b}$ and $\mathbf{u}$ are the fluctuating fields. Here, $\langle \underline{\mathbf{B}} \rangle = \mathbf{B}$, and $\langle \underline{\mathbf{U}} \rangle = \mathbf{U}$ meaning that the averaging procedure forces the rapidly varying fluctuating fields $\mathbf{b}$ and $\mathbf{u}$ to vanish. Furthermore, because of the solar cycle the time dependent mean fields fluctuate with time, and hence heliocentric distance [21].

The evolution and transport of fluctuations of velocity and magnetic fields can be described by the well-known two scale-separated incompressible MHD equations ([22 - 25]) in terms of the Elsässer variables as

$$\frac{\partial \mathbf{z}^{\pm}}{\partial t} + (\mathbf{U} \mp \mathbf{V}_A)\cdot\nabla\mathbf{z}^{\pm} + \frac{1}{2}\nabla\cdot(\mathbf{U}/2 \pm \mathbf{V}_A)\mathbf{z}^{\pm} + \mathbf{z}^{\mp}\cdot\left[\nabla\mathbf{U} \pm \frac{\nabla\mathbf{B}}{\sqrt{4\pi\rho}} - \frac{1}{2}I\nabla\cdot(\mathbf{U}/2 \pm \mathbf{V}_A)\right] = \mathbf{NL}_{\pm} + \mathbf{S}^{\pm}, \tag{1}$$

where I is the identity matrix, $\mathbf{NL}_{\pm}$ is a non-linear term that is responsible for the decay of turbulence, $\mathbf{S}^{\pm}$ is a source term, and $\mathbf{V_A}$ is the Alfvén velocity. Here, $\mathbf{z}^{\pm} = \mathbf{u} \pm \mathbf{b}/\sqrt{4\pi\rho}$ are the Elsässer variables ([26]), ρ is the solar wind density, and can be a function of both large-scales (e.g., solar wind scale) and small-scales (e.g., turbulence). Also, $\mathbf{z}^{+}$ ($\mathbf{z}^{-}$) represents the inward (outward) propagating modes with respect to the Sun.

By taking moments of equation (1) and imposing several assumptions (see [20] and [27]), Zank et al 1996 [27] developed a turbulent transport model that describes the transport and evolution of fluctuations of the fluctuating magnetic energy density (E_b) and correlation length (l) beyond 1-2 AU. However, [27] cannot describe the transport of turbulence within 1 AU since $\mathbf{V_A}$ was neglected and the cross helicity (E_c) was assumed to be zero, which is not observed to hold in the inner heliosphere ([28, 29]). The importance of [27] is that it reconciled the accepted turbulence description of fluctuations in the solar wind with the observed WKB-like (i.e., a linear wave description) decay of magnetic energy within a fully turbulence-based transport model. Zank et al 1996 [27] found good agreement with observations from ~ 1 AU to 40 AU. Later [7], and [8] used the formalism of [27] to show that the observed radial solar wind temperature profile could be explained by the dissipation of the fluctuating magnetic energy density.

Breech et al 2008 [12] used an approach similar to that initiated by Zank et al 1996 [27] to include cross-helicity in a turbulence transport model. The model compared well with observations from 0.3 AU to 100 AU. However, Breech et al 2008 [12] cannot address the transport of turbulence in complex sub-Alfvénic flows nor do they include the residual energy. To address these problems, Zank et al 2012 [20] developed a system of six coupled turbulence transport equations by taking moments of equation (1), and developing a closure that allows application to both sub-Alfvénic and super-Alfvénic flows. The six coupled turbulence transport equations of [20] in terms of $f =< {z^+}^2 >$ and $g =< {z^-}^2 >$ with $a = 1/2$ and $b = 0$, are

$$\frac{\partial f}{\partial t} + (\mathbf{U} - \mathbf{V}_A)\cdot\nabla f + \frac{1}{2}\nabla\cdot\mathbf{U}f + \nabla\cdot\mathbf{V}_A f + \frac{1}{2}\nabla\cdot\mathbf{U}E_D - \Gamma E_D - \nabla\cdot\mathbf{V}_A E_D = -2\frac{fg^{1/2}}{\lambda} + 2\langle \mathbf{S}^+\cdot\mathbf{z}^+\rangle; \tag{2}$$

$$\frac{\partial g}{\partial t} + (\mathbf{U} + \mathbf{V}_A)\cdot\nabla g + \frac{1}{2}\nabla\cdot\mathbf{U}g - \nabla\cdot\mathbf{V}_A g + \frac{1}{2}\nabla\cdot\mathbf{U}E_D - \Gamma E_D + \nabla\cdot\mathbf{V}_A E_D = -2\frac{gf^{1/2}}{\lambda} + 2\langle \mathbf{S}^-\cdot\mathbf{z}^-\rangle; \tag{3}$$

$$\frac{\partial E_D}{\partial t} + \mathbf{U}\cdot\nabla E_D + \frac{1}{2}\nabla\cdot\mathbf{U}E_D + \frac{1}{2\sqrt{fg}}\left[f\mathbf{V}_A\cdot\nabla g - g\mathbf{V}_A\cdot\nabla f\right] + \left[\frac{1}{2}\nabla\cdot\mathbf{U} - \Gamma\right]\frac{f+g}{2} + \nabla\cdot\mathbf{V}_A\frac{f-g}{2} = -E_D\left[\frac{f^{1/2}}{\lambda^-} + \frac{g^{1/2}}{\lambda^+}\right] + \langle \mathbf{S}^-\cdot\mathbf{z}^+\rangle + \langle \mathbf{S}^+\cdot\mathbf{z}^-\rangle; \tag{4}$$

$$\frac{\partial \lambda^+}{\partial t} + (\mathbf{U} - \mathbf{V}_A)\cdot\nabla\lambda^+ + \frac{E_D}{f}\left[\frac{1}{4}\nabla\cdot\mathbf{U} - \frac{1}{2}\nabla\cdot\mathbf{V}_A - \frac{\Gamma}{2}\right]\left[\lambda_D - \lambda^+\right] = 2g^{1/2} - 2\frac{\langle \mathbf{z}^+\cdot\mathbf{S}^+\rangle}{f}\lambda^+; \tag{5}$$

$$\frac{\partial \lambda^-}{\partial t} + (\mathbf{U} + \mathbf{V}_A)\cdot\nabla\lambda^- + \frac{E_D}{g}\left[\frac{1}{4}\nabla\cdot\mathbf{U} + \frac{1}{2}\nabla\cdot\mathbf{V}_A - \frac{\Gamma}{2}\right]\left[\lambda_D - \lambda^-\right] = 2f^{1/2} - 2\frac{\langle \mathbf{z}^-\cdot\mathbf{S}^-\rangle}{g}\lambda^-; \tag{6}$$

$$\frac{\partial \lambda_D}{\partial t} + \mathbf{U}\cdot\nabla\lambda_D + 2\left(\frac{1}{4}\nabla\cdot\mathbf{U} - \frac{\Gamma}{2}\right)\frac{f+g}{E_D}\left(\frac{f\lambda^+ + g\lambda^-}{f+g} - \frac{\lambda_D}{2}\right) + \nabla\cdot\mathbf{V}_A\left(\frac{f\lambda^+ - g\lambda^-}{f-g} - \frac{\lambda_D}{2}\right)\frac{f-g}{E_D} + \frac{1}{2E_D\sqrt{fg}}\left(f\mathbf{V}_A\cdot\nabla g - g\mathbf{V}_A\cdot\nabla f\right)\left[2\left(\lambda^+\lambda^-\right)^{1/2} - \lambda_D\right] + \frac{\sqrt{fg}}{E_D}\left[\left(\frac{\lambda^+}{\lambda^-}\right)^{1/2}\mathbf{V}_A\cdot\nabla\lambda^- - \left(\frac{\lambda^-}{\lambda^+}\right)^{1/2}\mathbf{V}_A\cdot\nabla\lambda^+\right] = \lambda_D\left[\frac{f^{1/2}}{\lambda^-} + \frac{g^{1/2}}{\lambda^+}\right] - \frac{\lambda_D}{E_D}\left[\langle \mathbf{z}^-\cdot\mathbf{S}^+\rangle + \langle \mathbf{z}^+\cdot\mathbf{S}^-\rangle\right], \tag{7}$$

where f is the energy corresponding to backward propagating modes, g is the energy corresponding to forward propagating modes, $E_D = \langle u^2\rangle - \langle b^2/4\pi\rho\rangle$ is the residual energy, λ^+ is the correlation length corresponding to backward propagating modes, λ^- the correlation length corresponding to forward propagating modes, and λ_D is the correlation length corresponding to the residual energy. Terms such as $\langle \mathbf{S}^\pm\cdot\mathbf{z}^\pm\rangle$ refer to sources of turbulence, and can be due to the stream-shear interaction between fast and slow wind shock waves, and pick up ions. The parameter $\Gamma(= n_i n_j \frac{\partial U_j}{\partial U_i})$ is the shear mixing term, which we neglect by choosing $\Gamma = 0$.

An important implication of these models is that backward propagating modes can be generated due to the reflection of forward propagating modes. In the absence of Alfvén velocity, mixing, dissipation and source terms, equations (2), (3), and (4) reduce to the well-known WKB Model (see Appendix A).

3. Sources of Turbulence

There are in principle three types of turbulence sources, and which source is important depends on location in the heliosphere. Firstly, stream shear is an important source of turbulence ([15, 30]). Secondly, shock waves can generate turbulence as they propagate through the solar wind ([31]). Finally, pickup ions created by charge exchange between the solar wind protons and interstellar neutral hydrogen is an important source of turbulence ([27, 32]). The first two sources are important within 4 - 5 AU, and the last source is important beyond the ionization cavity ($\gtrsim$ 6 - 10 AU) ([10, 11, 27, 32]). Here, we consider the region from 0.29 AU to 5 AU, and so only the stream-shear source of turbulence is necessary to include.

The source term to model turbulence generated by shear interactions can be written in a form similar to that introduced in ([27]),

$$\left\langle \mathbf{S}^{+} \cdot \mathbf{z}^{+} \right\rangle = C_{sh}(f)\frac{\Delta U_{shear}}{r}f; \tag{8}$$

$$\left\langle \mathbf{S}^{-} \cdot \mathbf{z}^{-} \right\rangle = C_{sh}(g)\frac{\Delta U_{shear}}{r}g; \tag{9}$$

$$\left\langle \mathbf{S}^{-} \cdot \mathbf{z}^{+} \right\rangle + \left\langle \mathbf{S}^{+} \cdot \mathbf{z}^{-} \right\rangle = [C_{sh}(f) + C_{sh}(g)]\frac{\Delta U_{shear}}{r}E_D, \tag{10}$$

where $C_{sh}(f)$ and $C_{sh}(g)$ are the strengths of shear interaction for the backward and forward propagating modes, respectively, and ΔU_{shear} (= 350 km/s) is characteristic difference between the fast and slow solar wind speeds. Equations (8), (9), and (10) describe the shear source of turbulence for the backward, forward, and residual energy, respectively. Shear driving generally supplies equal energy to forward and backward types of fluctuations ([29]) [i.e., $C_{sh}(f) \sim C_{sh}(g)$]. The derivation of equation (10) is shown in Appendix B. Similarly, the source terms due to the presence of shocks, following the approach of ([27]), are

$$\left\langle \mathbf{S}^{+} \cdot \mathbf{z}^{+} \right\rangle = C_{shock}(f)\frac{\Delta U_{shock}}{r}f; \tag{11}$$

$$\left\langle \mathbf{S}^{-} \cdot \mathbf{z}^{-} \right\rangle = C_{shock}(g)\frac{\Delta U_{shock}}{r}g; \tag{12}$$

$$\left\langle \mathbf{S}^{-} \cdot \mathbf{z}^{+} \right\rangle + \left\langle \mathbf{S}^{+} \cdot \mathbf{z}^{-} \right\rangle = [C_{shock}(f) + C_{shock}(g)]\frac{\Delta U_{shock}}{r}E_D, \tag{13}$$

where $C_{shock}(f)$ and $C_{shock}(g)$ parametrize the generation of energy for the backward and forward propagating modes, respectively, and ΔU_{shock} is the difference between the upstream and downstream speed of the shock.

4. Method

The steady state 1D coupled system of equations [(2)-(7)] is solved in spherical coordinates, neglecting the θ and ϕ components. We used a Runge-Kutta 45 method. The solar wind was assumed to be spherically symmetric i.e., $\mathbf{U} = U_0\hat{r}$, where U_0 = 400 km/s is the solar wind speed. The Alfvén velocity associated with the interplanetary magnetic field (Parker's model) is given by

$$V_A = \frac{B}{\sqrt{4\pi\rho}} = V_{A0}\left(\frac{R_0}{r}\right)\left[1 + \left(\frac{\omega_0 R_0}{U_0}\right)^2 \left(\frac{r}{R_0} - 1\right)^2 \sin^2\theta\right]^{1/2}, \tag{14}$$

where $R_0 = 10R_\odot$ ($R_\odot = 6.955 \times 10^5$ km is a solar radius), $V_{A0} = 400$ km/s is the Alfvén speed, $\omega_0 = 2.9 \times 10^{-6}$ rad/s is an angular speed of the Sun, and θ is colatitude with respect to solar rotation. We chose $\theta = \pi/2$ i.e., the ecliptic plane. The suffix 0 represents a reference point. The total turbulent energy E_T and the cross-helicity E_C can be calculated by using the following relations [20],

$$E_T = \frac{f+g}{2} = \langle u^2 \rangle + \langle b^2/4\pi\rho \rangle; \tag{15}$$

$$E_C = \frac{g-f}{2} = 2\left\langle \mathbf{u}\cdot\mathbf{b}/\sqrt{4\pi\rho}\right\rangle, \tag{16}$$

The inner boundary conditions for the turbulence variables at 0.29 AU are $f_0 = 753$ (km/s)^2, $g_0 = 13,515$ (km/s)^2, $E_{D0} = -57.07$ (km/s)^2, $\lambda_0^+ = 0.00143$ AU, $\lambda_0^- = 0.000779$ AU and $\lambda_{D0} = 0.00286$ AU. Also, we used $C_{sh}(f) \sim C_{sh}(g) = 16.35$. The suffix 0 represents a reference point which in our case is 0.29 AU.

5. Discussion

Here, we present solutions to the steady-state equations (2)-(7) for an expanding spherically symmetric flow in the presence of stream-shear driving. Theoretical results are shown in Figure 1 and a comparison of theoretically computed correlation lengths with observations from 0.29 AU to 5 AU is shown in Figure 2.

Figure (1) shows the numerical solution of our model [(2)-(7)] from 0.29 AU to 5 AU. The first column from top to bottom shows the total turbulent energy E_T, the energy corresponding to backward propagating modes f, and the correlation lengths (the dashed curve identifies λ^+, the dashed-dotted-dashed curve $\lambda_D/2$, and the solid curve λ^-), respectively. Similarly, the second column from top to bottom shows the energy corresponding to forward propagating modes g, the normalized residual energy $\bar{E}_D$ ($= E_D/E_T$), and the normalized cross-helicity $\bar{E}_C$ ($= E_C/E_T$), respectively. Fig.1 shows that the total energy in fluctuations $E_T = (< z^{+2} > + < z^{-2} >)/2$ decays uniformly from 0.29 AU to 5 AU despite the stream-shear source of turbulence. In our simple model of the stream-shear source, we have assumed that the source itself weakens with increasing heliocentric distance. Consequently, dissipation of $< z^{+2} >$, $< z^{-2} >$, and E_D ensure the overall decay of E_T. Note that we had introduced source terms into the WKB model, E_T would increase monotonically. Notice that g also decays monotonically, but f does not. An initial decrease in f is followed by an increase and then plateau followed by a slow decay with increasing heliocentric distance. The increase in f is due simply to the generation of backward propagating modes. The residual energy decays monotonically with radial distance reaching an approximately constant negative value beyond ~ 2.5 AU, showing that fluctuations in outer heliosphere are magnetically dominated. The cross-helicity also decays from a value of ~ 1 at 0.29 AU to 0.5 at 5 AU, reflecting the increase in f. Finally, all three of the correlation lengths $\lambda^\pm$, and λ_D increase with increasing heliocentric distance, despite the driving by shear. That the correlation lengths do not decrease reflects the weakening of the turbulence driving i.e., the dissipation proceeds more rapidly than the injection of energy.

Figure (2) is a comparison of the computed correlation lengths and those observed from 0.29 AU to 5 AU. The scattered "△"and "+"diagram show the observational correlation lengths for the backward and forward propagating modes, respectively. The dotted curve shows the correlation length corresponding to backward propagating modes, the solid curve

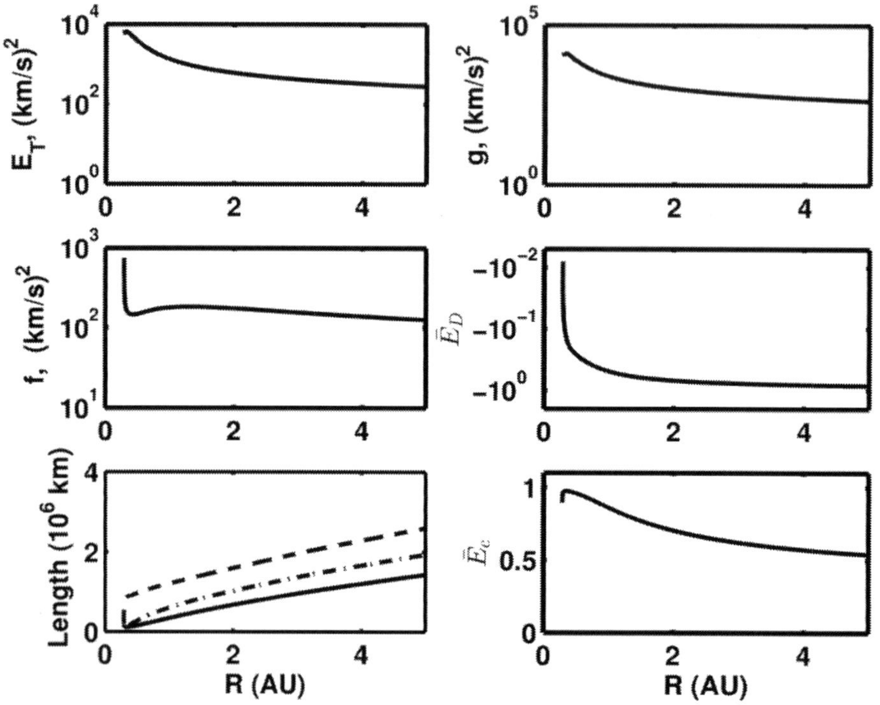

Figure 1. Solutions of the coupled turbulent transport equations [(2)-(7)] from 0.29 AU to 5 AU with turbulence driven by stream-shear. See text for details.

the correlation length corresponding to forward propagating modes, and the dashed-dotted-dashed curve denotes the (correlation length)/2 corresponding to the residual energy. Figures 3 and 4 correspond to Figures 1 and 2 except that shear driving has now has been neglected. The same format has been adopted.

The differences between Fig. 1 and Fig. 3 are apparent. All solutions decay but, in the absence of driving, the decay is more rapid. Two primary differences between the driven and undriven cases exist in the behavior of f and the correlation lengths. Without shear driving, although backward modes are generated, their energy nonetheless decays monotonically, unlike the driving case. In both cases, the correlation length increases with increasing heliocentric distance, but the scale is much larger for the undriven case. In the absence of driving, the dissipation rate does not have to increase as it does when driving present (this because the dissipation rate has to increase to balance the rate of energy injections.) Finally, the behavior of the residual energy and cross-helicity, while different in detail, like the driven case, decays with increasing heliocentric distance. As before, fluctuations become increasingly magnetic in character with distance from the Sun and the ratio between the energy in inward and outward propagating modes slowly tends to equalization.

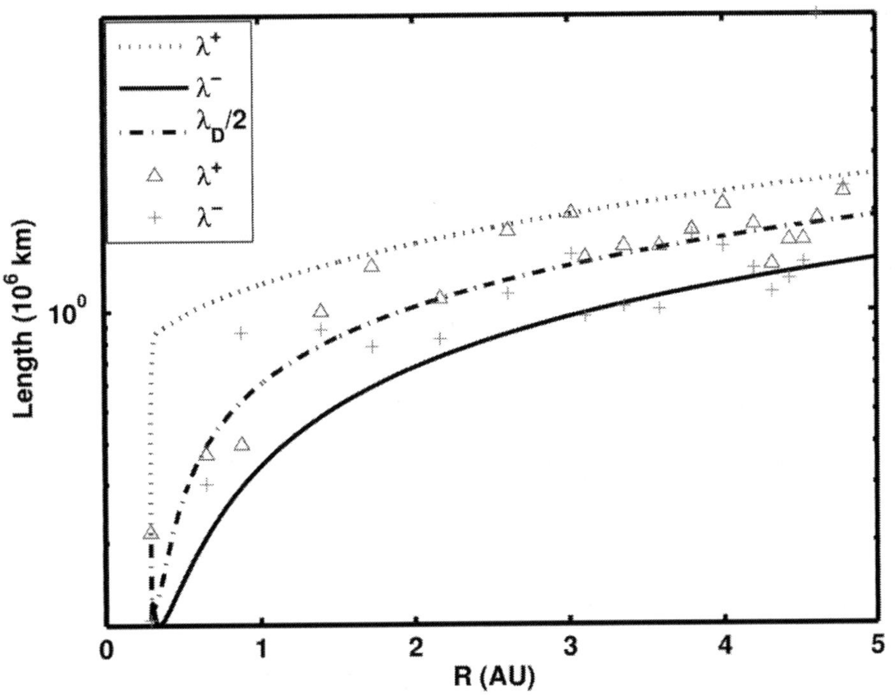

Figure 2. Comparison of the computed correlation lengths with those observed from 0.29 AU to 5 AU, assuming driving by stream-shear.

6. Conclusion

The turbulence transport models solved in this paper are more complex than previous models ([12, 27]), and they contain more information. The dependence on heliocentric distance for virtually all (non-spectral) quantities of interest to turbulence studies can be predicted on the basis of the Zank et al 2012 [20] model. We present the first numerical solution of this model, applicable to the region from 0.29 AU $\sim$ 5 AU. Furthermore, this model can be used to investigate turbulence in both sub-Aflvénic and super-Alfvénic flows. It is of course critical to validate complex models against observations. We make a start by comparing our computed correlation length with those observed between 0.29 AU and 5 AU. We find that the correlation lengths λ^+, λ^-, and $\lambda_D/2$ have some variation, but they show good agreement with the observation of the correlation length from 0.29 AU to 5 AU.

Acknowledgments

We acknowledge the partial support of NASA grants NNX14AC08G and NNX11AO64G.

Appendix A

Neglecting the Alfvén velocity, mixing, dissipation, and source terms, equations (2), (3), and (4) can be written as

$$\frac{\partial f}{\partial t} + \mathbf{U} \cdot \nabla f + \frac{1}{2} \nabla \cdot \mathbf{U} f = 0; \qquad (17)$$

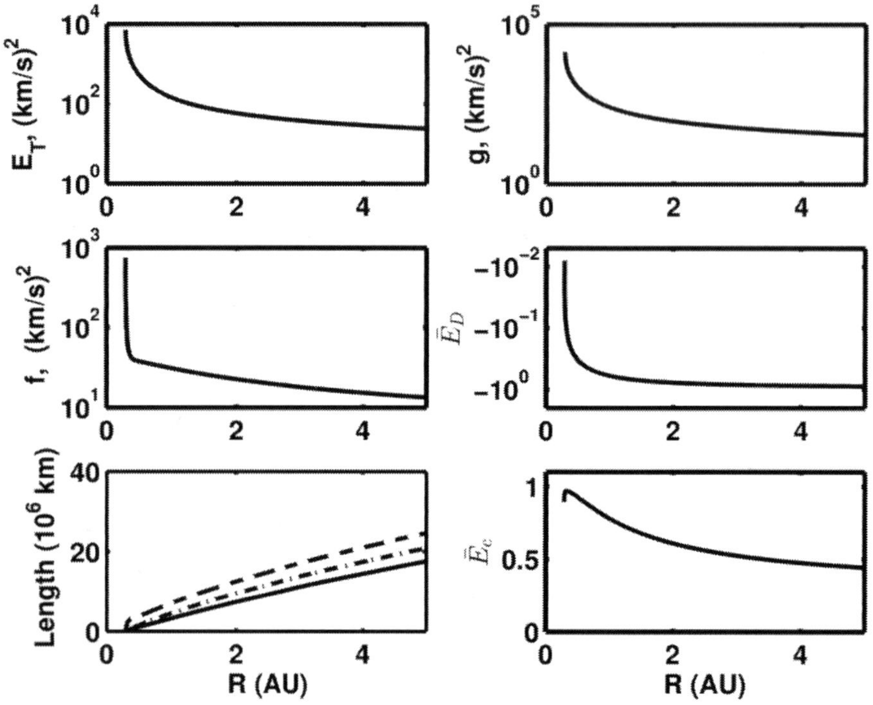

Figure 3. Solutions of the coupled turbulent transport equations [(2)-(7)] from 0.29 AU to 5 AU (without a source term).

$$\frac{\partial g}{\partial t} + \mathbf{U} \cdot \nabla g + \frac{1}{2} \nabla \cdot \mathbf{U} g = 0; \tag{18}$$

$$\frac{\partial E_D}{\partial t} + \mathbf{U} \cdot \nabla E_D + \frac{1}{2} \nabla \cdot \mathbf{U} E_D = 0. \tag{19}$$

Multiplying equation (19) by 2 and subtracting it from equations (17) and (18), we get

$$\frac{\partial (f + g - 2E_D)}{\partial t} + \mathbf{U} \cdot \nabla (f + g - 2E_D) + \frac{1}{2} \nabla \cdot \mathbf{U} (f + g - 2E_D) = 0, \tag{20}$$

which can be written in terms of the fluctuating magnetic energy density $E_b = (f + g - 2E_D)/4$ as

$$\frac{\partial E_b}{\partial t} + U \cdot \nabla E_b + \frac{1}{2} \nabla \cdot U E_b = 0, \tag{21}$$

Equation (21) is the well-known WKB model.

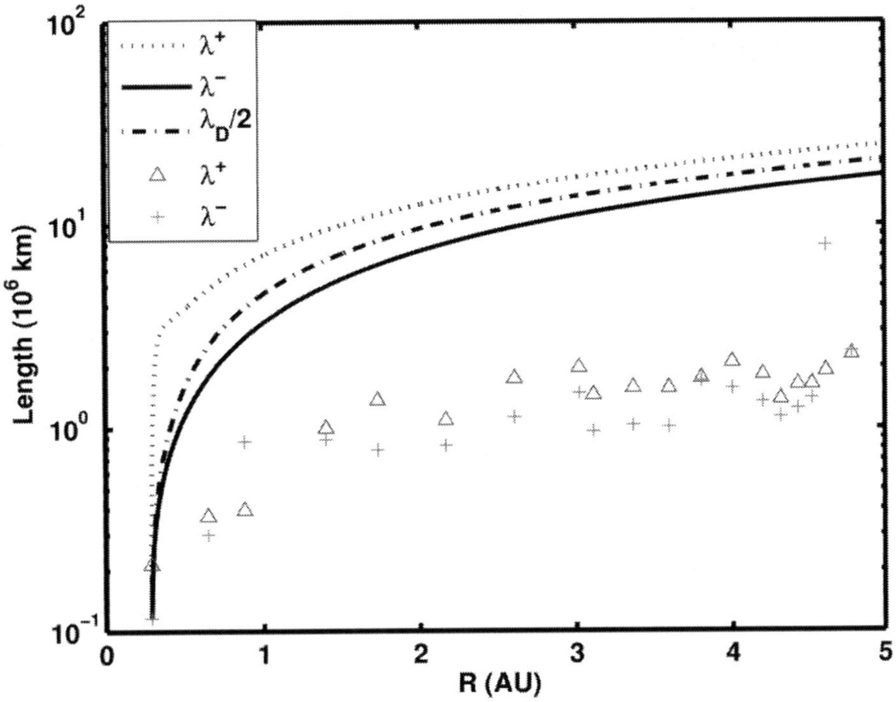

Figure 4. Comparison of the computed correlation lengths with those observed from 0.29 AU to 5 AU (without a source term).

Appendix B

The derivation of the source term for stream-shear interaction is straight forward,

$$\begin{aligned}\langle \mathbf{S}^- \cdot \mathbf{z}^+ \rangle + \langle \mathbf{S}^+ \cdot \mathbf{z}^- \rangle &= \left\langle C_{sh}(g)\frac{\Delta U}{r}\mathbf{z}^- \cdot \mathbf{z}^+ \right\rangle + \left\langle C_{sh}(f)\frac{\Delta U}{r}\mathbf{z}^+ \cdot \mathbf{z}^- \right\rangle \\ &= C_{sh}(g)\frac{\Delta U}{r}\langle \mathbf{z}^- \cdot \mathbf{z}^+ \rangle + C_{sh}(f)\frac{\Delta U}{r}\langle \mathbf{z}^+ \cdot \mathbf{z}^- \rangle \\ &= [C_{sh}(g) + C_{sh}(f)]\,\frac{\Delta U}{r}E_D\end{aligned}$$

7. References

[1] Li G, Zank G P, & Rice W K M 2003 *J. Geophys. Res. (Space Physics)* **108** 1082
[2] Zank G P, Li G & Verkhoglyadova O 2007 *Space Sci. Rev.* **130** 255
[3] Leer E, Holzer T E & Fla T 1982 *Space Sci. Rev.* **92** 11041
[4] Oughton S, Matthaeus W H, Dmitruk P 2001 *ApJ* **551** 565
[5] Verdini A, Velli M, Matthaeus W H, Oughton S & Dmitruk P 2010 *ApJ* **708** L116
[6] Williams L L, Zank G P, & Matthaeus W H 1995 *J. Geophys. Res.* **100** 17059
[7] Matthaeus W H, Zank G P, Smith C W and Oughton S 1999 Phys. Rev. Lett. **82** 3444
[8] Smith C W, Matthaeus W H, Zank G P 2001 *J. Geophys. Res.* **106** 8253
[9] Smith C W, Isenberg P A, Matthaeus W H, & Richardson J D 2006 *ApJ* **638** 508
[10] Isenberg P A, Smith C W, and Matthaeus W H 2003 *ApJ* **592** 564
[11] Isenberg P A 2005 *ApJ* **623** 502
[12] Breech B, Matthaeus W H, Minnie J 2008 *J. Geophys. Res.* **113** 8105

[13] Bavassano B, Dobrowolny M, Mariani F & Gosling N F 1982 *J. Geophys. Res.* **87** 3617
[14] Bruno R & Carbone V 2013 *Living Rev. in Solar Phys.* **10** 2
[15] Coleman Jr P J 1968 *ApJ* **153** 371
[16] Belcher J W & Davis Jr L 1971 *J. Geophys. Res.* **76** 3534
[17] Matthaeus W H & Goldstein M L 1982 *J. Geophys. Res.* **87** 6011
[18] Iroshnikov P S 1964 *Soviet Ast.* **7** 566
[19] Podesta J J, Roberts D A & Goldstein M L 2007 *ApJ* **664** 543
[20] Zank G P, Dosch A, Hunana P 2012 *ApJ* **745** 35
[21] Adhikari L, Zank G P, Hu Q, & Dosch A 2014 (Submitted to *ApJ*)
[22] Zhou Y & Matthaeus W H 1990 *J. Geophys. Res.* **95** 10291
[23] Zhou Y & Matthaeus W H 1990 *J. Geophys. Res.* **95** 14863
[24] Marsch T & Tu C 1989 *J. Geophys. Res.* **95** 8211
[25] Marsch T & Tu C 1989 *J. Geophys. Res.* **95** 11945
[26] Elsässer W M 1950 *Phys. Rev.* **79** 183
[27] Zank G P, Matthaeus W H and Smith C W 1996 *J. Geophys. Res.* **101** A8 17093-17107
[28] Matthaeus W H, Minnie J, Breech B 2004 *Geophys. Res. Lett.* **31** 12803
[29] Breech B, Matthaeus W H, Minnie J 2005 *Geophys. Res. Lett.* **32** 6103
[30] Roberts D A, Goldstein M L, Matthaeus W H, & Ghosh S 1992 *J. Geophys. Res.* **97** 17115
[31] Whang Y C 1991 *Space Sci. Rev.* **57** 339
[32] Williams L L & Zank G P 1994 *J. Geophys. Res.* **99** 19229

Instability of the Heliopause

K Avinash[1,2], G P Zank[2,3], B Dasgupta[2], and Shikha Bhadoria[1]

[1]Department of Physics and Astrophysics, University of Delhi, Delhi, 110007, India
[2]Center for Space Plasma and Aeronomic Research (CSPAR), University of Alabama in Huntsville, Huntsville, AL 35805, USA
[3]Department of Space Science, University of Alabama in Huntsville, Huntsville, AL 35805, USA

E-mail: garyp.zank@gmail.com

Abstract. The heliopause (HP) separates the tenuous hot heliosheath plasma from the relatively dense cool magnetized plasma of the local interstellar medium (LISM). Fluid acceleration in the HP region can therefore drive Rayleigh-Taylor-like and Kelvin-Helmholtz-like instabilities. Charge exchange coupling of plasma ions and primary interstellar neutral atoms provides an effective gravity, suggesting the possibility of Rayleigh Taylor-like (RT-like) instabilities. Shear flow due to the velocity difference between the heliosheath and the interstellar flows drives Kelvin Helmholtz-like (KH-like) modes on the heliopause. Magnetic fields damp the classical KH instability. However, we show that energetic neutral atoms (ENAs) destabilize KH-modes,even in the presence of interplanetary and interstellar magnetic fields. We consider a model that includes a number of effects that are important in the heliosphere such as resonant change exchange between the primary neutrals and the solar wind plasma, ENAs from the inner heliosheath, plasma flows along the heliopause and magnetic fields in the inner and outer heliosheath. We find that the nose region is unstable to RT-like modes for HP parameters, while the shoulder region is unstable to a new instability that has the characteristics of a mixed RT-KH-like mode. These instabilities are not stabilized by typical values of the magnetic fields in the inner and outer heliosheath close to the nose and shoulder regions. Whereas ENAs have a stabilizing influence on the RT instability in the vicinity of the nose region (due to counter streaming), they have a destabilizing influence on the KH instability in the vicinity of the flanks. We find that even in the presence of interplanetary and interstellar magnetic fields, ENAs can drive a new form of KH-like instability on the flanks. An analysis of the collisional and anomalous magnetic field diffusion time scales shows that ideal MHD is an appropriate model at the HP. The interstellar magnetic field therefore drapes over the HP and does not diffuse into the inner heliosheath (IHS). However, RT-like, RT-KH-like, and KH-like instabilities serve to drag outer heliosheath (OHS)/interstellar magnetic field into the IHS, allowing for local reconnection of interplanetary and interstellar magnetic field. Such reconnection may 1) enhance the mixing of plasmas across the heliopause, and 2) provide open magnetic field lines that allow easy ingress of galactic cosmic rays into the heliosphere and easy loss of anomalous cosmic rays.

1. Introduction

The Heliopause (HP), a tangential discontinuity within a magnetohydrodynamic (MHD) description, separates the high temperature, tenuous solar wind (SW) plasma in the inner heliosheath (IHS) from the relatively dense, cooler, plasma of the local interstellar medium (LISM) in the outer heliosheath (OHS). Across the HP the pressure of the plasma within the heliosphere balances the pressure of the surrounding interstellar medium [1]. The HP has been crossed by Voyager 1 [2,3] at $\sim$ 125 AU and the magnetic fields on either side are approximately parallel along the surface of the HP.

Neutral interstellar hydrogen drifts through the HP, coupled weakly via resonant charge exchange with the plasma. The energetic neutrals in the IHS created through charge exchange are called secondary neutrals or energetic neutral atoms (ENAs) [4,5] and propagate upstream into the OHS across the HP. These secondary neutrals or ENAs experience charge exchange with ions in the OHS leading to significant heating of the OHS plasma [6,7,8].

Multi-fluid simulations by Zank et al. 1996 [6] and Liewer et al. 1996 [9] and later by Florinski et al. 2005 [10] and Borovikov et al. 2008 [11] showed that the HP was unstable to large scale modes. Zank 1999a [12] showed that frictional drag between the plasma and neutral H in the vicinity of the HP drives a Rayleigh-Taylor-like (RT-like) instability. Subsequent simulations confirmed the existence of a RT-like instability caused by charge exchange between the primary neutrals from the LISM (or OHS) and SW ions [10]. In particular, high resolution simulations by Borovikov et al. 2008 [11] identified two types of unstable modes in the vicinity of HP; a RT-like mode driven mainly by the charge exchange between neutrals and plasma ions and a RT-KH-like mixed mode, driven by a combination of charge exchange and plasma flow shear across HP. The charge exchange interaction introduces an effective gravitational force acting on the plasma in the direction of the flow of the LISM neutrals [12], which flow from the OHS to the IHS. The plasma density gradient is in the opposite direction as the LISM plasma is much denser than the SW plasmas. Hence RT-like modes are expected to be unstable in the neighborhood of the nose region of the HP, where the effective gravitational force and the plasma density gradient are nearly in the opposite direction. The plasma flows on the other hand, which are weak in the vicinity of the nose region, are strong on the flanks with significant shear across the HP. Hence the Kelvin-Helmholtz (KH) modes might be unstable on the flanks of the HP.

The charge exchange interaction introduces momentum and energy sources in the plasma dynamics and thus introduces significant changes in the dynamics of the system. Zank 1999a [12] derived a modification of the plasma incompressibility condition due to charge exchange in the heliosphere and formulated the problem of the stability of the HP in terms of the charge exchange modified plasma incompressibility condition. Later, a more detailed numerical and analytic analysis of the RT instability of HP of axisymmetric heliosphere in the vicinity of nose region, using the charge exchange modified plasma incompressibility condition was given by Florinski et al. 2005 [10].

To study the stability of modes in the vicinity of HP, we use a model that includes resonant charge exchange with primary neutrals from LISM, magnetic fields, plasma flows and ENAs. Plasma flows and flow shear are important on the flanks of HP. Magnetic fields are important since they can significantly change the stability thresholds of the RT and KH instabilities [Chandrasekhar, 1981]. There is uncertainty in the magnitude and direction of the magnetic field in the vicinity of HP. The results of Burlaga et al. 2013 [3], when interpreted as the crossing of the HP [2], suggest that the magnetic fields on either side of the HP are approximately parallel, with perhaps a change in the orientation. We present a fully, general analysis below and then consider the magnetic field parameters and densities suggested by the Voyager 1 magnetometer and wave instrument measurements. One of the main effects of the interstellar magnetic field is that it introduces a north-south asymmetry in the heliosphere as is evident by the Voyager crossings of the TS. Estimates [13] based on the numerical modeling of the IBEX observation of the enhanced energetic neutral atom flux, constrain the field to be $< 3\ \mu$G (see also [14,8,15]). As the heliosphere is asymmetric, we carry out a stability analysis in a local reference frame located on the HP. This model is applicable to the entire HP i.e., close to the nose, flanks and the shoulder region in between. In the neighborhood of the nose where plasma flows are weak, it generalizes the earlier analyses [12,10] to include the effect of the magnetic field in the IHS and OHS.

2. Model equations

In the vicinity of the HP, the plasma and its interaction with neutral hydrogen via resonant charge exchange is described by the following equations,

$$\frac{\partial \rho}{\partial t} + \nabla \cdot (\rho \mathbf{u}) = 0; \quad (1)$$

$$\rho \left(\frac{\partial \mathbf{u}}{\partial t} + \mathbf{u} \cdot \nabla \mathbf{u} \right) + \nabla p = \frac{1}{c} \mathbf{J} \times \mathbf{B} - \nu_p \rho (\mathbf{u} - \mathbf{V}_p) - \nu_s \rho (\mathbf{u} - \mathbf{V}_s); \quad (2)$$

$$\frac{\partial p}{\partial t} + \mathbf{u} \cdot \nabla p + \gamma p \nabla \cdot \mathbf{u} = \nu_p \rho \left(\frac{\gamma - 1}{2} (\mathbf{u} - \mathbf{V}_p)^2 + \frac{kT_p}{m_H} - \frac{p}{2\rho} \right) + \nu_s \rho \left(\frac{\gamma - 1}{2} (\mathbf{u} - \mathbf{V}_s)^2 + \frac{kT_s}{m_H} - \frac{p}{2\rho} \right); \quad (3)$$

$$\frac{\partial N_\alpha}{\partial t} + \nabla \cdot (N_\alpha \mathbf{V}_\alpha) = -\nu_\alpha \frac{\rho}{m_H}; \quad (4)$$

$$\nabla \times \mathbf{B} = \frac{4\pi}{c} \mathbf{J}. \quad (5)$$

Here, ρ, p and J are the plasma mass density, velocity, pressure and current density respectively, $\mathbf{B}$ is the magnetic field and $\gamma = 5/3$ is the specific heat ratio. The charge exchange interaction introduces sources and sinks in the momentum and energy equations which appear on the RHS of Eq. (2) and (3) [16,17]. In our analysis we consider two populations of neutrals; primary neutrals from the LISM, which are relatively cooler and denser ($T_p \sim 10^4$ K), and ENAs or secondary hot and rarer neutrals ($T_s \sim 10^6$ K) from the IHS, created by the charge exchange of primary neutrals with the hot plasma in the IHS [6,8,18]. The density, velocity and charge exchange frequency of these two populations of neutrals is denoted by N_α, $\mathbf{V}_\alpha$, and ν_α respectively (where $\alpha = p, s$). The charge exchange frequency is given by the expression $\nu = \sigma U^* N$ where $\sigma = 3 \times 10^{-15}$ cm^2 is the charge exchange interaction cross section [19] and the speed U^* is the characteristic interaction speed between protons and neutrals. U^* is typically 1.5 times the thermal speed of hot neutrals [10]. The neutral velocity is taken constant since we consider scale lengths less than the charge exchange mean free path.

The relative importance of various terms on the RHS of Eq. (3) in the IHS and OHS is important in simplifying equations (1) - (5). Typically in the neighborhood of the HP, the plasma temperature varies from $\sim 5 \times 10^5$ K along the flanks to $\sim 10^6$ K at the nose in the IHS [11]. The temperature of ENAs also lies in a similar range. When hot neutrals propagate upstream into the OHS, they cause significant heating of the plasma in the OHS due to secondary charge exchange with LISM protons [6,8]. The plasma temperature is typically $\sim 5 \times 10^4$ K. The primary neutral density in the OHS is $N_p \sim 0.2$ cm^{-3} while the ratio of primary to secondary hot neutral H atoms is typically $N_s/N_p \sim 0.01$ [6,18]. On assigning the indices h (heliospheric) and i (interstellar) for IHS and OHS variables, respectively we have in the first parenthesis on the RHS of Eq. (3), $(p/2\rho)_h \simeq 10^6$ K, $T_p \simeq 10^4$ K and $V_p \sim 25$ km/s, hence the third term is the largest, as first pointed out by Zank 1999a [12]. This is also bigger than all the terms in the second parenthesis as $\nu_{sh}/\nu_{ph} < 1$. In the OHS, the plasma term is 5×10^4 K, hence it is still the largest term in the first parenthesis [10]. In the second parenthesis, the second term proportional to $T_s \sim 5 \times 10^5 - 10^6$ K is the largest, but, since $(kT_s/m_H)/(p/2\rho)_i \sim 10$ to 20 and $\nu_{si}/\nu_{pi} \sim 0.01$ to 0.05, $\nu_p(p/2\rho)_i$ is still greater than $\nu_s(kT_s/m_H)$ and hence it is the largest term on the RHS of Eq. (3). Thus, in both the IHS and OHS, the charge exchange modified incompressibility condition becomes

$$\nabla \cdot \mathbf{u} = -\frac{\nu_p}{2\gamma}. \quad (6)$$

This is consistent with the original condition suggested by Zank 1999a [12] and is different from Florinski et al. 2005 [10] and Dasgupta et al. 2006 [20] who retain the hot neutral thermal velocity term in the OHS. The effect of ENAs is still retained in the momentum conservation equation in Eq. (2).

The HP boundary has roughly three regions: The nose region where the interstellar plasma and the solar wind plasma flows each stagnate with weak plasma flows; the flanks, well separate from the nose region, with significant plasma flow and flow shear across the HP, and finally the shoulder region between the flanks and nose. From the nose region, the plasma flow on either side of HP is diverted towards the flanks. Thus the plasma flow is along the HP surface while the flow shear and the plasma density gradient are perpendicular to it. The primary neutrals from the LISM flow unhindered across the HP. In the nose region, the neutral flow is in the direction opposite to that of the solar wind plasma flow. The angle θ between the two flows gradually decreases from 180° along the HP to nearly 0° on the flanks where both flows are nearly in the same direction. The ENAs on the other hand, which are generated in the heliosheath, propagate upstream from the IHS, in a direction roughly perpendicular to the HP everywhere.

The analysis is done in a local reference frame, assuming a planar geometry where the z-axis is perpendicular and the x-y plane is tangential to the HP surface. The solar wind flow is in the region $z < 0$ and the interstellar plasma flow in $z > 0$. The x-axis is aligned along the equilibrium plasma flow velocity $\mathbf{u}_0 = (u, 0, 0)$ while the flow shear and plasma density gradient are along the z-axis across the HP. The primary neutrals move downstream across the HP boundary in the x-z plane with a constant speed $\mathbf{V}_p$. Let θ be the angle between the z-axis and $\mathbf{V}_p$. Along the HP boundary, this angle changes, being close to 180° in the nose region and nearly 90° in the flank region. The ENAs, on the other hand, move upstream across the HP at an angle Ψ from the local z-axis, which remains close to zero everywhere along the HP boundary.

Avinash et al. 2014 [21] show that the plasma in the vicinity of the HP is nearly ideal and hence the magnetic field will not diffuse into the heliosphere but instead will drape over it. In this case, the magnetic field $\mathbf{B}$ lies on the HP surface in the x-y plane. Since the plasma in the vicinity of HP is nearly ideal, the evolution of $\mathbf{B}$ is governed by

$$\frac{\partial \mathbf{B}}{\partial t} = \nabla \times \mathbf{u} \times \mathbf{B}. \tag{7}$$

The equation describing the zeroth-order equilibrium heliopause is given by

$$\frac{\partial}{\partial z}\left(p + \frac{B^2}{8\pi}\right) = \left(\nu_p V_{pz} + \nu_s V_{sz}\right)\rho, \tag{8}$$

where the zeroth-order pressure p and magnetic field $\mathbf{B}$ depend only on the z-coordinate.

3. Stability of the Heliopause

The plasma continuity equation is given by

$$\frac{\partial \rho}{\partial t} + \mathbf{u}\cdot\nabla\rho = \frac{\nu_p}{2\gamma}\rho, \tag{9}$$

after eliminating the divergence of the flow velocity through Eq. (6).We use the incompressibility condition given in Eq. (6) as the equation of state to calculate the pressure.

Since the HP may be regarded as a tangential discontinuity (TD) at $z = 0$ in our coordinate system, the stability problem constitutes an expansion in terms of singular perturbations. The equations are solved across and away from the singular surface and the solutions are asymptotically matched to obtain the eigenvalue. For the stability analysis, we consider a simple plasma density and flow profile, which are distinct constants in the IHS and OHS i.e., $\rho_h =$ const., $\mathbf{u}_h =$ const. in $z < 0$ and $\rho_i =$ const., $\mathbf{u}_i =$ const. in $z > 0$, with a jump $z = 0$.

The density, pressure, magnetic field, and velocity are perturbed as $\rho + \delta\rho$, $p + \delta p$, $\mathbf{B} + \delta\mathbf{B}$, and $\mathbf{u} + \delta\mathbf{u}$ respectively. The perturbed linearized equations are then

$$\frac{\partial \delta\rho}{\partial t} + \mathbf{u}\cdot\nabla\delta\rho + \delta w\frac{d\rho}{dz} = \frac{\nu_p}{2\gamma}\delta\rho; \tag{10}$$

$$\rho\frac{\partial \delta\mathbf{u}}{\partial t} + \rho\delta\mathbf{u}\cdot\nabla\mathbf{u} + \delta\rho\mathbf{u}\cdot\nabla\mathbf{u} + \rho\mathbf{u}\cdot\nabla\delta\mathbf{u} + \nabla\left(\delta p + \frac{\delta\mathbf{B}\cdot\mathbf{B}}{4\pi}\right)$$
$$= \frac{\mathbf{B}\cdot\nabla\delta\mathbf{B}}{4\pi} - \nu_p\left(\mathbf{u} - \mathbf{V}_p\right)\delta\rho - \nu_p\rho\delta\mathbf{u} - \nu_s\left(\mathbf{u} - \mathbf{V}_s\right)\delta\rho - \nu_s\rho\delta\mathbf{u}; \tag{11}$$

$$\nabla\cdot\delta\mathbf{u} = 0 = \nabla\cdot\delta\mathbf{B}; \tag{12}$$

$$\frac{\partial\delta\mathbf{B}}{\partial t} = \nabla\times\delta\mathbf{u}\times\mathbf{B} + \nabla\times\mathbf{u}\times\delta\mathbf{B}, \tag{13}$$

where δw is the z component of the perturbed plasma velocity. The eigenfunction $f(z)\exp[nt + i\mathbf{k}\cdot\mathbf{r}]$ is used to describe the perturbed quantities, and represents waves propagating on the singular surface in the x-y plane $\mathbf{k} = (k_x, k_y)$. Matching the solution across the singular surface with the solution away from the singular surface finally gives the dispersion relation (DR) as [21]

$$\left[\rho_h\left(n + \nu_{ph} + \nu_{sh} + i\mathbf{k}\cdot\mathbf{u}_h\right)\left(n + i\mathbf{k}\cdot\mathbf{u}_h\right) + \frac{(\mathbf{k}\cdot\mathbf{B}_h)^2}{4\pi}\right]$$
$$+\left[\rho_i\left(n + \nu_{pi} + \nu_{si} + i\mathbf{k}\cdot\mathbf{u}_i\right)\left(n + i\mathbf{k}\cdot\mathbf{u}_i\right) + \frac{(\mathbf{k}\cdot\mathbf{B}_i)^2}{4\pi}\right]$$
$$= \frac{(\rho_i - \rho_h)k}{2}\left[\frac{(-\nu_{pi}V_p\cos\theta - \nu_{si}V_s\cos\Psi)(n + i\mathbf{k}\cdot\mathbf{u}_i)}{n - \nu_{pi}/2\gamma + i\mathbf{k}\cdot\mathbf{u}_i}\right.$$
$$\left. + \frac{(-\nu_{ph}V_p\cos\theta - \nu_{sh}V_s\cos\Psi)(n + i\mathbf{k}\cdot\mathbf{u}_h)}{n - \nu_{ph}/2\gamma + i\mathbf{k}\cdot\mathbf{u}_h}\right]. \tag{14}$$

The dispersion relation Eq. (14) takes into account charge exchange interactions with primary neutrals from the LISM and ENAs from the heliosheath, as well as plasma flows and magnetic fields in the IHS and OHS. Eq. (14) is quartic in n and possesses either four real roots or two pairs of complex roots. Instability occurs if the roots have positive real parts.

We use (14) to study the stability of a particular region of the HP by appropriately choosing the value of θ and Ψ. Typical values of the plasma density and temperature in the IHS and OHS and neutral density in the nose and flanks region have been given above. Typical values of the neutral flow velocity, charge exchange frequencies, and magnetic fields and ratio of plasma densities in the OHS and IHS are $V_p \simeq 20$ km/s, $V_s \simeq 100$ km/s, $\nu_{pi} \simeq 10^{-9}$ s^{-1}, $\nu_{ph} \simeq 10\nu_{pi} \simeq 10^{-8}$ s^{-1}, $\nu_{si} = 10^{-11} \simeq \nu_{sh}$, $B_i \simeq 3$ μG, $B_h \simeq 5$ μG and $\alpha \simeq 20$ [11,8].

(i) Nose region of the HP: This region is characterized by weak plasma flows $U \equiv |v_i - v_h| \simeq 0$ km/s, LISM neutral flow along z ($\cos\theta \sim -1$) and ENAs that are counter streaming along $-z$, hence $\cos\Psi \simeq 1$. In this case, the dispersion relation is fourth order in n. Because the plasma density gradient and the effective gravity due to primary LISM neutrals are in opposite directions, the RT mode dominates. The counter streaming ENAs, magnetic field, and the ion-neutral collisions provide a stabilizing influence on the RT instability. In the absence of a magnetic field and ENAs, the dispersion relation Eq. (14) reduces to the cubic dispersion relation obtained originally by Zank 1999a [12] and later by Florinski et al. 2005 [10].

Fig. 1 shows the four roots for the dispersion equation in the presence of magnetic fields in the IHS and OHS. Compared to the no magnetic field case (not shown), the plot shows a marginal reduction of the growth rate even when the unstable mode is propagating in the

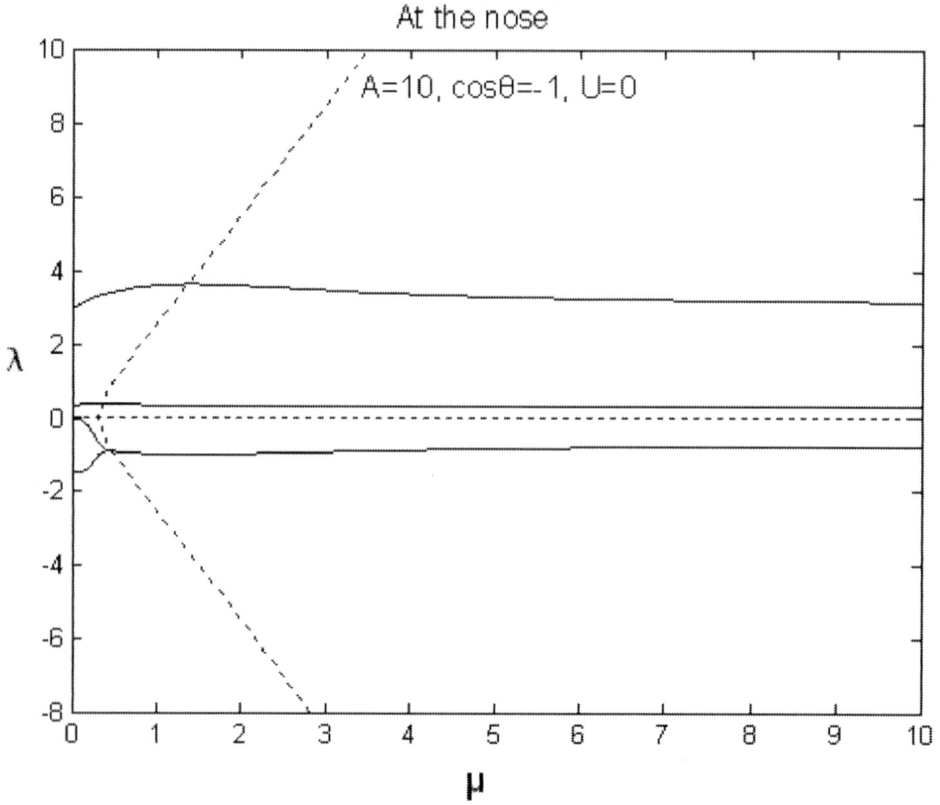

Figure 1. The roots of the dispersion relation near the nose with magnetic field. With a non-zero magnetic field, there are four roots instead of three. The stabilization of the unstable roots due to the magnetic field is marginal. Here the label axes are normalized as $\lambda \equiv n/\nu_{pi}$ and $\mu \equiv \mathbf{k}V_p/\nu_{pi}$.

direction of the magnetic field, which is when stabilization by a magnetic field is expected to be strongest. Stabilization by the magnetic field for other angles is weaker and zero for perpendicular propagation. Most importantly, we find that stabilization by the magnetic field is not complete even for very large values of the field. Typical values of the growth rates are $\lambda \equiv n/\nu_{pi} \simeq 1$ - 3, which for heliospheric parameters corresponds to a growth time of 50 to 100 years. This time scale agrees with that of quasi-periodic oscillations of the nose seen in numerical simulations of the HP [10]. Our results thus show that the nose region of the HP is unstable to RT-like modes with a growth time of roughly 50-100 years with wavelengths ranging from 100 to 500 AU. The stabilizing influence of magnetic fields in the IHS and OHS and of ENAs is weak in the nose region.

(ii) Shoulder region of the HP: This region is characterized by the simultaneous presence of a KH-like instability due to flow shear and the RT-like instability, though the latter is somewhat weakened as compared to the nose region because $|\cos\theta| < 1$. For our calculation we choose $\theta = 135°$, $\cos\theta = -0.7$ and $\cos\Psi \simeq 1$. A typical value for the flow shear across HP is $U \equiv |v_i - v_h| \simeq 2$. Values for the other parameters are the same as given above. The dispersion relation has four roots for these parameters, which are shown in Fig. 2. Two roots have positive

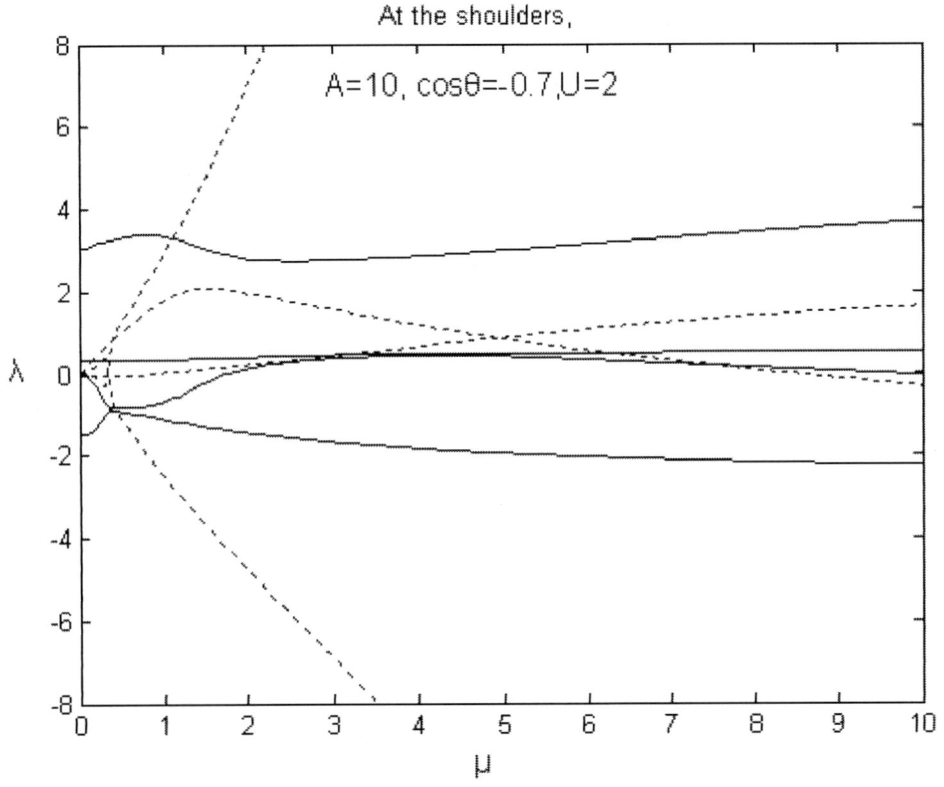

Figure 2. The roots of the dispersion relation in the shoulder region. The unstable roots correspond to mixed RT-KH modes. The dotted lines identify the imaginary part of the solution to the dispersion relation. The imaginary part of the solution shows that these are propagating modes. The axes are described in the caption of Fig. 1.

real parts thus indicating instability. The mixed KH-RT mode is a new form of instability driven by a combination of ENAs and primary neutrals, and is seen in simulations [11].

(iii) Flank region of the HP: In the flanks, the LISM neutral flow (as well as the plasma flow) is along the x direction, hence $\cos\theta = 0$. Since the plasma density gradient is along z, the RT mode is absent and LISM neutrals play no role in destabilizing the flanks (and in fact can act to damp a KH instability [12]). However, ENAs now play an important role. The ENAs travel upstream nearly along z, hence $\cos\Psi \simeq 1$. Consequently, energetic neutrals influence the KH instability on the flanks, which is caused by flow shear along z. In the absence of ENAs, the dispersion relation for modes propagating parallel to the magnetic field is a quadratic equation with two roots, both of which are stable. This is the KH instability that is stabilized by magnetic fields. However, when ENAs are retained, the dispersion relation now admits roots with positive real parts, i.e., unstable roots, which show that ENAs revive the instability on the flanks. Thus, in the presence of a magnetic field, ENAs are essential for driving an instability of the HP flanks.

4. Conclusions

With the crossing of the HP by Voyager 1, its structure and stability have assumed great interest. In an idealized view, the HP is a tangential discontinuity that separates the tenuous and hot

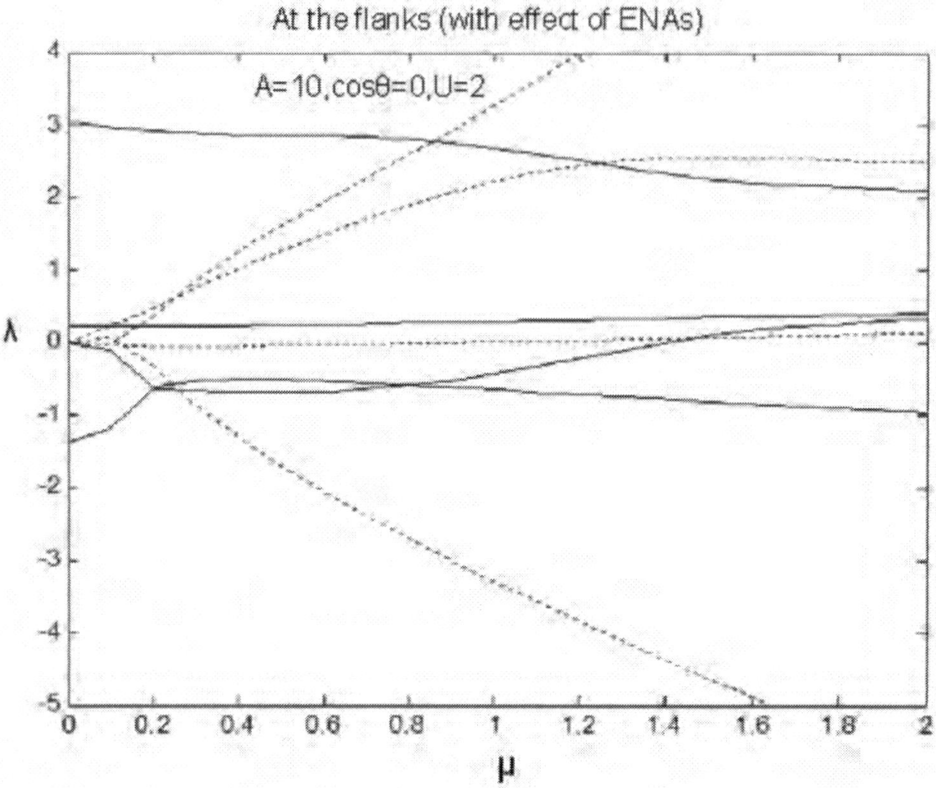

Figure 3. The roots of the dispersion relation in the flank region with magnetic field and ENAs. The plot illustrates three unstable roots showing the destabilizing influence of ENAs on the KH modes. The axes are described in the caption of Fig. 1.

heliosheath plasma from the relatively dense and cool plasma of the local interstellar medium. We have explored the effects of resonant change exchange between primary interstellar neutrals atoms and the solar wind plasma, ENAs, interstellar and heliosheath magnetic fields, and plasma flows on the stability of the HP. We use Zank's charge exchange modified incompressibility condition [12] as the equation of state in our model. Under typical conditions, the plasma resistivity due to electron-neutral collisions, electron-ion collisions, or turbulence (anomalous) is found to be small hence the interstellar magnetic field drapes around the heliopause causing a north-south asymmetry. The stability analysis is performed in a local reference frame located on the HP, and we consider three regions. The nose region where plasma flows are weak is unstable to RT-like modes; the shoulder region is unstable to mixed RT-KH-like modes, and the flank region (with strong flow and flow shear) is unstable to a KH-like instability. The dispersion relation shows that, in the relevant parameter space, magnetic fields cannot stabilize any of the instabilities. However, ENAs are found to be essential for destabilizing the flanks when magnetic fields are included.

Let us consider what the effect might be of an instability on the HP that drags a loop of interstellar magnetic field (colored blue) into the inner heliosheath (IHS) – Figures 4 A and B. At the nose and shoulder regions of the HP, the latter being the current location of Voyager 1,

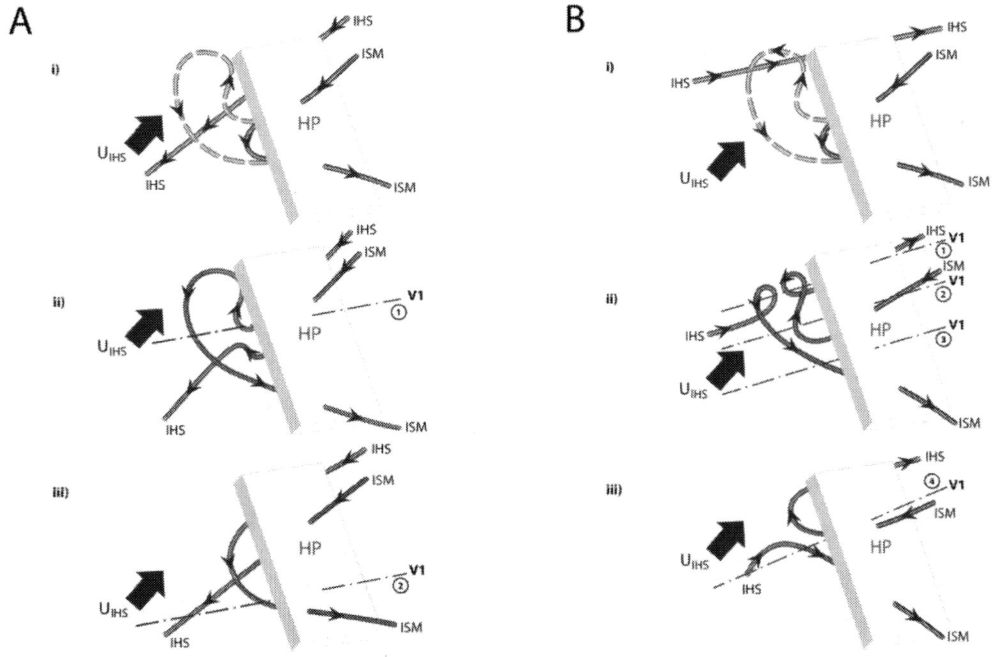

Figure 4. A. Cartoon showing the possible evolution of an OHS/VLISM magnetic field line dragged into the IHS by either a RT-like or mixed KH-RT-like instability on the HP. For convenience, we have assumed that we are in the stationary frame of the OHS flow. The red line shows the IHS magnetic field. In this example, we suppose that the IHS and OHS magnetic field lines are approximately parallel to one another on either side of the HP (sketched as the shaded block, and labeled HP). The heavy black arrow depicts the IHS plasma flow direction as measured in the frame of the OHS flow. (i) The solid blue line depicts the initial incursion of the OHS magnetic field dragged into the IHS by a RT-like instability. The dashed blue line shows the later OHS magnetic field line after being dragged northward and westward (assuming we are considering the approximate location of Voyager 1) by the IHS plasma flow. The IHS magnetic field (red line) is convected northward and westward by the IHS plasma flow. (ii) As the IHS magnetic field line is advected in the IHS plasma flow, it will eventually be forced against the intruding OHS magnetic field line, which in the OHS plasma flow frame of reference, is anchored to the HP. Because the OHS magnetic field has been dragged as illustrated in (i), there will always be a region that is anti-parallel to the impinging IHS magnetic field. In this region, as illustrated, the IHS and OHS magnetic field lines can reconnect. In so doing, the entering OHS magnetic field line is connected to an IHS magnetic field section, thereby allowing direct access of OHS plasma into the IHS. Conversely, the other segment of the IHS magnetic field is now connected to the OHS magnetic field, thereby allowing direct access of IHS plasma into the OHS. (iii) After some time, the magnetic field connecting the IHS to the OHS and vice versa will be dragged into the parallel configuration illustrated here. The dash-dotted lines, labeled A1 and A2, depict possible Voyager 1 trajectories through the magnetic field line topologies and HP. B. Same as A except that the IHS is now oriented anti-parallel to the OHS magnetic field. Four possible Voyager 1 trajectories have been sketched, B1 - B4. The net result is that again the IHS and OHS magnetic fields reconnect and allow the exchange of plasma from IHS to OHS and vice versa.

the RT element of the instability described here will drag a loop of magnetic field through the HP into the IHS. If we adopt a frame of reference that is co-moving with the interstellar flow adjacent to the local HP, then the magnetic field lines on the interstellar side of the HP are fixed. The initial loop is illustrated in Fig. 4 (i). The flow in the IHS is faster than in the very local ISM (VLISM) and has both a northward and westward component as viewed from the Sun. As the VLISM magnetic field protrudes into the IHS, it is dragged in a corresponding direction (relative to the VLISM flow) with the footpoints remaining anchored to the same relative locations on the HP. The dragged and distorted interstellar magnetic field is shown in Fig. 4A (i) by the dashed blue line. The interplanetary field (colored red) in the inner heliosheath can be oriented in one or the other direction as illustrated in Fig. 4A or B, depending on the sector boundary structure ahead of the HP. Interplanetary magnetic field that was below the interstellar loop protruding into the IHS will be pushed up against the loop as it is dragged (more slowly) by the IHS flow. As illustrated in Fig. 4A (i), part of the dragged loop of VLISM magnetic field will be anti-parallel to the IHS magnetic field that is pushing up against it. These sections of the VLISM and IHS magnetic fields can experience reconnection, with the resulting re-configured field lines (now red and blue) illustrated in Fig. 4. No island-like structures are expected to form but instead, one part of the IHS magnetic field connects to the VLISM magnetic field that returns to the VLISM. The other section of the IHS magnetic field connects to the VLISM magnetic field that is entering the IHS from the ISM. The IHS flow will tend to straighten the recently reconnected magnetic field, yielding a final configuration similar to that shown in Fig. 4 (iii). The key points are that i) one of the field lines allows direct access from the IHS to the VLISM, and thus allows direct access of energetic heliospheric particles to the ISM, and the other magnetic field line allows direct access of energetic galactic particles from the VLISM into the IHS; and ii) the magnetic field structure on either side of the HP is unlikely to be very complicated.

There is an important implication of the analysis presented here. As discussed in detail in Avinash et al. 2014 [21], the classical collisional and anomalous magnetic field diffusion time scales show that the VLISM magnetic field may be described using ideal MHD. The OHS magnetic field therefore drapes over the HP and thus should not be able to reconnect with IHS magnetic field (reconnection found in the vicinity of the HP in global numerical simulations is due to numerical diffusion and not due to the inclusion of a physical magnetic field diffusion coefficient). However, the instabilities described here, as illustrated in the cartoons of Fig. 4 (and possibly other forms of instability, e.g., [1]), provide a critical mechanism that allows IHS interplanetary and OHS interstellar magnetic field to reconnect. The form of the reconnected IHS and VLISM magnetic field described above might be an explanation for the observed galactic and anomalous cosmic ray behavior seen by Voyager 1. Furthermore, since the magnetic field configuration holds for either orientation of the IHS magnetic field, we believe that this explanation is quite robust. We note that Borovikov & Pogorelov 2014 [22] suggest that solar cycle variation might provide the perturbation to initiate the Zank instability. It is also possible that that the highly dynamical state of the IHS [23,24] will perturb the HP and thus drive instabilities, which can then lead to magnetic field reconnection. Such reconnection may 1) enhance the mixing of plasmas across the heliopause, and 2) provide open magnetic field lines that allow easy ingress of galactic cosmic rays into the heliosphere and easy loss of anomalous cosmic rays.

Acknowledgments

This work was supported by NASA grant NNX10AC17G and IBEX NASA/SwRI award NNG05EC85C, subcontract A99132BT. Discussions with V. Florinski, J. Heerikhuisen and N. Pogorelov are gratefully acknowledged. GPZ appreciates discussions at the team meeting "Heliosheath Processes and Structure of the Heliopause: Modeling Energetic Particles, Cosmic

Rays, and Magnetic Fields" supported by the International Space Science Institute in Bern, Switzerland.

References

[1] Fahr, H. J., Neutsch, W., Grzcdzielski, S., Macek, W., and Ratkiewicz, R., 1986, Space Sci. Rev., 43, 329
[2] Gurnett D. A., Kurth W.S., Burlaga L. F., and Ness N. F., 2013 Science 341, 1489
[3] Burlaga L. F., Ness N. F., Gurnett D.A., and Kurth W.S., 2013 ApJ, 778:L3
[4] Gruntman, M., 1992, Planet. Space Sci., 40, 439
[5] Gruntman, M., Roelof, E. C., Mitchell, D. G., Fahr, H. J., Funsten, H. O., and McComas, D.J., 2001, J. Geophys. Res., 106, 15,767
[6] Zank, G. P., Pauls, H. L., Williams, L. L., and Hall, D. T. 1996, J. Geophys. Res., 101, 21, 639
[7] Zank, G. P., Muller, H.-R., and Wood, B. E. 2001, Phys. Plasmas, 8, 2385
[8] Zank G. P., Heerikhuisen, J., Wood B. E., Pogorelov, N. V, Zirnstein E., and McComas D. J., 2013, ApJ, 763, 20
[9] Liewer, P. C., Karmesin, S. R., and Brackbill, J. U. 1996, J. Geophys. Res., 101, 17,119
[10] Florinski, V., Zank, G. P., and Pogorelov, N. V. 2005, J. Geophys. Res., 110, A07104
[11] Borovikov, S. N., Pogorelov, N. V., Zank, G. P., and Kryukov, I. A. 2008, ApJ, 682, 1404
[12] Zank, G. P. 1999a, in AIP Conf. Proc. 471, Solar Wind Nine, ed. S. R. Habbal et al. (Melville, NY: AIP) 783
[13] Heerikhuisen J. and Pogorelov N. V. 2011 ApJ, 738:29
[14] Strumik, M, Ben-Jaffel, L., Ratkiewicz, R., Grygorczuk, J., 2011, ApJL, 741:L6
[15] Ben-Jaffel, Lotfi, Strumik, M., Ratkiewicz, R., Grygorczuk, J., 2013, ApJ, 779, 130
[16] Holzer, T. E. 1972, J. Geophys. Res., 77, 5407
[17] Zank, G. P. 1999b, Space Sci. Rev., 89, 413
[18] Heerikhuisen, J., Florinski, V., and Zank, G. P. 2006, J. Geophys. Res., 111, A06110
[19] Lindsay, B. G., and Stebbings, R. F. 2005, J. Geophys. Res., 110, A12213
[20] Dasgupta, B., Florinski, V., Heerikhuisen, J., and Zank, G. P. 2006, in AIP Conf. Proc. 858, Physics of the Inner Heliosphere, ed. J. Heerikhuisen et al. (Melville, NY: AIP), 51
[21] Avinash, K., Zank, G.P., Dasgupta, B., and Bhadoria, S., 2014, ApJ, in press
[22] Borovikov, S. N., Pogorelov, N. V., 2014, ApJ, 783, L16
[23] Zank, G. P., and Muller, H.-R. 2003, J. Geophys. Res., 108, 1240
[24] Washimi, H., Zank, G.P., Hu, Q., Tanaka, T., Munakata, K., and Shinagawa, H., 2011, MNRAS, 416, 1475

Journal of Physics: Conference Series **577** (2015) 012003 doi:10.1088/1742-6596/577/1/012003

Helium abundance in giant planets and the local interstellar medium

L. Ben-Jaffel[1,2] and I. Abbes[1,2]

[1] UPMC Univ. Paris 06, UMR7095, Institut d'Astrophysique de Paris, F-75014, Paris, France
[2] CNRS, UMR7095, Institut d'Astrophysique de Paris, F-75014, Paris, France

E-mail: bjaffel@iap.fr

Abstract. The sun and giant planets are generally thought to have the same helium abundance as that in the solar nebula from which they were formed 4.6 billion years ago. In contrast, the interstellar medium reflects current galactic conditions. The departure of current abundances from the primordial and protosolar values may help trace the processes that drive the nucleosynthesis evolution of the galaxy and planetary interior formation and evolution. The Galileo probe measured the He abundance *in situ* the atmosphere of Jupiter, showing that He is only slightly depleted compared to the solar value. For Saturn, contradictory estimates from past Voyager observations make its He abundance very uncertain. Here, we use He 58.4 nm dayglow measured from the outer planets by the Voyager ultraviolet spectrometers to derive the He abundance in the atmosphere of Jupiter and Saturn. With the He abundance reported here, credible models become possible for the interior of Jupiter, Saturn, and extrasolar planets. We also use the solar He 58.4 nm line measured by the Solar Heliospheric Observatory to derive the He abundance inside the focusing cone. Finally, we compare He abundances derived here with primordial and protosolar values, stressing the unique opportunity offered by inner heliosphere observations and future Voyager *in situ* local interstellar medium measurements to derive the He abundance in the very interstellar cloud in which we reside.

1. Introduction

According to the standard Big Bang Nuclueosynthesis (BBN) model, the universe was dense and hot at the beginning of its expansion phase. During four long minutes, nuclear reactions produced a significant fraction of the baryonic mass of the universe in the form of helium-4 (~25%) and traces of deuterium and lithium. At the end of the expansion period, the temperature of the universe dropped to a level that does not allow the cascade of nuclear reactions required to produce the other heavy elements of the periodic table. The mass embedded in those few initial species forms the so-called primordial baryonic mass. A few million years later, stars started forming with those primordial light elements (H, He, etc.). With the extreme density and temperature conditions in the interior of stars, nuclear reactions then produced new He and heavy species, which contaminated the interstellar medium through a variety of stellar mass loss processes.

One of the most intriguing results of the BBN model is that it allows one to predict a large primordial He mass fraction $Y_{BBN} = 0.2485 \pm 0.0002$ at the end of the expansion phase of the universe

Journal of Physics: Conference Series **577** (2015) 012003 doi:10.1088/1742-6596/577/1/012003

(based on the recent Planck determination of the baryon density) [1]. Because non-primordial He and heavy metals could only be produced in stars in contrast to D that is consumed, regions of poor metallicity could reflect the He condition not far from that which prevailed at the end of the expansion phase. This simple idea fueled the study of a collection of metal-poor extragalactic H II targets to obtain the He abundance versus the gas metallicity, usually expressed as the O I/H I ratio [2]. When extrapolated to zero metallicity, the technique should provide a measurement of the primordial He mass fraction Y_p before the emergence of stars in the observed region. Such was the concept without taking into account systematic errors. Indeed, the application of this technique showed how the inferred Y_p was sensitive to our level of knowledge of the driving processes in the gas phase of the studied targets. The continuous effort of the scientific community is illustrated in Figure 1 in terms of the temporal evolution of their results.

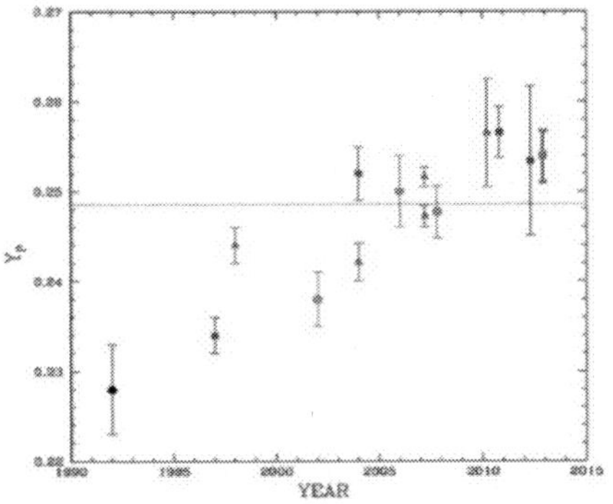

Figure 1. Primordial He mass fraction derived from extragalactic H II regions studies over time. Plot adapted from [1] with the newest value obtained by [2] added. For reference, the dotted line shows the Y_{BBN} value.

Besides H II regions, the paradigm described above has been applied to the protosolar nebula, the formation and evolution of the sun, and the interstellar medium, which represent targets of high interest. For example, the comparison between the He mass fraction in the protosolar nebula and the photosphere of the sun revealed a diffusive settling that reduces the He and heavy species abundances relative to H I [3]. For giant planets, the Galileo probe *in situ* measurement also revealed a He mass fraction smaller in the outer layers of Jupiter compared to the protosolar value [4]. In the local interstellar medium, the ionized gas near early-B stars and in the Orient nebula indeed have a poor metallicity level, yet the exact He and metal abundances are also uncertain because gas depletion into dust grains complicates the data interpretation. In addition, the actual local interstellar medium (LISM) properties around the solar system may be different from the conditions that prevailed at the sun's birthplace in the galaxy because it may have migrated over its lifetime.

In the following, we revisit giant planets' He abundance using He 58.4 nm dayglow measured with Voyagers 1 and 2 ultraviolet spectrometers (UVS) during their encounters with the outer planets (Section 2). In Section 3, we also use the solar He 58.4 nm line but as measured by Solar Ultraviolet Measurements of Emitted Radiation (SUMER) instrument onboard the Solar Heliospheric Observatory (SoHO) to constrain the He abundance inside the gravitational focusing cone of the interstellar gas flow, which reflects the local interstellar cloud (LIC) composition. Finally, we compare our results to the protosolar and primordial He values in an attempt to connect heliospheric physics with the larger cosmological picture of the origin and evolution of the interstellar medium.

2 Giant planets' helium abundance

An accurate measurement of the helium abundance in the atmosphere of giant planets is a key step in understanding the fundamental problem of the formation and evolution of giant planets in solar and extrasolar systems. In the past, using Voyagers 1 and 2 radio occultation observations to derive a pressure-temperature altitude profile, along with infrared emission spectra from most of the outer planets, a sophisticated approach was implemented to derive the He abundance by adjusting the atmospheric He /H_2 mixing ratio until the infrared spectra are best recovered [5]. However, the first *in situ* He mass fraction value $Y_{Galileo} = 0.234 \pm 0.005$ for Jupiter [4] provided by the Galileo Entry probe does not agree with the indirect Voyager value $Y = 0.18 \pm 0.04$ [5], casting doubt on the indirect method. Furthermore, contradictory estimates reported for the He abundance in Saturn using similar remote sensing methods with values ranging from $Y_{sat} = 0.06 \pm 0.05$ (pre-Galileo value [6]) to $Y_{sat} \sim 0.18\text{-}0.25$ (post-Galileo value [7]) added to the confusion and the lack of confidence the community has in the proposed values. In contrast to Jupiter, no space probe is actually planned to measure the Saturn He abundance *in situ*, thus making the diagnostic of this primordial parameter obtained with the Galileo probe incomplete.

2.1 Voyager UVS observations & data analysis

To derive the helium abundance, we use the first resonance line of He as emitted by the upper atmosphere of most planets at 58.4nm. The He 58.4nm dayglow thus far observed during the successive Voyager encounters with the outer planets corresponds to the solar He emission line back-scattered by the He atoms confined in their upper atmospheres (e.g., Table 1). The observed dayglow mainly depends on the He atmospheric total content, the eddy mixing power of the atmosphere below the homopause level, and the temperature profile above the homopause level.

The bulk of the atmosphere of the two giant planets is assumed to be composed mainly of H_2, He, and CH_4 (Figure 2). We solved the continuity equation for He and CH_4 as minor constituents, including molecular diffusion and eddy mixing parameterized by the eddy diffusion coefficient K_H. The background gas is assumed to be in hydrostatic equilibrium with a varying mean molecular mass.

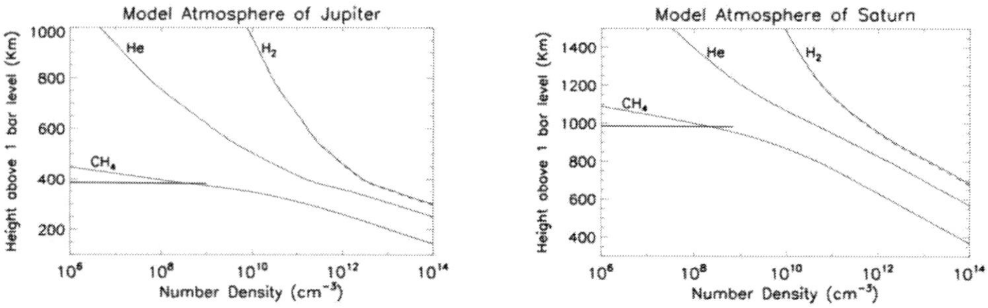

Figure 2. Atmospheric model derived in this study. A short horizontal line segment indicates the homopause level. (Left) Jupiter. (Right) Saturn.

To estimate a disk average range of K_H, we use stellar occultation observations made with the UVS instrument during spacecraft encounters with each of the planets. Of particular interest in our case is the light curve trend obtained for wavelength ranges that are sensitive to the absorption by CH_4. Because methane is the third-most abundant component of a giant planet's atmosphere, the corresponding light curve in the far ultraviolet shows a clear cut-off at the homopause level, offering the cleanest direct technique for locating its level and the corresponding eddy diffusion strength [8]. The K_H values derived here (e.g., Table 1) are consistent with past estimates [8, 9].

The next parameter of importance for the He dayglow modeling is a good estimate of the temperature profile in the upper atmosphere, preferentially obtained for each planet during the same period of the airglow observation. As shown in Figure 3, the different temperature profiles so far reported for Jupiter and Saturn do not show a noticeable variation (less than ~10%) around the homopause level [9, 10, 11]. The profiles shown in Figure 3 are explored as potential distributions of temperature versus altitude, including estimated measurement and model uncertainties.

To derive the He abundance, we adjust the atmospheric He mixing ratio f_{He} until our radiation transfer model (RT) of the He 58.4nm sun-reflected emission line of the planet matches the corresponding observed brightness. The RT model includes photon absorption by molecular and atomic hydrogen and resonance scattering by He with partial frequency redistribution [12, 13].

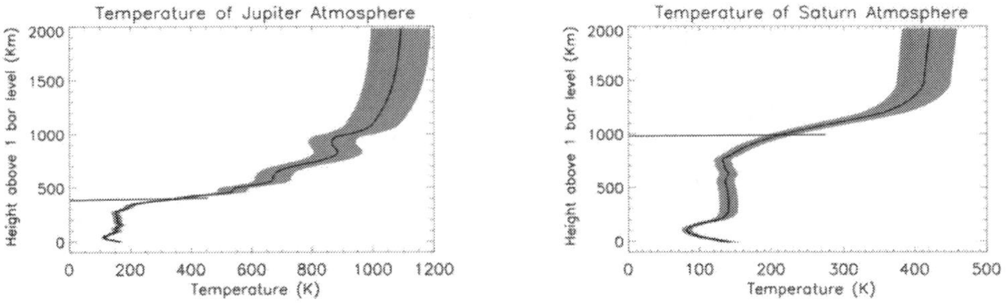

Figure 3. Temperature profile values (shaded) used in He 58.4nm Brightness RT model. The horizontal line segment indicates the homopause altitude level. (Left) Jupiter. (Right) Saturn.

Because the temperature depends on the mean molecular mass, its altitude profile and species distribution are updated iteratively for each mixing ratio (f_{He}) assumed deep in the atmosphere.

Table 1. Key input parameters for the He 58.4nm brightness radiation transfer calculations.

Planet	Eddy diffusion K_H ($cm^2\ s^{-1}$)	1 bar-level Temperature (°K)	He 58.4nm Brightness (Rayleigh)	He 58.4nm solar flux @ 1AU[a] (Photons $cm^{-2}\ s^{-1}$)
Jupiter	$(2.5\text{-}4.0)\ 10^6$	165	6.5±1.1 (V1) & 5.7±1.2 (V2)	$3.832\ 10^9$
Saturn	$(1.5\text{-}2.5)\ 10^7$	135	3.1±0.4 (V1) & 4.2±0.5 (V2)	$4.694\ 10^9$

[a] Integrated solar flux at the time of the V1 observation.

2.2 Helium abundance in Jupiter and Saturn: results

The case of Jupiter is very interesting because the Galileo probe already measured the helium abundance with a high degree of accuracy, thus offering a key test for our He 58.4nm airglow-based technique. For the eddy diffusion coefficient range shown in Table 1 (from occultation observations), and for the temperature profiles shown in Figure 2, we derive a He mixing ratio in the range $f_{He} \sim 0.159 \pm 0.034$ to fit both V1 and V2 UVS He 58.4nm airglow observations. The result corresponds to a He mole fraction $q_{He} \sim 0.137 \pm 0.025$ and mass fraction $Y_{Jup} \sim 0.237 \pm 0.038$. This abundance is nicely consistent with the Galileo probe *in situ* values $q_{He} \sim 0.1359 \pm 0.0027$ and $Y_{Jup} \sim 0.234 \pm 0.005$ measured in the atmosphere of the planet [4]. The agreement between our result and the Galileo probe measurement successfully validates our technique.

The comparison of our model calculations with Saturn's He 58.4nm intensity from the Voyager UVS brightness observations reveals a surprising result. For the eddy diffusion coefficient values derived from occultation observations (e.g., Table 1) and the temperature profile shown in Figure 2 with the attached error bars, we derive f_{He}~ 0.151 ± 0.025 in the atmosphere of Saturn. This mixing ratio corresponds to a mole fraction q_{He}~ 0.131 ± 0.02 and mass fraction Y_{Sat} ~ 0.226 ± 0.03, as summarized in Table 2. This is a higher abundance for Saturn than the value Y_{sat} = 0.06 ± 0.05 found earlier by [6]. As a consequence, the mean molecular mass should be updated in the continuity equation [6]. In order to fit the background density with the He mixing ratio derived here, the 1-bar level temperature in Saturn's atmosphere should be ~149 °K, a value larger than previously derived by [10]. Surprisingly, the high 1 bar level temperature supports the finding of [14], who suggested that a warmer atmospheric layer is required in their model of Saturn's formation.

Table 2. Helium abundance and temperature at the 1 bar level that best fit the He 58.4nm brightness observed by V1 & V2 for Jupiter & Saturn. Galileo probe results are shown for reference [4, 11].

Planet	1 bar-level Temperature (°K)	Mixing ratio (f_{He}=He /H_2)	He mole fraction (q_{He})	He mass fraction (Y_{planet})
Jupiter (Galileo)	166	0.156 ± 0.006	0.136 ± 0.003	0.234 ± 0.005
Jupiter (UVS)	165	0.16 ± 0.03	0.137 ± 0.025	0.237 ± 0.038
Saturn (UVS)	149	0.15 ± 0.025	0.131 ± 0.02	0.226 ± 0.030

3 Local interstellar medium helium abundance

The helium abundance in the local interstellar medium remains particularly uncertain because different LISM regions may have different metallicity and probably different He abundances. Those differences are intimately related to the chemical evolution of the galaxy. To measure the LISM He abundance, three main regions have been considered in the past: the protosolar nebula, the Orient nebula, and the gas phase around early-B stars.

For the protosolar nebula, the sun's photosphere is usually used to derive the He abundance. Using a sophisticated model of solar oscillation, the present-day He abundance in the sun's photosphere is derived as $Y_{photosphere}$ ~ 0.2485 [3]. From the photosphere He mass fraction, sophisticated 3D hydrodynamics and radiation transfer modeling is then used to derive the protosolar He and metals mass fractions that prevailed 4.6 Gy ago: Y_{proto} ~ 0.2703 and Z_{proto} ~ 0.0142 [15]. It is important to stress that the values are very model-dependent. Usually the difference between the photosphere and protosolar values is explained by diffusive settling that brings He and heavy species deeper toward the sun's core. In addition, the protosolar nebula value represents an indication of the galactic composition at the time and birthplace of the sun, which may be different for other regions of the galaxy and the actual epoch.

The second method for estimating the He mass fraction in the LISM is to use the nearby bright Orient nebula. Traditionally, this nebula is considered a standard source for the retrieval of the ionized gas chemical composition in the galactic environment around the sun. Using recombination and collision excited emission lines from ionized species, the conclusion was that the He mass fraction and the metallicity of the Orient nebula are slightly larger than in the sun's photosphere with Y_{Orient} ~ 0.278 and Z_{Orient} ~ 0.0137 [16]. However, neither mass fraction is independent of the other, and each hinges on the degree of success in properly including depletion into dust grains.

A third method for deriving He and heavy elements mass fractions in the LISM is to probe the ionized gas around early B-type stars [17]. Those short-lived stars have the interesting property of not having had time to migrate from their birthplace, in contrast to old stars like the sun. In addition, the dust-grain phase seems nearly nonexistent for those objects, which provides a good opportunity to derive a present-day cosmic chemical composition reference. Using a few targets, [17] showed that the technique is very promising and derived $Y_{E\text{-}B\text{-}stars}$ ~ 0.276 and $Z_{E\text{-}B\text{-}stars}$ ~ 0.014 that are close to mass fractions derived from other techniques; however, sophisticated models are needed to reduce systematic uncertainties.

A common feature that we noticed for the three different targets considered above for the retrieval of cosmic abundances is that the ability to retrieve those abundances requires first a deep and sophisticated description of the physics of the region under study. Therefore, an accurate determination of He and metal mass fractions is closely related to our ability to model the physics in those astrophysical sites, much like the requirements in comparable studies for retrieving the primordial He mass fraction from extragalactic H II regions. In the following, we propose to apply the same paradigm to derive the He abundance inside the local interstellar cloud (LIC).

3.1 Helium abundance in the LIC

The LIC can be described as the ISM cloud through which our solar system is actually moving. Whether described as a separate object or a piece of a much larger structure [18, 19], the pristine composition of the LIC remains a topic of vigorous discussion [20, 21]. Following the same technique of using the imprint of the LIC gas on nearby stellar emission lines, Extreme Ultraviolet Explorer (EUVE) spectra of six white dwarfs were used to derive H I and He I neutral density columns versus distance from the sun, revealing an interesting finding of a rather constant ratio H I/He I ~ 14 ± 2 (2σ error bar) that could be used for the immediate neighborhood of the sun [22]. However, using the same EUVE instrument, another white dwarf (REJ 1032+532) showed a quite different ratio H I/He I ~ 7.2 [23], a finding that casts doubt on the uniformity of the H I/He I ratio initially derived when using only six target stars [22]. We believe the main problem lies in the short list of targets used in that it does not allow derivation of reliable statistics.

Another tentative effort to derive the He abundance in the LIC utilized the He ionization degree versus distance from the sun, using local thermodynamic equilibrium (LTE) models applied on an extended set of seventeen hot white dwarfs observed by EUVE [24]. Besides the limitation in using LTE models for the selected white dwarfs, the large scattering that appears in the data makes the lack of gradient thus far reported for the He ionization degree (~40%) uncertain, particularly the extrapolation to short distances from the solar system. Here, it is important to note that in contrast to many prior attempts to derive the He abundance in different regions of the galaxy, the techniques thus far applied to the LIC make the assumption that the He mass fraction is cosmic (He/H =0.1), and they return only partial information (ionization degree, H I/He I ratio, etc.). For reference, sophisticated modeling appeared in the literature, including radiation transfer effect for the photoionization modeling inside the LIC, and an attempt to make the link between the inner heliosphere *in situ* measurement and the distant LISM observations [25]. However, all those studies assume a cosmic value for the LISM He mass fraction and a constant H I/He I ratio in the LISM as derived from EUVE observations, and thus those assumptions are questionable, as discussed above.

Following the same line of thinking that prevailed in the retrieval of most cosmic chemical elements, we propose in the following a new and simple approach, principally based on *in situ* measurement obtained or planned for the near future, to derive an accurate and consistent He mass fraction in the LIC. This should ultimately form a link between heliospheric physics and cosmology in a rather elegant way.

3.2 Helium abundance in LIC: back to the basics

Because the solar system is moving inside the LIC, any sensor at any distance from the sun does provide *in situ* measurements. Over the last few decades, with space missions probing few yet varying neighborhoods around the sun, plasma, fields, and radiations *in situ* measurements became available over time [20, 21, 26]. Rather than enumerate a detailed description of available data, we prefer to list the ingredients required to derive the helium abundance and assess their accuracy. Basically, the density of five species is required: neutral hydrogen (H I), neutral helium (He I), ionized hydrogen (H II), ionized helium (He II), and the double ionized helium (He III). In the event that one of the three ionized species is not available, one may include the electron number density Ne, using the property of plasma neutrality.

3.2.1 Hydrogen & helium neutral components in the LIC

The interaction between the solar wind and the LIC produces a specific configuration with the heliopause interface that deflects all impinging ionized species but allows the neutral components to penetrate while suffering few charge exchanges on their way. Charge exchanges produce pick-up ions, which ultimately produce a population of energetic neutral atoms. The important distinction here is that the interaction region between the solar wind and the LIC strongly modifies the incoming neutral population with the setup of many sub-population species (charge exchange history for neutrals and pick-up ions) that have specific kinetic properties, yet are fully related to the primary LIC population. To access the undisturbed H I population, a sophisticated description of the interaction between the solar wind and the LIC flow is required, including state-of-the-art MHD modeling for plasma and a kinetic description for the neutrals [27, 28].

One of the pioneer techniques for estimating the LIC neutral density is to use the sky background glow produced by the solar radiation (H I Ly-α, Ly-β, He Ly-α) backscattered by interplanetary neutrals [29, 30]. The main strength of this local remote sensing technique is that the observed emissions are directly related to the unknown neutral population. However, opacity effects in a moving medium induce multiple scatterings that require state-of-the-art radiation transfer 3D models in addition to the 3D kinetic description of the H I neutrals themselves [31, 32, 33]. The technique provides valuable insight about the inner heliosphere, but unfortunately, many problems still persist in adequately and consistently describing the outer heliosphere sky background observations made by the Voyagers spacecraft (e.g., poorly constrained LIC parameters that are key inputs in all models, data calibration, missing atoms' recoil during Ly-α scattering, etc.). As a consequence, the retrieved neutral H density remains poorly constrained (~ 0.1 - 0.25 cm^{-3}) despite an apparent consensus (Figure 4).

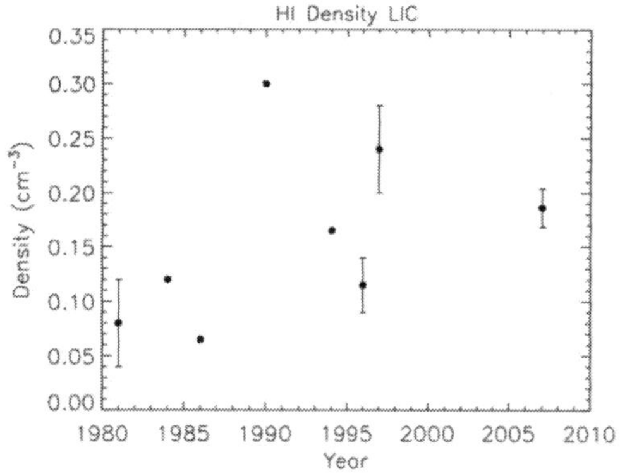

Figure 4. Neutral H number density estimated for the LIC by different techniques over time. Despite claims of its exact value, this quantity remains uncertain.

Nevertheless, the progress made by the community since the first discovery of the sky background glow is tremendous. It shows that 3D models should be improved to include most missing effects (see above), yet the data inversion should be implemented following sensitivity studies that scan all parameters of the problem (LIC & SW parameters, etc.). Such a technique requires expensive computer simulations [34], yet it should be but one piece of the global approach proposed in this study.

A second technique that may give access to neutral number densities is to use pick-up ions *in situ* measurements made by several spacecraft inside the heliosphere (ACE, Messenger, Ulysses,

Journal of Physics: Conference Series **577** (2015) 012003 doi:10.1088/1742-6596/577/1/012003

Voyagers, etc.) [20]. Because pick-ions have the primary neutral population as their intrinsic source and are intimately related to the filtration processes at work near the heliopause, they offer the opportunity to strongly constrain any model of interaction between the solar wind and the LIC plasma. Usually, obtained measurements have been used efficiently to derive the required neutral density level at the termination shock (TS) position [20]. Obtaining the TS density level is a good start, yet sophisticated 3D plasma-neutral models are required to derive the undisturbed LIC neutral density. Because the phase space of unknown parameters is large (LIC & SW parameters), sensitivity studies are also required, yet they must be within the framework of a global approach that includes both pick-up ions and sky background radiation observations.

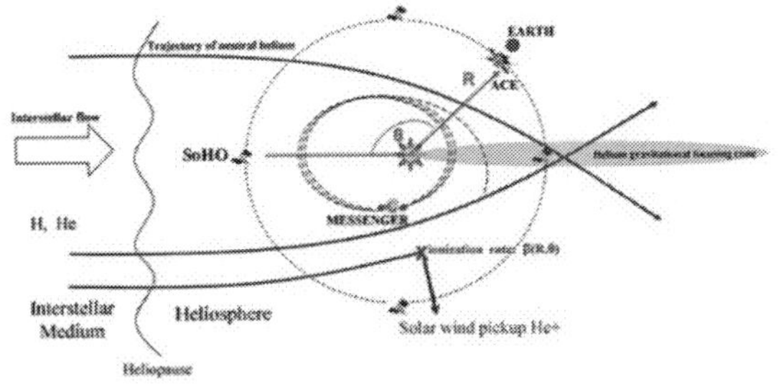

Figure 5. Sketch showing the He focusing cone and key space missions that provide *in situ* or remote observations of He atoms in the inner heliosphere. Reproduced from [35] where we added a reference to SoHO.

A third technique is fully related to the He intrinsic distribution, particularly the formation of the focusing cone on the downstream of the flow (e.g., Figure 5). Because it has a specific spatial distribution, the He cone was used in the past specifically to derive the LIC properties from different data sets (see [21] for a review). In that frame, during the two first years of the SoHO mission, we observed the sun at the He resonance line (58.4nm) with the SUMER instrument along the orbit of SoHO inside and outside the He focusing cone (Figure 5). The aim was to detect the absorption feature that should appear in the solar He 58.4nm line profile due to the opacity of He atoms that fill the space between the sun and the spacecraft. In contrast to classical backscattering emission observations, this technique is free of solar flux and instrument calibration uncertainties [33]. Within the planned program, we were twice able to sample the angular region of the He cone, which allows checking and validating any spectral feature detection. In addition to the yearly repetition of the program, we also obtained time series of different quiet regions on the solar disk or off-disk in order to monitor any potential time variation of the source.

The SUMER instrument description and its performances before and in-flight are well described in [36]. We use the 1x300" slit pointing at either the quiet sun or off-disk from June 1996 to December 1997 in order to obtain time series of the He 58.4 line profile in the second order of the SUMER detectors. Because absorption by He atoms is almost negligible (less than 1 %) for the region outside the focusing cone, line profiles obtained around June of each year (when the SoHO orbital position is near the apex direction) are used to obtain a good reference for the undisturbed solar line profile. The selected reference line profile is obtained by iteration over available time series spectra in which instant line profiles do not depart more than a fraction of percent from the average line profile.

For the line profile inside the focusing cone, we apply the same statistical selection as for the undisturbed solar line profile but for time series obtained during the December period (SoHO in downwind direction). The comparison between the line profiles inside to outside the focusing cone clearly shows a blue-shifted feature that corresponds to the expected absorption by the intervening He atoms in the cone zone (Figure 6). We verified and obtained similar results using the two available data sets recorded successively in 1996 and 1997.

Journal of Physics: Conference Series **577** (2015) 012003 doi:10.1088/1742-6596/577/1/012003

To obtain the kinetic properties of the He atoms, we consider the so-called hot model, which includes solar gravity, radiation pressure, and photon and electron ionization [37]. To analyze the observed spectra, we use the solar line profile obtained outside the focusing cone as our reference and apply the He absorption provided by the hot model before comparing to the line profile observed inside the cone. Our first quick analysis shows a very extended absorption that cannot be fit with classical temperature levels, casting doubt on the assumptions made in the classical hot model.

It is important to stress that this difficulty of fitting observations inside the cone is not new as similar problems appeared for the interpretation of pick-up He ions distributions obtained by both ACE and MESSENGER spacecraft [35]. Despite the large number of unknowns compared to available data, we could obtain the following LIC parameters N(He I) ~0.015 cm^{-3} and V(He I) ~ 26.2±2 km s^{-1}, yet the temperature T(He I) remains poorly constrained. It is important to stress that the indicated values are uncertain and are still under investigation by incorporating ACE, MESSENGER, and Ulysses pick-up ion *in situ* measurements [35]. The interpretation of this kind of data is complex, particularly for regions close to the sun [21], yet we believe the new SoHO/SUMER observations, together with He pick-up ions measurements at different distances from the sun, will provide new and strong constraints on both the He LIC parameters and the processes that define the He atoms distribution.

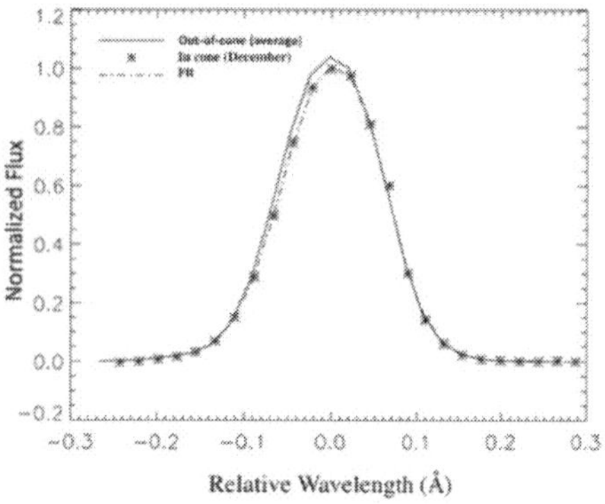

Figure 6. Solar He 58.4nm emission line observed by SoHO/SUMER in and out of the He focusing cone. The best fit to the in-cone line profile is shown (dashed-dotted line).

3.2.2 *Hydrogen, helium, and electron plasma component in the LIC*

The heliosphere acts as a barrier that deflects most LISM electrons and ionized species. This means that we have no direct access to those species from inside the heliosphere. Fortunately, with Voyagers 1 and 2 ranging outside the heliopause, we have a real and unique opportunity not only to measure charged components that could be accessed directly from plasma detectors, but also to obtain combined radio waves and magnetometer measurements [26]. For the recent and exceptionally low-activity solar minimum, the heliopause position was close enough, which allowed the Voyager 1 radio instrument to obtain the first indications of the LIC plasma density level (Ne ~ 0.05-0.08 cm^{-3}) [26]. Because Voyagers 1 and 2 operations will continue up to year ~2025, we have a rare opportunity to directly measure those key plasma parameters in the near future [38].

3.2.3 *How to derive the helium abundance: implementing the global approach*

In situ measurements outside the heliopause are fundamentally important for the retrieval of the helium abundance. When the exciting moment arrives in which one will be able to see Voyagers 1 and 2 probes reach the LISM, a proper description from the inner boundary at the solar corona up to the outer boundary at the heliopause should be obtained using high-resolution MHD-kinetic models. This should be achieved following sensitivity studies that, while expensive, are efficient in capturing the best space parameters that are consistent with available observations (plasma and pick-up ions *in situ* measurements, ENAs observations, and Lyman-α data). Because the unknown parameters space is large, one should favor multiple-observations data analysis rather than focusing on a single dataset.

This global approach showed its strength recently when Voyagers 1 and 2 plasma measurements associated with the IBEX ribbon and a full sensitivity study were used accurately to constrain the LISM magnetic field [34].

4 Conclusions

We reviewed the helium abundance at different sites of the universe starting with the Big Bang era, considered the protosolar nebula and giant planet composition, and examined the current local interstellar medium. From the studies thus far conducted on different astrophysical sites [1, 2, 3], the first lesson learned is that an accurate determination of the He abundance strongly constrains the key processes that control the formation of the astrophysical objects from the primordial universe down to the current sun and giant planets. The second important lesson is that the accuracy of the He abundance is strongly related to the accuracy of the description of the physics of the object under study.

Here, we considered two targets—the giant planets of the solar system and the local interstellar cloud— to which we attempted to apply the initial steps of a global approach, following the path of prior studies conducted on extragalactic H II regions and the sun. For the giant planets, we proposed a comparison of sophisticated radiation transfer models with the He 58.4nm airglow as observed during the encounters of the Voyagers spacecraft with the planets. In order to fit the planetary He 58.4nm airglow, our finding is that the He mass fraction should be $Y_{Jup} \sim 0.237 \pm 0.038$ and $Y_{Sat} \sim 0.226 \pm 0.030$, respectively, for Jupiter and Saturn. Our results are consistent with the Galileo probe measurements for Jupiter, yet they stand in contrast with past findings for Saturn. For the higher He abundance obtained for Saturn, the stratospheric temperature should be raised to ~149 °K in order to fit the background gas density thus far observed. With the He abundance reported here, credible models become possible for the interior of Jupiter, Saturn, and extrasolar planets [39, 40].

For the local interstellar cloud, we found that the many observations and *in situ* measurements obtained by different techniques lacked any strategy for linking the different pieces of the puzzle in order to obtain the He mass fraction. Here, we proposed a new strategy based on a global approach. Starting from the inner boundary of the heliosphere at the position of the solar corona, we showed how different data sets should be associated together to constrain the radial distribution of the He ionization rates, which should provide a very accurate number density of the He neutral component. Also, *in situ* measurements by Voyagers 1 and 2 plasma and/or radio emissions detectors in the near future, combined with magnetometer measurements, should provide the pristine density level of key ionized species in the LISM. When combined with sophisticated MHD-kinetic description of the heliosphere, those measurements should allow us to derive the density level of H I. The proposed approach should ultimately provide the He abundance in the very interstellar cloud in which we reside. This would build a bridge between heliosphere physics and cosmology in a rather elegant way by showing that all past efforts so patiently made by the heliospheric community were not done merely for the narrow view of a specialized group, but rather as a piece in the big picture of the existence of the universe and its evolution.

Acknowledgements

Authors acknowledge support from CNES, CNRS and UPMC. They thank the referee for very useful comments that helped clarify the manuscript.

References

[1] Aver, E., K. A. Olive, R. L. Porter, and E. D. Skillman 2013 *J Cosm and Astroparticles* **11** 017; Steigman, G. 2012 *Advances in High Energy Physics* **2012** ID 268321
[2] Izotov, Y. J., G. Stasinska, and G. Guseva 2013 *Astron & Astrophys* **558** A57
[3] Basu, S. and H. M. Antia 2008 *Physics Reports* **457** 217

[4] Von Zahn, U. and D. M. Hunten 1995 *Science* **272** 849
[5] Gautier, D., B. Conrath, M. Flasar, R. Hanel, and V. Kunde 1981 *J Geoph Res* **86** 8713
[6] Conrath, B. J., D. Gautier, R. A. Hanel, and J. S. Hornstein 1983 *Astrophys J* **282** 807
[7] Conrath, B. J. and D. Gautier 2000 *Icarus* **144** 124
[8] Smith, G. R., D. E. Shemansky, J. B. Holberg, and B. R. Sandel 1983 *J Geoph Res* **88** 8867
[9] Yelle, R. V., A. Y. Leslie, R. J. Vervack, and B. R. Sandel 1996 *J Geoph Res* **101** 2149
[10] Hubbard, W. B., C. C. Porco, D. M. Hunten, and J. B. Holberg 1997 *Icarus* **130** 404
[11] Niemann, H. B., S. K. Atreya, G. R. Carignan, T. M. Donahue, and S. H. Way 1996 *Science* **272** 846
[12] Parkinson, C. D., A. I. Stewart, A. S. Wong, Y. L. Yung, and J. M. Ajello 2006 *J Geoph Res* **111** E02002
[13] Ben-Jaffel, L, Y. J. Kim, and J. T. Clarke 2007 *Icarus* **190** 504
[14] Guillot, T. 1999 *Planet & Space Sci* **47** 1183
[15] Asplund, M., N. Grevesse, A. J. Sauval, and P. Scott 2009 *Ann Rev Astron Astrophys* **47** 481; Lodders K 2013 *Astrophys J* **591** 1220
[16] Esteban, C,. M. Peimbert, J. Garcia-Royas, and M. Rodriguez 2004 *Mon Not R Astron Soc* **355** 229
[17] Nieva, M. F. and N. Przybilla 2012 *Astron & Astrophys* **539** 143
[18] Redfield, S. and J. I. Linsky 2008 *Astrophys J* **673** 283
[19] Gry, C. and E. B. Jenkins 2014 *Astron & Astrophys* in press. *ArXiv*: 1404.0326
[20] Gloeckler, G. and J. Geiss 2004 *Adv in Space Res* **34** 53
[21] Möbius, E. et al. 2004 *Astron & Astrophys* **426** 897
[22] Dupuis, J, S. Vennes, and S. Browyer 1995 *Astrophys J* 455 574
[23] Holberg, J. B., F. C. Bruhweiller, M. A. Barstow, and P. D. Dobbie *Astrophys J* **517** 841
[24] Wolff, B., K. Koester, and R. Lallement 1999 *Astron & Astrophys* **346** 969
[25] Slavin, J. D. and P. C. Frisch 2002 *Astrophys J* **565** 364
[26] Gurnett, D. A., W. S. Kurt, L. F. Burlaga, and N. F. Ness 2013 *Science* **341** 1489
[27] Izmodenov, V. V., Y. G. Malama, and M. S. Ruderman 2009 *Space Sci Rev* **146** 329
[28] Pogorelov, N. V., J. Heerkhuisen and G. P. Zank 2008 *Astrophys J* **675** L41
[29] Bertaux, J. L. and J. E. Blamont 1971 *Astron & Astrophys* **11** 200
[30] Thomas, G. E. and R. F. Krassa 1971 *Astron & Astrophys* **11** 218
[31] Fayock, B., G. P. Zank, and J. Heerkhuisen 2013 *Astrophys J* **775** L4
[32] Katushkina, O. A. and V. V. Izmodenov 2011 *Adv in Space Res* **48** 1967
[33] Puyoo, O., L. Ben-Jaffel, and C. Emerich 1997 *Astrophys J* **480** 262
[34] Ben-Jaffel, L., M. Strumik, R. Ratkiewicz, and J. Grygorczuk 2013 *Astrophys J* **779** 130
[35] Gershman, D.J., G. Gloeckler, J. A. Gilbert, J. M. Raines, L. A. Fisk, S. C. Solomon, E. C. Stone, and T. H. Zurbuchen 2013 *J Geophys Res* **118** 1389
[36] Wilhelm, K. et al. 1995 *Sol Phys* **162** 189; Lemaire, P. et al. 1997 *Sol Phys* **170** 105
[37] Bzowski, M., H. J. Fahr, and D. Rucinski 1997 *Astron & Astrophys* **326** 396
[38] Stone, E. C., A. C. Cumming, F. B. McDonnald, B. C. Heikkila, and W. R. Webber 2008 *Nature* **454** 71
[39] Leconte, J. and G. Chabrier 2012 *Astron & Astrophys* **540** 20L
[40] Guillot, T. and D. Gautier 2014 *ArXiv*:1405.3752 to appear in *Treatise on Geophysics*

13th Annual International Astrophysics Conference: Voyager, IBEX, and the Interstellar Medium IOP Publishing
Journal of Physics: Conference Series **577** (2015) 012004 doi:10.1088/1742-6596/577/1/012004

On the Stability of the Interstellar Wind through the Solar System

Jean-Loup Bertaux (1) and Rosine Lallement (2)

1. LATMOS/IPSL, Université de Versailles Saint Quentin, INSU/CNRS, 11 Bd D' Alembert, 78200 Guyancourt, France 2. GEPI Observatoire de Paris, CNRS, Université Paris Diderot, Place Jules Janssen 92190 Meudon, France

jean-loup.bertaux@latmos.ipsl.fr

Abstract. As a follow-up of a recent study [1], we challenge the claim that the flow of interstellar helium through the solar system has changed substantially over the last decades. We argue that only the IBEX-Lo 2009/2010 measurements are discrepant with older consensus values. Then we show that the probability of the claimed variations of longitude and velocity are highly unlikely (≈1%), in view of the absence of change in latitude and absence of change in the (V, λ_∞) relation, while random values would be expected. Finally, we report other independent studies showing the stability of Helium flow and the Hydrogen flow over the years 1996-2012, consistent with the 70's earlier determinations of the interstellar flow.

1. Introduction

Since the discovery of the interstellar wind (the flow of neutral H and He atoms through the solar system) with OGO-5 spacecraft in 1970 [2,3], a number of other space investigations have been used to determine the characteristics of this flow (direction, velocity, temperature). It was recently suggested [4] that the direction of interstellar helium flow has changed by ≈ 5.5° within the last ≈ 40 years. At a velocity of 25 km/s, this corresponds to 210 AU, much smaller than the mean free path between collisions for an interstellar He atom (≈1,000 AU). From the point of view of the physics of rarefied gas dynamics, this is questionable, since several collisions are needed to reach a Maxwell-Boltzmann distribution which can be defined by macroscopic parameters as the temperature and the bulk velocity. Rather than discussing this aspect, several comments are now developed concerning a number of He wind direction measurements assembled in [4] to claim a secular change of the He flow direction. Within the frame of an ISSI workshop in 2003-2004 discussing UV measurements of resonance fluorescence at 58.4 nm of Helium atoms, *in-situ* He atoms and pick-up He^+ ion measurements, a set of Helium flow parameters was obtained that we call here "consensus value" [5]: velocity at infinity V=26. 24 ± 0.45 km/s, ecliptic longitude of downwind flow (J2000) λ_∞=74.68 ± 0.56°, latitude β_∞= -5.31 ± 0.28°, temperature T= 6306 ± 390 K. Figure 1 is extracted from [4], where we have added an index corresponding to each of the following comments.

Journal of Physics: Conference Series **577** (2015) 012004 doi:10.1088/1742-6596/577/1/012004

1. The pick-up Ions (PUI) measured by PLASTIC instrument on board STEREO mission of He^+, O^+, Ne^+ are forming a crescent in the upstream direction [6] with a very large angular width (160°) inadequate for a precise determination of the longitude, and also there are transport effects which may deviate the PUI away from the trajectory of the interstellar neutrals.

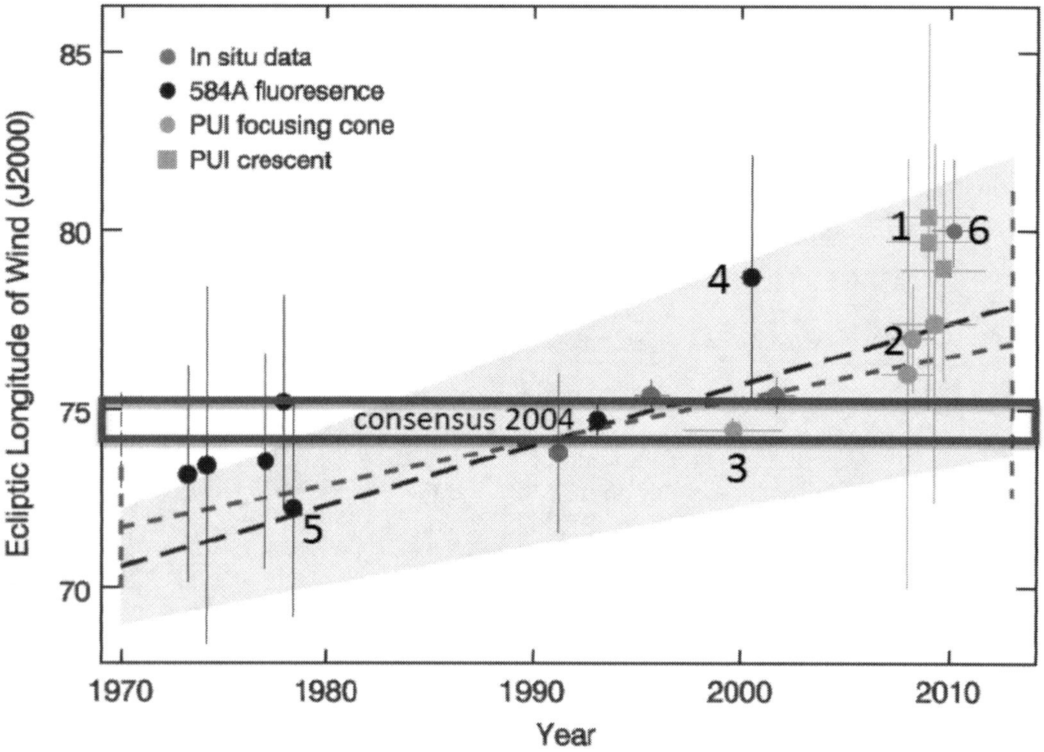

Figure 1. Adapted from [4]. This is the ecliptic longitude of the Helium downwind flow as a function of time as determined by a number of space experiments and techniques. For details, see [1,4]. Numbers from 1 to 6 are attached for reference in the discussion. The blue rectangle materializes the longitude and its error bar as discussed in [5], a consensus value combining EUV observations at 58.4 nm of He resonance, and GAS/Ulysses *in situ* measurements [15] up to 2002.

2. PLASTIC/STEREO PUI He^+ measured in the focusing cone [6]. While the He^+ *in situ* measurements [6] display a maximum at a longitude of 75.0 ± 0.3° over several years, the authors [6] use a complicated method combining by pair the Earth's orbits, with the surprising result that the He cone would be at a longitude of 77.4 ± 1.9° as plotted on figure 1. This was recently criticized in [1], from which is based the analysis of the present presentation. However, assuming that the analysis of [6] is correct, it has been demonstrated by Chalov and Fahr [7] and more recently Chalov [8] through detailed modeling that the He^+ maximum measured in the cone may be deflected from the He cone to larger longitudes by solar wind transport effects.

3. SWICS/ACE PUI He^+ measured in the focusing cone [9] in the period 1998-2002 are giving a downwind direction of λ_∞= 75.13 ± 0.33°. It is shown in [8] through modeling that "the angular displacement of the cone in ecliptic longitude depends on the ion mean free path. When the mean free path is small (high level of turbulence), the ion velocity distribution function is close to isotropy and the displacement is small or absent at all. " This was the case for 1998-2002 conditions of

SWICS/ACE, but not for PLASTIC/STEREO. All in all, the longitude of the maxima of He^+ should be considered as upper limits only.

4. The error bar given for the Nozomi observations [10] of He I 58.4 nm in 2001 is wrongly displayed as being ± 3.4°. We quote the paper of Nagakawa et al. [10]: « Observations suggest that the helium flow arrives from the ecliptic coordinates of (258.7 °± 3.4°), which within error bars is consistent with the primary helium flow direction derived from direct observations (e.g. Witte et al. 2004) ». There is an unfortunate typographic error in this sentence. Nagakawa et al. [10] meant: "...arrives from the ecliptic coordinates of (258.7°, + 3.4°) ». Otherwise, it would be written «...arrives from the ecliptic ***longitude*** of (258.7 °± 3.4°) ». Actually, there are no error bar indicated for the He flow longitude in the Nagakawa et al. paper, and the error is likely much larger than erroneously quoted by [4], probably 6° as discussed in [1].

5. This point at 72.2° is taken from the analysis of Prognoz 6 58.4 nm photometric measurements (Dalaudier et al., 1984, [11]). However, the authors [11] quote the value of 72.2° as "an eye fit preliminary determination" based on a fraction of the data, and it is clear from the whole article that it can by far not compete with the second result, based on the whole dataset and a careful data-model adjustment. The abstract of [11] contains only the value of 74.5 ± 3°, correctly corrected into 75.2 ± 3° and plotted on figure 1 by [4] to account for the change between J1950 and J2000 (+0.7° increase of ecliptic longitudes). We conclude that the use of the first value 72.2° to determine a linear time trend is not justified.

6. The only solid point away from the consensus value is the determination of the longitude in 2009-2010 from *in situ* measurements of neutral He atom from the IBEX-Lo instrument, analyzed very seriously and thoroughly in Möbius et al [12] and Bzowski et al. [13]. From essentially the same data sets, a longitude of 79.0° +3.0°, (− 3.5°) is found in [12], while a longitude of 79.2° is found in [9], but with no error bar quoted. In McComas et al [14], a combination of IBEX values is claimed to give (their Table 1) λ_∞=79.00° and 1σ uncertainty of 0.47°, with no justification of this lower error bar. Note that this new value of 79.0° is called "consensus value" in [14] inappropriately, since it results from two IBEX studies of the same data set.

Strangely enough, the IBEX red point 6 on figure 1 is at a longitude λ_∞=80.0° instead of the original IBEX value of 79.0°. In order to understand this suspicious upward shift of 1°, one has to read carefully a footnote in Table S1 of the Supplementary On line Material of [4]. We quote:

"This He flow direction [IBEX, 79.0°+3.0, -3.5] is further constrained to be 80.0°+2.0, -1.0 degrees by restricting the longitude range using the independent LIC temperature obtained from Sirius data (SM-S4) combined with the longitude parameter range shown in Fig. 1 of [our reference [14]]". The argumentation in [4, Supplement Online Material] is the following: the width of interstellar absorption lines to star Sirius (at a distance of 2.7 parsec) of Fe^+, Mg^+, and Ca^+ ions indicate a temperature of 5800 +300 -700K. When these values are fed into the IBEX relationship (longitude, T) (Fig. 1 of [14]), it converts into a longitude and its error bar which then is combined with the IBEX longitude of 79.0° +3.0, -3.5° into a new value of 80.0°+2.0, -1.0°. We disagree totally with this approach that is not consistent because the interstellar temperature parameter to Sirius is averaged over 2.67x2.06 10^5 =5.5 10^5 Astronomical Units, while the whole paper [4] seeks to demonstrate changes at a scale of 200 AU! This average value should not be combined with local values. At the same time the IBEX point is moved upward by 1°, the original error bar (+3.0°, -3.5°) is also strongly reduced without any justification. Being at the extreme boundary of the examined time and longitude intervals, this manipulation of the original data increases the weight of this point by a factor of 4, which has certainly consequences in the linear least square fit of the series of data points and its confidence interval.

In addition, there is an issue to use Sirius as a proxy of ISM local temperature, since it has multiple components and no usable D I line. There are much better determinations of the temperature of the LIC that suggest significantly higher T~7500 K [15] and according to B. Wood (private communication) the best single line of sight to represent the LIC at the solar system would probably be Eps Eri (d=3.3 pc, T=7800 ± 800 K, in [16]). If these higher temperatures were considered, it would push down the longitude instead of increasing it.

Journal of Physics: Conference Series **577** (2015) 012004 doi:10.1088/1742-6596/577/1/012004

2. Likelihood of the claimed He changes

The IBEX-Lo longitude value is significantly different (by 4.3°) from the old consensus value. It may be either that there is a historical change, as claimed in [4], or one (or two) of the longitude determination(s) (consensus or IBEX) is (are) wrong. We show that the former situation is very unlikely. Indeed, the ecliptic latitude of the flow measured by IBEX is basically unchanged (at -4.9 ± 0.2°, point A on figure 2a), from the previous consensus value (β_∞= -5.31 ± 0.28), while the longitude has changed (point B). The change in latitude is 0.41 ± 0.34°.

On figure 2a is illustrated the situation in the sky plane. With a motion of 4.3° in a random direction from the old consensus value (marked A) to point B, it could have moved anywhere along a circle of 4.3° radius centered on point A. The probability that the latitude of a random change is within 0.4° from 0 in the motion from A to B is 0.8°/4.3°*π= 6 %, if one considers the fact that it could also fall on the other side of A at the same latitude.

On the other hand, the IBEX show a change of the velocity of He flow at infinity, previously at V=26. 24 ± 0.45 km/s and now at 23 km/s. On figure 2b are plotted the positions of point A and B in a graph (V, λ_∞). If the wind is changing, changes in V and λ_∞ should be de-correlated: the new velocity V(B) could have any new value within ± 5 km/s from V(A) (this value of a velocity change ± 5 km/s is rather arbitrary, but not very different from the claimed V(A)-V(B)). However, the new couple V(B), λ_∞(B) falls exactly on a well known curve V(λ_∞), which is the locus of all couples (V, λ_∞) giving hyperbolic trajectories tangential to the same Earth longitude position λ^E_{max}= 130° (figure 3). Therefore, this is where an instrument like IBEX-Lo, in the configuration of rotating about the sun direction, looking at 90° from sun direction, would measure a maximum of the He flux. The old couple V(A), λ_∞(A) is also exactly on the same locus. The probability of claimed observed change of λ_∞ while staying on the same locus (V, λ_∞) is the ratio of the thickness ΔV(locus)≈1 km/s of the locus to the (somewhat arbitrary) extent of the possible change of V, ≈10 km/s. The ratio and then the probability of the observed change is therefore ΔV(locus)/ 10 km/s = 0.1.

Therefore, the combined probability of observing the particular values of changes as claimed by IBEX is of the order of 0.6%, which is very unlikely. What is more likely is that one of the Helium parameters measurements (IBEX or old consensus) is wrong.

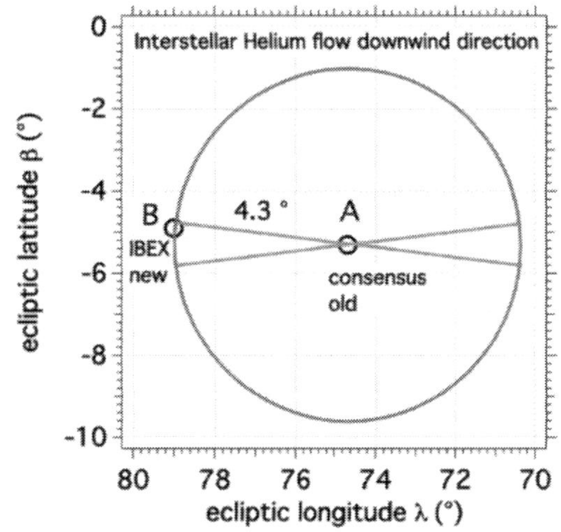

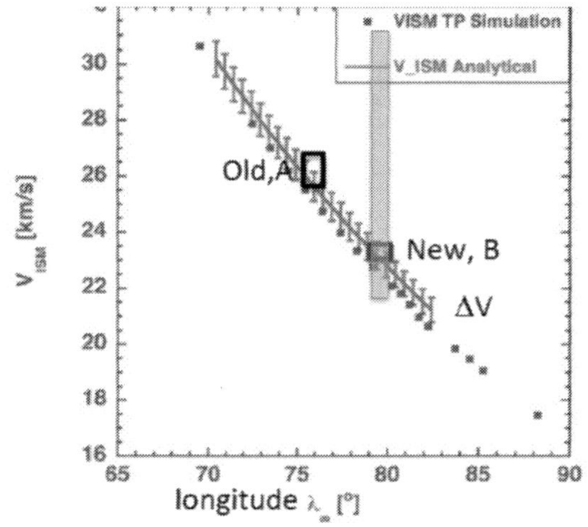

Figure 2a (left): The positions in the sky plane of He downwind direction flow, as determined earlier (A, old consensus) and IBEX, point B. The probability for a random change to stay within the portion of circle limited by the blue lines (essentially no latitude change) when there is a change of longitude of 4.3° is ≈6%. Figure 2b (right): adapted from figure 23 of [13]. There is a relationship between the downwind longitude λ_∞ and the velocity V imposed by the Earth longitude at which is measured the maximum He flux. Blue points are the analytical approach of [12]; red bars are the model results of [13]. A marks the old consensus values and B the new IBEX value. If there were a random change of the interstellar wind parameters, there would be no reason why the B point would be on the same locus as point A. It could be anywhere along the grey rectangle.

3. Non-Linearity / Dead Time correction effect

It was noted by Lallement and Bertaux [1] that the two coincidences mentioned above have a common particularity. When IBEX (in terrestrial orbit) is not far from longitude position λ^E_{max}= 130°, the latitude of incoming flow is easily determined from the latitude where is measured the maximum He count rate, along one scan of the rotation around the sun pointing axis. And the particular locus (V, λ_∞) is determined by the Earth longitude λ^E_{max} at which the He count rate is maximum, looking at 90° from sun direction. Therefore, IBEX-Lo finds angular positions of maximum flux where they should be if the old values are valid. However, what imposes new values for the longitude and velocity is the way the signal decreases when IBEX is further away from λ^E_{max}= 130°. The counted maxima are decreasing more slowly than would be predicted by a model with the old values. An instrumental effect like a non-linearity could mimic such a behavior, if higher fluxes are producing relatively smaller counting rates, as suggested by [1]. A problem of dead time in the counting chain of IBEX-Lo do present this feature, and this idea was explored more thoroughly from modeling [1]. Actually, this effect is known to exist [12], and an estimate (not very accurate) is given "…up to 5 ms". In addition to He atoms impact on the detector, there are some electrons which participate as an additional background to this dead time. Assuming a background of 22 electrons/s as discussed in [12], a simulation was performed for various putative values of the dead time. The IBEX counting rates were taken from [13] (blue circles on figure 3, the maximum count rate was 25.3 counts s^{-1}, then normalized to the IBEX model, red line) and corrected for the dead time. It was found that if the dead time is of 7 ms, the maximum true counting rate (corrected from dead time) would turned to be 43 count s^{-1}.

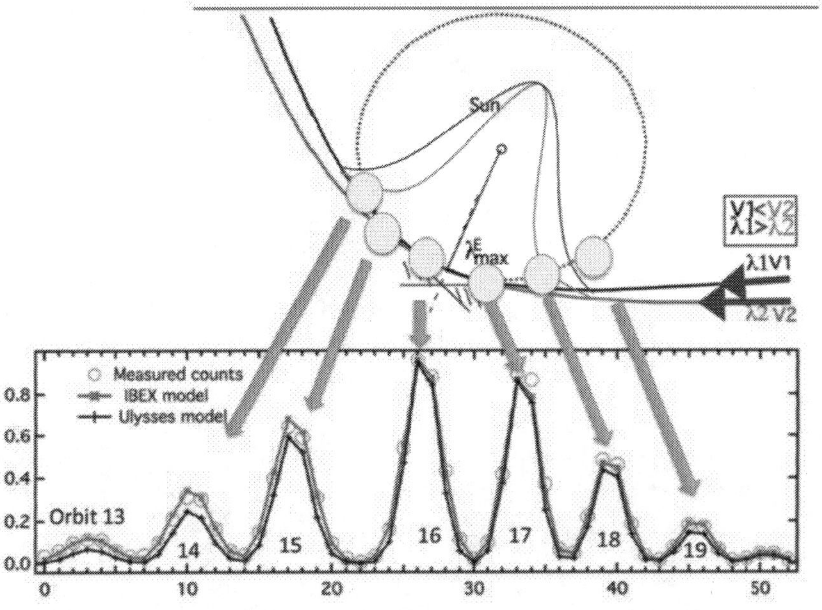

Figure 3 Top: Green circles indicate the positions on Earth's orbit where were collected the IBEX measurements, with the corresponding peak along an IBEX scan at 90° from sun plotted on bottom panel (blue circles, normalized to the best IBEX model of [13]). The data of successive orbits are concatenated for clarity. On top is illustrated that a family of hyperbolic trajectories are tangent to the Earth's orbit at the same point, giving the relationship (V, λ_{∞}) of figure 2b. Bottom: the black curve (taken from [13]) is the Ulysses model [17] (basically the old consensus model). All data and model are normalized to 1 at the maximum. This Ulysses model is distinct from the IBEX model in that successive peaks are decreasing faster with distance from the maximum (*adapted from* [1]).

On figure 4, the Ulysses model (also labelled Witte et al. model [17]), which relative values were taken from [13], is normalized to this maximum counting rate. It is clear that now the corrected counting rates follows much better the overall behaviour of the model computed by [13] from Witte et al parameters. Indeed, when varying the value of the dead time, 7 ms was the value giving the best fit to this model; at bottom of Figure 4 is plotted the difference model–data (residual), showing fluctuations, but also a trend that suggests that a slightly lower longitude could significantly improve the data-model agreement.

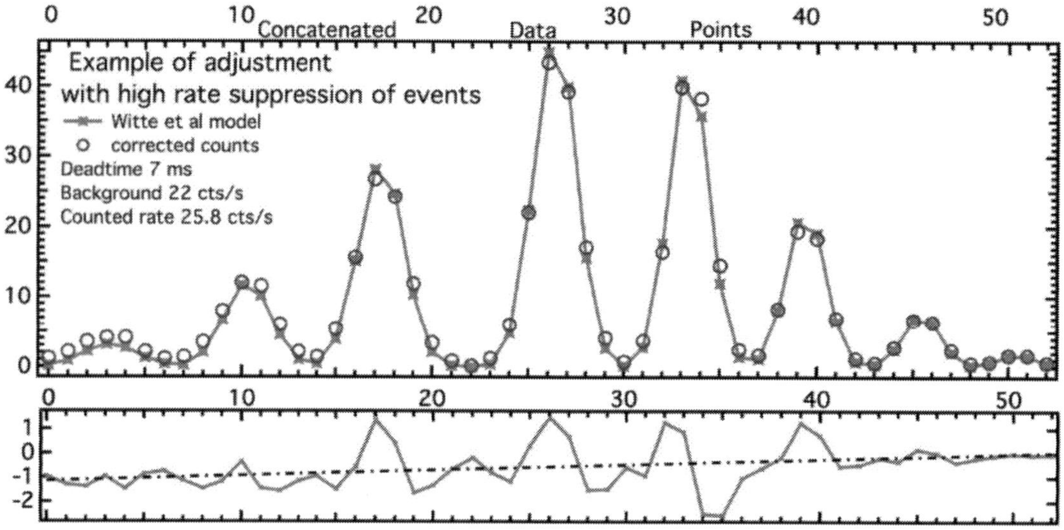

Figure 4 Blue circles: the IBEX counting rates were corrected assuming a dead-time of 7 ms, a maximum count rate of 25.8 counts/s due to He atoms and an electron background of 22 counts/s. Units are counts/s. The red curve labeled Witte et al is the same as the black curve of figure 3, labeled Ulysses with parameters from [17]. The sum of squared residuals after normalization is only slightly higher than for the [13] parameters. *Bottom*: a small asymmetry is found in all residuals, that we attribute to the fact that the Witte et al longitude is slightly too high compared to the actual flow longitude (*adapted from* [1]).

4. The spade of Damocles

There are a number of instruments and observations that have provided, or will provide, indications about the changes of the interstellar wind direction, or its stability. Clearly, if the Helium flow has changed, the Hydrogen flow should also have changed.

4.1. High spectral Resolution of interplanetary Hydrogen emission with HST and SWAN.

High spectral resolution of the interplanetary Lyman-α emission give also an information on the velocity of the H interstellar flow through the solar system. Some data coming from HST and from

SWAN instrument on board SOHO (using the absorption hydrogen cell [18,19,20,21,22]) were recently [23] compared to model predictions with two different sets of parameters, labelled "Ulysses" and "IBEX" (figure 5). Along their travel through the solar system, H atoms are suffering ionization (which kills more slow atoms than fast atoms) and Lyman-α radiation pressure. The observed velocity (measured by a Doppler shift) of the H Lα line is different from the H velocity at infinity. Both effects are modulated by the solar cycle, since ionization and solar Lα both depend on the solar cycle. These effects are small at solar cycle minima (the observed velocity is near the velocity at infinity). It is clear on figure 5 that both SWAN data and HST data are more consistent with the Ulysses parameters than with the IBEX parameters. In particular, the velocity at infinity is greater than the value of 23 km/s invoked by [14] to claim the absence of bow shock outside the heliosphere. This putative bow shock is back again.

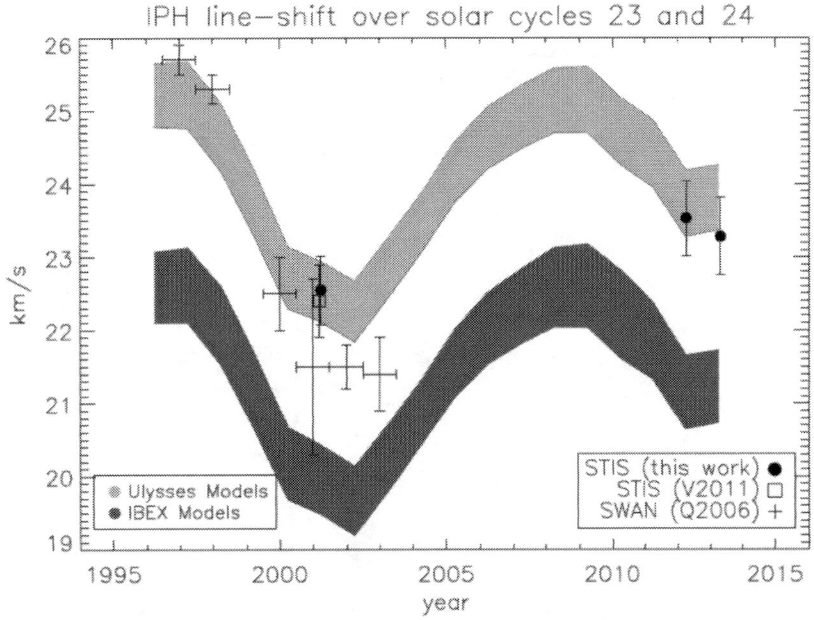

Figure 5. Taken from [21]. The observed Doppler shift of the interplanetary H Lα line was measured by HST and SWAN/SOHO several times. They are compared to a model which takes into account the effect of the solar cycle (radiation pressure, ionization) with the two sets of parameters, IBEX or Ulysses. Clearly the Ulysses parameters are favored.

4.2. SWAN photometric Lα data in the ecliptic.

In the solar system, the S plane, perpendicular to the ecliptic (Fig.6), containing the sun and the interstellar wind vector is a symmetry plane for H and He distributions. Therefore, particular sets of photometric data may be used to determine accurately the position of this plane. For instance, looking at right angle from the sun, in the ecliptic, but forward (F) or backward (B) (w.r.t. the Earth's motion), the intensities should be equal when the Earth is in upwind position (U) or in the downwind position (D). One complication is that the illuminating sun in Lα is not spherically symmetric because of spots, which introduces some noise. It remains to be seen if the direction is stable, and if this photometric value is consistent with the H upwind determination from SWAN Hydrogen cell absorption measurements, sensitive to the velocity of H atoms, analyzed by Lallement et al. [20].

4.3. SWAN photometric Lα data to the North pole.

Another set of SWAN intensities have been analyzed, all in the direction of the North ecliptic pole [24]. They are displayed on figure 7. Obviously, the intensity is maximum at position (U), and minimum at position (D). There is a strong modulation associated to the solar cycle, with less intensity at solar max, because of the stronger Lα radiation pressure and increased ionization which increase the size of the ionization cavity carved into the H flow. Figure 7 shows that intensity variations cannot be simulated by a stationary model H distribution. A much better fit is obtained with a complete 3D, time-dependent model [25]. This data set is yet to be analyzed with the prospect of determining the position of U and D. One advantage of this viewing geometry is that the intensities to the North ecliptic pole (or the South) are less sensitive to variations of the solar illumination, seeing (more or less) a complete North (or South) solar hemisphere).

4.4 Messenger UV spectrometer

Interplanetary Lα photometric data collected by the UV spectrometer instrument on board Messenger in the ecliptic plane are presently analyzed (Quémerais, private communication) to search for the incoming longitude of the hydrogen flow.

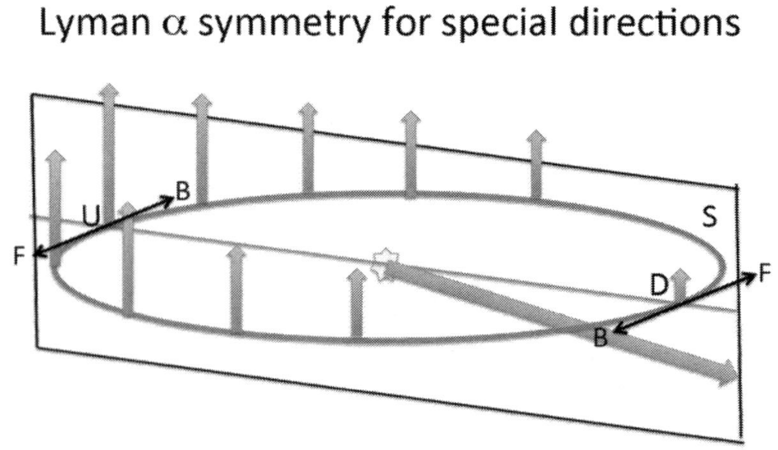

Figure 6. The S plane, perpendicular to the ecliptic containing the sun and the interstellar wind vector is a symmetry plane for H and He distributions. Looking at right angle from the sun, in the ecliptic, but forward (F) or backward (B) (w.r.t. the Earth's motion), the Lα intensities should be equal when the Earth is in upwind position (U) or in the downwind position (D). Also, when looking always to the North (or South) pole, the Lα intensity (mimicked by the length of blue arrows) should be maximum at U and minimum at D, because of the progressive destruction of H atoms by solar ionization along the flow.

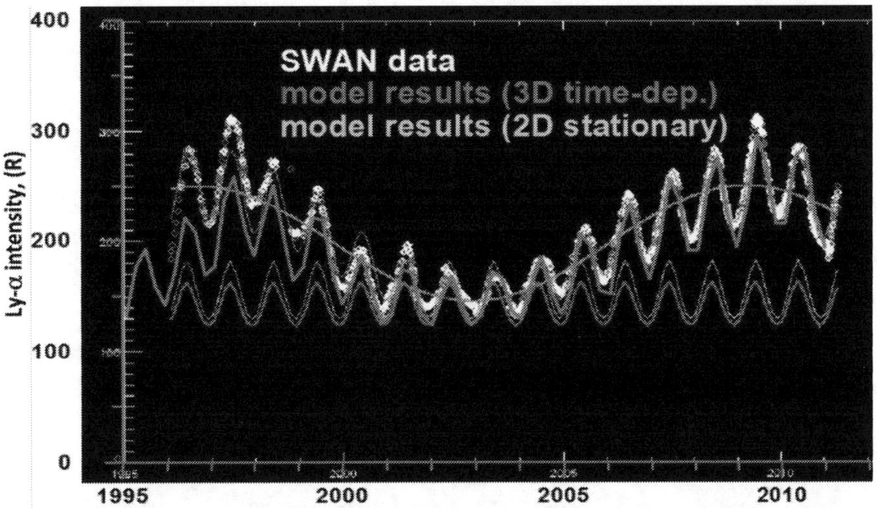

Figure 7. Evolution of the Lα intensity in the direction of the North Ecliptic Pole from SWAN/SOHO over 17 years (white diamonds). There is a solar cycle influence which is not accounted for in the 2D stationary models (green), while it is well reproduced by a time dependent/3D model (from O. Katushkina, [24]).

4.4 Gas/Ulysses Helium flow reanalysis.

In view of the IBEX Helium interpretations [12,13,14], three different teams have re-analyzed the Gas/Ulysses *in-situ* data, including now the 2007 data set which was not existing in the earlier analysis [15]. Wood [24] (this conference) concluded that his re-analysis was giving results similar to the original analysis. Katushkina et al. [27] did a similar work. Finally, Bzowski et al. [28] made also a re-analysis, using the same model as was used in [13] and the same fitting techniques for the analysis of IBEX helium data. They analyzed three epochs 1996, 2002, 2007 (only the fast North South orbital scan of Ulysses provide useful measurements). They find as best fit parameters for the upwind direction $\lambda = 255.3°$, $\beta = 6°$ (J2000), V = 26.0 km/s, T = 7500 K. The direction and velocity are in good agreement with older findings (our consensus values) [5] while the temperature is somewhat higher.

4.5 Angular distance between the IBEX ribbon center and the Hydrogen Deviation Plane (HDP).

Perhaps the most spectacular IBEX result is the discovery of the so-called "ribbon", the locus on the sky from where are coming most of high energy ENAs. This structure is forming a (sometimes incomplete) circular structure around one point, the center of the ribbon. The position on the sky of this center is [29]: ecliptic longitude 219.2 ± 1.3°, latitude: + 40 ± 2.3°. On the other hand, it was discovered [22] that the incoming of Hydrogen atoms was deviated from the incoming of He atoms, because some of the interstellar H^+ are deviated by the heliopause obstacle, kinked by the interstellar magnetic field, and then produce H neutrals by charge exchange. Therefore, the plane containing both H and He vectors was dubbed the Hydrogen Deviation Plane (HDP), containing the interstellar magnetic field outside of the heliopause. The normal to this plane re-determined in 2010 [30] is at $\lambda = -10.6°$, $\beta = 37.5°$ (ecliptic J2000), and the angle between this normal and the direction of ribbon center is 90.0°: the ribbon center is right on the HDP as determined in 2010 [30]. If the new IBEX value of He direction is taken, it gives another normal to HDP: $\lambda = -3.2°$, $\beta = +57.5°$, making an angle of 76° with the ribbon center, which would then be at 14 ° from this new HDP, a substantial deviation from the HDP.

5. Conclusions

The IBEX helium direction and velocity [12,13,14] are in marked discrepancy with earlier determinations of the Helium flow, either from optical measurements of He 58.4 nm resonance, or from Gas/Ulysses [17, 26, 27, 28] *in-situ* measurements. This discrepancy was interpreted as the result of a secular variation of the interstellar wind direction [4], an interpretation vigorously challenged [1]. Here we show first that the claimed variations of longitude and velocity are highly unlikely (with a probability <1%), in view of the absence of change both in latitude and in the coupling (V, λ_∞). Therefore either the IBEX value, or the old one is true. We have then proposed an instrumental effect for IBEX-Lo sensor that could bias the fit to a model, the inability to count all He events when their rate is too high. As reported in [1], the IBEX-Lo data could be as well fitted by the old parameters, when corrected for a dead time of 7 ms, not far from a quoted possible value of 5 ms [12]. Other instrumental effects could also be investigated, as for instance the detection efficiency of Helium atoms as a function of their kinetic energy, which is assumed to be constant in the IBEX analyses [12,13].

In parallel, several independent new analyses or re-analyses of He and Hydrogen flow indicate a long term stability of the direction and velocity of the interstellar flow through the solar system. Additional analyses of other data sets are expected in the near future, telling on whose head will fall the spade of Damocles.

6. Acknowledgements

We thank Gary Zank for the invitation to the 13th Annual International Astrophysics conference in Myrtle Beach, during March 2014. We also wish to thank Dimitra Koutroumpa, Eric Quémerais, and Olga Katushkina for communication of their most recent Lα results on SWAN. We wish to thank the referee for useful comments and Brian Wood for pointing to astronomical measurements of LIC temperature. This work was supported by CNES (Centre National des Etudes Spatiales) and CNRS (Centre National de la Recherche Scientifique) in France.

References

[1] Lallement R. and Bertaux J.L. 2014 A & A, 565, 41
[2] Bertaux J.L., and Blamont J.E. 1971 A & A , 11, 200
[3] Thomas G.E., and Krassa R.F. 1971 A & A , 11, 218
[4] Frisch P.C., Bzowski M., Livadiotis G., et al. 2013, Science, 341, 1080
[5] Möbius E., Bzowski M., Chalov S., et al. 2004, A&A, 426, 897
[6] Drews C. et al. 2012, J. Geophys. Res.,117, A09106
[7] ChalovS.V. and Fahr H. J. 2006, Astron. Lett., 32, 487
[8] Chalov S. 2014, A & A, in print
[9] Gloeckler G., Möbius E., Geiss J. et al. 2004, A&A, 426, 845
[10] Nakagawa H, et al., 2008, A&A , 491, 29
[11] Dalaudier F., Bertaux J. L., Kurt V. G., & Mironova E. N. 1984, A&A 134, 171
[12] Möbius E., Bochsler P., Bzowski M., et al. 2012, Ap.J. supp.series
[13] Bzowski, M. Kubiak M. A., Möbius E., et al. 2012, Ap.J. supp.series
[14] McComas D.J. et al. 2012, Science, 336, p.1291
[15] Redfield, S. & Linsky, J.L., 2008, Ap.J. 673, pp. 283-314.
[16] Dring,A.R., et al., 1997, Ap.J. 488, pp. 760-775
[17] Witte, M. 2004, A&A, 426, 835
[18] Bertaux J.L., Kyrölä E., Quémerais E., Lallement R. et al. 1995 Solar Physics, Volume 162, pp. 403-439
[19] Bertaux J.L., Quémerais E., Lallement R., Kyrölä E., Schmidt W. 1997 Solar Physics, Volume 175, Issue 2, pp.737-770
[20] Costa J. et al. 1999 A&A, 349, 660

[21] Quémerais E. et al. 1999 JGR, 104 (A6), 12585
[22] Lallement R. et al. 2005 Science 307, 5714, p1447
[23] Vincent F.E. et al. 2014, Ap.J. L., 788, L25, 6 pp.
[24] Katushkina O. 2014 private communication.
[25] Izmodenov V. et al. 2013, ISSI book and workshop.
[26] Wood, B.E., Mueller, H.-R., & Witte, M. 2014, Ap.J., submitted; (also presented at the ICNS conference in March 2014)
[27] Katushkina O.A. , V. V. Izmodenov, B. E. Wood , D. R. McMullin, Neutral Interstellar Helium Parameters Based on Ulysses/GAS and IBEX-Lo Observations: What are the Reasons for the Differences, 2014, Ap.J., vol.789, p.80
[28] Bzowski M. et al., 2014 submitted to A & A, Astro.Ph., available (arXiv1405.0623v1).
[29] Funsten H.O. et al. 2013 Ap.J., Volume 776, 15 pp.
[30] Lallement R. et al. 12th Solar Wind Conference, 2010 AIP Conference Proceedings, Volume 1216, pp. 555-558

13th Annual International Astrophysics Conference: Voyager, IBEX, and the Interstellar Medium IOP Publishing
Journal of Physics: Conference Series **577** (2015) 012005 doi:10.1088/1742-6596/577/1/012005

Voyager 1 Observations of the Interstellar Magnetic Field and the Transition from the Heliosheath

L Burlaga[1]
NASA Goddard Space Flight Center, Greenbelt, Maryland, Code 673, Greenbelt, Maryland 20771, USA

N Ness
Institute for Astrophysics and Computational Sciences, Catholic University of America, Washington DC 20064, USA

[1]E-mail: lburlagahsp@verizon.net

Abstract. Voyager 1 (V1) has been observing interstellar magnetic fields (ISMF) for more than one year, from 2012/209 to at least 2013.6. From 2013.0 to 2013.6 the difference between the azimuthal angle of the ISMF and the Parker spiral angle at the latitude 34.6° of V1 was (22 ± 3)° and the corresponding difference of the elevation angle was (0 ± 8)°. During 2012 the deviation from the Parker spiral angle was somewhat smaller. The interstellar magnetic field has a West to East polarity, opposite to the direction of planetary motions. The magnitude of the ISMF varied smoothly in the range 0.38 nT to 0.59 nT with an average strength 0.49nT. The strongest interstellar fields were observed behind a shock at 2012/297 that was preceded by 2.2 KHz plasma oscillations, which implies an interstellar electron density $n_e = 0.05/cm^{-3}$. The ISMF was observed after V1 crossed a current sheet CS0 having the structure of a tangential discontinuity. The inclination of this current sheet is consistent with an interstellar magnetic field draped on a blunt heliopause. Two other current sheets (sector boundaries) were observed earlier in the heliosheath at 2012/167 and 2011/276 with high inclinations (99 +/-10)° and (89 ± 10)°, respectively). The transition from heliosheath to interstellar magnetic fields is related to a two-step increase in the cosmic ray intensity observed by V1 from 2012.30 to 2012.65. The first step-increase began near the end of an unusual away-polarity sector, and it reached a plateau when V1 moved into a toward-polarity sector that ended at CS0. The second step-increase began slowly after V1 crossed CS0, and it ended abruptly at 2012/237.728.

Introduction

Voyager 1 (*V1*) encountered a boundary across which the flux of low-energy termination shock particles (TSPs) and anomalous cosmic rays (ACRs) accelerated in the heliosphere dropped to very low values compared to heliosheath values. The spacecraft crossed this "particle boundary" five times, first on 2012/210.6 and last on 2012/237.7 [1]. Initially, it was widely thought that the particle boundary might be the heliopause, which separates the heliosheath from the interstellar medium; although an alternative explanation had been suggested by McComas and Schwadron [2]. It is expected that the magnetic field direction should be different in the interstellar medium than in the heliosheath, since heliosheath magnetic fields are produced by the rotation of the sun, whereas interstellar magnetic fields have a very different origin. However, Burlaga et al. [3] found that the magnetic field direction did not change significantly across the particle boundaries. Since small differences in the B can be measured accurately (± 2°), it was concluded that the particle boundaries were not multiple crossings of the heliopause.

The purpose of this paper is a) to study the nature and structure of the heliosheath magnetic fields before and after the last polarity boundary CS0 observed by *V1* during ≈2012/208-209, b) discuss the structure of this boundary, c) show how this boundary relates to the heliopause, and d), describe the nature and properties of **B** beyond CS0, including the observation of a shock or pressure wave. We use *V1* magnetic field observations [4] to study **B** within CS0, prior to CS0 and after V1 crossed CS0.

2. Heliosheath and interstellar magnetic fields

Voyager 1 observations from 2012.5 to 2013.6 are shown in figure 1. Prior to crossing a current sheet CS0, V1 was observing heliosheath magnetic fields directed along the Parker spiral direction [5, 6] pointing toward the

[1]lburlagahsp@verizon.net.

sun ($\lambda_p = 90°$). In this paper, we use an RTN coordinate system whose origin is at the spacecraft. Immediately after crossing the current sheet, V1 observed magnetic fields significantly different from the Parker spiral field direction pointing away from the sun ($\lambda_P = 270°$). The difference between λ and λ_p increased following the crossing of CS0, reaching a plateau after ≈ 2013.2 when ($<\lambda> - \lambda_p$) = 22° ± 3° and ($<\delta> - \delta_P$) = 23° ± 8°. The sustained difference between the observed magnetic field direction and the Parker spiral magnetic field direction and absence of sectors following the crossing of CS0 at ≈ 2012.5691 (≈2012/209.3) ≈ July, 27, 2012 and continuing to 2013.6, as shown in figure 1, strongly suggests that V1 was observing interstellar magnetic fields throughout the interval. The direction of this interstellar magnetic field is close to the –T direction, pointing from west to east when looking from Voyager to the sun.

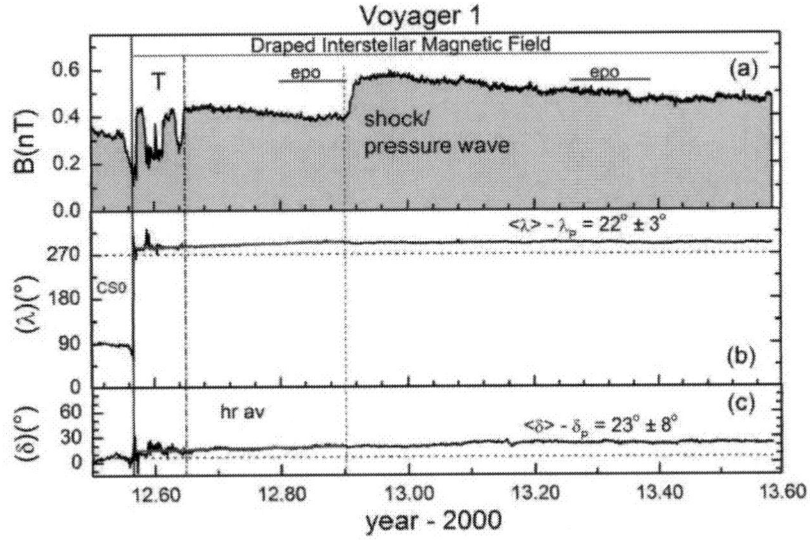

Figure 1. The interstellar magnetic fields following the crossing of a current sheet CS0, which may be the heliopause, by V1. Panel (a) shows the magnetic field strength profile, while (b) and (c) should the azimuthal angle λ and elevation angle δ. A region containing both interstellar and heliosheath magnetic fields in flux tubes on interstellar magnetic field lines is labelled "T".

The small difference between the interstellar magnetic field direction and the heliosheath magnetic field direction observed by V1 was initially surprising. However, a re-examination of the results of Pogorelov et al. [7] , plotting angles instead of components of **B**, showed that they actually predicted that the angle between the draped interstellar magnetic field in the heliosheath fields should be $\delta \approx 25$ ° and $\lambda \approx 290°$ ($<\lambda> - \lambda_P$) ≈ 20°, consistent with the observations shown in figure 1, within the uncertainties of the observations. The simulations of Borovikov and Pogorelov [8] based on exact fitting of the HP do not support the idea of Opher and Drake [9] that the ISMF vector becomes nearly parallel to the solar equatorial plane regardless of its direction in the unperturbed LISM.

The magnitude of **B** shown in Figure 1a is stronger than the magnetic fields observed in the heliosheath prior to the crossing of CS0 by V1, except in the region labelled T in figure 1 a. In this region, Burlaga et al., [3] showed that, there were 5 boundaries at which the strength of **B**, changed abruptly in association with changes in the energetic particles and cosmic rays [1]. Some people suggested that V1 crossed the heliopause moving from the heliosheath to the LISM three times before it remained in the LISM after August 25. However, Burlaga et al. [3] showed that the magnetic field direction did not change significantly (< 2°) across any of these boundaries, and argued that V1 was not crossing the heliopause at these times, because one does not expect the interstellar magnetic field (ISMF) to have the same direction as the heliosheath magnetic field.

The ISMF has been observed continually since V1 crossed CS0, and the five boundaries observed previously separated interstellar magnetic flux tubes that alternately contained heliosheath plasma and the plasma of the LISM as shown by Burlaga and Ness [10]. It is possible that some of the ISMF magnetic field lines and region T reconnected to interstellar magnetic field lines beyond the location of V1, allowing heliosheath plasma and energetic particles to move to the position of V1. Magnetic reconnection has been suggested by Swisdak et al. [11] and Strumik et al. [12]. Alternatively, configurations analogous to flux transfer events in the earth's heliosheath might be responsible for the structure and region T [13].

The magnetic field strength following region T, from 2012.5691 to at least 2013.6, is stronger than the magnetic fields observed in the heliosheath, varying from 0.38 nT to 0.59 nT, with an average $\langle B \rangle = 0.49$ nT. The fluctuations of 48 s averages of B during each hour were extremely small, at the limit of measurement (0.005 nT). Thus, the extended observations of the magnetic fields following the passage of CS0 clearly indicate that V1 is in the new region, with properties expected from magnetic fields of interstellar origin.

Gurnett et al. [14] showed that *Voyager 1 (V1)* observed electron plasma oscillations from 2013/99 to 2013/142 at a frequency of 2.6 kHz, corresponding to a very high electron density of 0.8 cm^{-3}, during the intervals indicated by the horizontal line to the right in figure 1a. They also noted an earlier electron plasma oscillations event at 2.2 kHz in October-November, 2012 shown by the horizontal line to the left in figure 1a. Since such densities are characteristic of interstellar plasma and are at least 50 times larger than expected for heliosheath plasma, Gurnett et al. concluded that *V1* was immersed in the interstellar plasma. Gurnett et al. [14] suggested that the increase in density from 0.05 cm^{-3} to 0.08 cm^{-3}, from the time of the first burst of electron plasma oscillations to the second, is an indication of a density ramp like that inferred by Gurnett et al. [15] from remote observations of radio waves by V1, which were assumed to be radio waves generated by a shock in interstellar plasma just beyond the heliopause.

Figure 1 shows that **B** varies little before and after each of these two electron plasma oscillation events, apart from a slowly monotonically increasing trend with time. Thus, it appears that V1 was immersed in interstellar magnetic fields from the time that it crossed CS0, on day 219, 2012 to at least 2013.6.

3. CS0 and heliopause

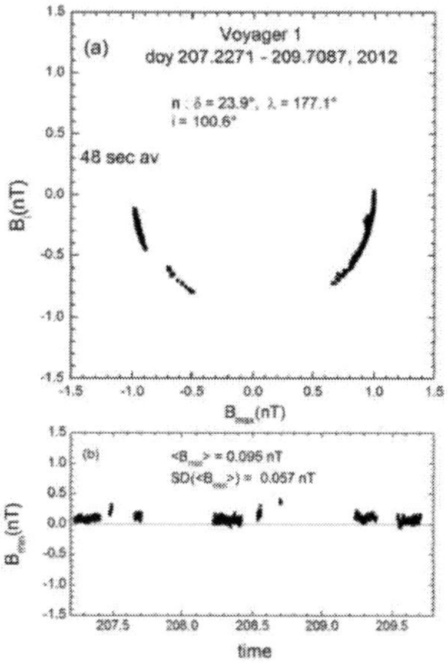

Figure 2. Minimum variance analysis of CS0. (a) The normalized component B_i in the intermediate variance direction versus B_{max} in the maximum variance direction. (b) the normalized component B_{min} in the minimum variance direction versus time in days.

In this section we consider the structure of **B** in the current sheet CS0 and its possible relationship to the heliopause. The direction of **B** rotated within CS0 during the interval from approximately day 207.2271 - 209.7087, 2012. We examine the structure of this rotation using the minimum variance method. A well-defined minimum variance direction was found in the direction **n** = (-0.91, 0.05, 0.415) in the RTN coordinate system whose origin is at V1. The elevation angle $\delta = 23.9°$ and the azimuthal angle $\lambda = 177.1°$. The normalized components of **B** in the intermediate variance direction versus the components of **B** in the maximum variance direction, respectively, are plotted in figure 2(a), where one sees that **B** rotates through a large angle in the plane normal to **n.** A geometrical construction shows that this minimum variance plane is inclined by an angle i = 100.6° with respect to the solar equatorial plane. The magnitude of the component of **B** in the minimum variance direction B_{min} is shown in figure 2(b) as a function of time in doy. If CS0 were the current sheet associated with a tangential discontinuity, the average of B_{min} would be 0. The observed value of $< B_{min} > =$ 0.095 nT with a standard deviation of 0.057 nT. Thus, the observations of CS0 are consistent with the tangential discontinuity, but one cannot exclude very small nonzero component.

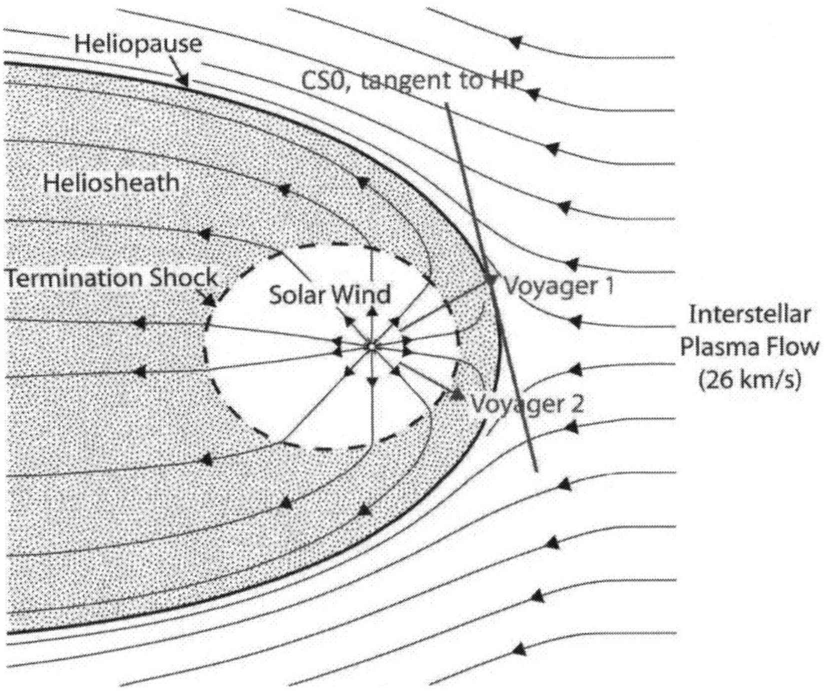

Figure 3. Schematic of the heliopause interacting with interstellar plasma flow. The tangent to the heliopause at the position of V1 is qualitatively consistent with the orientation of CS0

Figure 3 shows a schematic of the heliosphere. The solar wind velocity is directed radially away from the sun and moves toward the termination shock, where it is diverted tailward and moves through the heliosheath.

4. Interstellar shock or pressure wave

Figure 4 shows that a large smooth jump in B on 2012/333 followed the interval containing electron plasma oscillations from 2012 day 297 to 2012 day 332, which was analysed by Burlaga et al. [16]. This jump resembles a shock with $B_2/B_1 = 1.4$. The jump in B was associated with a small but significant change in λ. From the change in the magnetic field direction one can calculate that the normal to the surface corresponding to the jump in B is directed 27° above the radial direction in the meridional plane. Knowing direction of B prior to the jump, we calculate that the angle between the ambient magnetic field and the shock normal is 85°, indicating a quasi-perpendicular shock. Knowing the density of the upstream medium from the plasma wave oscillations, one can calculate the density behind the shock, since the shock is quasi-perpendicular with $N_2/N_1 = B_2/B_1 = 1.4$. Assuming an interstellar temperature of 20,000° K just beyond the heliopause as suggested by Zank et al. [17] the upstream Alfvén speed is $V_{A1} = 38$ km/s and the sound speed is $V_{S1} = 17$ km/s, which gives a magnetoacoustic speed of 40 km/s. Thus, the shock was moving at greater than 40 km/s, consistent with the shock speed beyond the heliopause calculated by Washimi et al. [18]. The ratio $(V_{AS}/V_s)^2 = 5$.

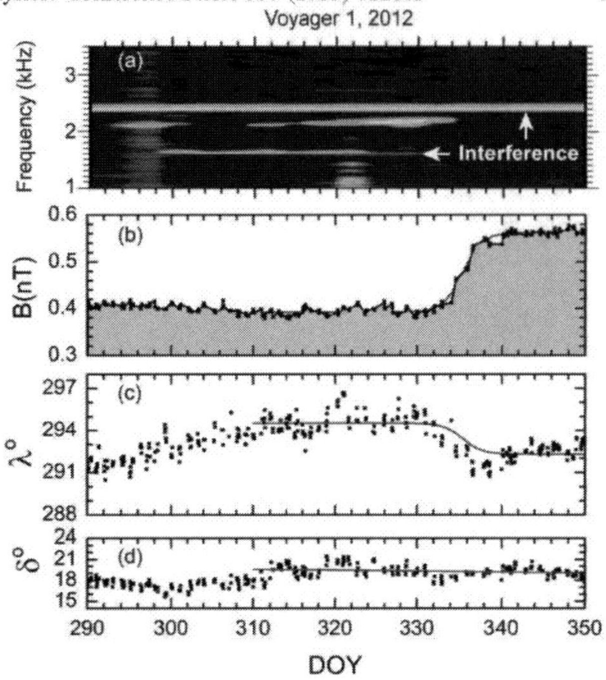

Figure 4. A shock or pressure wave the LISM. (a) The band between the two interference signals represents the frequency of electron plasma oscillations. (b) The jump in the magnetic field corresponds to a shock or pressure wave. (c) λ changes across the jump in B, but there is no change in δ (d)

By analogy with shocks observed at 1 AU (where pickup protons, and neutral particles do not affect the dynamics) this shock in interstellar plasma appears to be a sub-critical, laminar, quasi-perpendicular, shock. However, the shock seems to be much thicker than those observed at 1 AU. Further theoretical studies are needed to understand the nature of such shocks or pressure waves in the interstellar plasma and magnetic field.

5. Transition from the heliosheath to the LISM

We have presented evidence for a distinct boundary CS0 after which interstellar magnetic fields were observed, which has the orientation expected for the heliopause. However, observations of the heliosheath plasma on

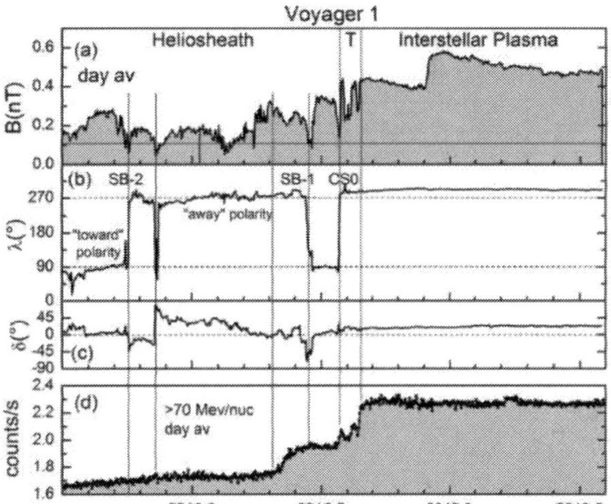

Figure 5. The transition from the heliosheath to the interstellar plasma. The magnetic field strength (a), the azimuthal angle (b), and elevation angle (c), and the counting rate of >75 MeV/nuc cosmic rays.

interstellar magnetic field lines in region T in figure 1 indicate that the transition from the heliosheath to the interstellar plasma is not simply an isolated tangential discontinuity. Figure 5 above, from Burlaga & Ness [10], shows that the transition occurs over a broader region, beginning prior to the crossing of SB1 by V1. This transition is most evident in the 2-step increase in the counting rate of cosmic rays >70 MeV/nuc [19] shown in figure 5 (d). The first step, beginning on ≈2012.30 within an away sector in which B points toward the sun

along the Parker field direction, and it ends at the boundary of that sector, SB-1. This step-increase is followed by a brief plateau in the cosmic ray counting rate when V1 observed a short sector in which B points toward the sun. The second step-increase in the cosmic ray counting rate begins at the time V1 crosses CS0 and it ends when V1 crosses region T and remains in the interstellar magnetic field until 2013.6.
The two-step profile of the cosmic ray counting rate has not yet been explained. Although it has been proposed that magnetic bubbles may exist in the region observed prior to the crossing of CS0 [11, 12], there is no clear evidence for a series of nested magnetic bubbles. However, Burlaga and Ness [10] show that the sector boundaries, SB-2 and SB-1 are nearly parallel to CS0, and the away polarity sector within these boundaries has an unusual jump in δ extending from near 0° to in nearly 90°, declining to 0° prior to the onset of the first step-increase in the cosmic ray counting rate. The magnetic field strength varies within the sector between SB-1 and SB-2, and there are large dips in B at the sector boundaries, as well as at the jump in δ between the boundaries. Such decreases in B are often observed at sector boundaries, but they are occasionally observed at structures called D-sheets by Burlaga and Ness [20] which could be associated with magnetic reconnection. The structure of the region observed prior to and following the passage of CS0 shown in figure 5 is not understood.

There are many theories of various types of instabilities that might occur on the heliopause [21]. A large-scale instability [8, 22] driven by a charge exchange instability [23, 24, 25] can explain why V1 observed interstellar plasma, even though theories predict that the heliopause of a stationary heliosphere should be well beyond the position of V1.

Acknowledgments T. McClanahan and S. Kramer provided support in the processing of the data. D. Berdichevsky computed correction tables for the three sensors on each of the two magnetometers. N. F. Ness was supported by NASA Grant NNX12A63G3G to the Catholic University of America. L. F. Burlaga was supported by NASA Contract NNG14PN24P.

References
[1] Stone E C, Cummings A C, McDonald F B, et al. 2013, *Science* **341** 150; Krimigis S M, Decker, R B, Roelof E C et al. 2013 *Sci* **341**, 144
[2] McComas D A and Schwadron N A 2012 *ApJL*, **758**, 19 doi:10.1088/0004-637X/758/1/19
[3] Burlaga L F, Ness N F and Stone, E C 2013 *Sci*, **341**, 147, doi:10.1126/science.123545
[4] Behannon K W, Acuña M H, Burlaga L F, et al. 1977 *SSRv*, **21**, 235
[5] Parker E N 1958 *ApJ* **128**, 664
[6] Parker E N 1963, Interplanetary Dynamical Processes (Interscience Publishers, New York)
[7] Pogorelov N V, Borovikov S N, Zank G P and Ogino T 2009 *ApJ* **696** 1478
[8] Borovikov S N, and Pogorelov N V 2014 *ApJ* **783** L16 doi10.1088/2041-8205/783/1/L16
[9] Opher M and Drake J F 2013 *ApJ*, **778** L26 doi:10.1088/2041-8205/778/2/L26
[10] Burlaga L F and Ness N F 2014 *ApJ* **784** 146 doi:10.1088/0004-637X/784/2/146
[11] Swisdak M, Drake J F and Opher M 2013 *ApJ* **774**, L8
[12] Strumik M A, Czechowski, Grzedzielski S, Macek W M and Ratkiewicz R 2013 *ApJL* **773**:L23, doi:10.1088/2041-8205/773/2/L23
[13] Schwadron N A and McComas D *APJ*, **599** 1395
[14] Gurnett D A, Kurth W S, Burlaga L F and Ness N F 2013 *Sci* **2025** doi:10.1126/science.124168115
[15] Gurnett D A, Kurth W S, Allendorf S C and Poynter R L 1993 *Sci* **262** 199
[16] Burlaga L F, Ness N F, Gurnett D A and Kurth W S 2013 *ApJL* **778**: L3, doi: 10.1088/2041-8205/778/1/L3
[17] Zank G P and Müller H R 2003 *JGR* **108** A6, 1240 doi:10.1029/2002JA009689
[18] Washimi H, Zank G P, Hu Q, et al. 2011 *MNRAS* **416** 1475-1485
[19] Webber W and McDonald F 2013 *GeoRL* **40** 166520
[20] Burlaga L B and Ness N F 1968 *Canadian Journal of Physics*, **46**, S962; Burlaga L F 1968 *Solar Physics* **4**, 67
[21] Zank G P 1999 *SSRv* **89** 413; Pogorelov N V, Borovikov S N, Zank G P et al. 2012 *ApJ* **750** L4; Krimigis S M, Decker, R B, Roelof E C et al. 2013 *Sci* **341**, 144
[22] Borovikov S N, Pogorelov N V, and Ebert R W 2012 *ApJ* **750**:42, doi:10.1088/0004-637X/750/1/42
[23] Liewer P C, Karmesin S R and Brackbill J U 1996 *JGR* **101** 17119
[24] Zank G P, Pauls, H L, Williams, L L and Brackbill D T 1996 *JGR* **101**, 21,639 , doi: 10.1029/96JA02127
[25] Zank G P 1999 *Solar Wind 9 Conf, AIP Conf Proc.* **471**, 783, doi: 10.1063/1.58660

Recent Particle Measurements from Voyagers 1 and 2

R B Decker[1], S M Krimigis[1,2], E C Roelof[1] and M E Hill[1]
[1] Johns Hopkins University, Applied Physics Laboratory, Laurel, MD 20723, USA
[2] Office of Space Research and Technology, Academy of Athens, Athens, Greece

E-mail: robert.decker@jhuapl.edu

Abstract. We summarize recent measurements made by the LECP instrument on Voyager 1 near its crossing of the heliopause and entry into the interstellar medium on or about day 238 of 2012 at 121.6 AU, and on Voyager 2 mainly during the period 2012-2014.4 characterized by large variations in the intensities and angular distributions of low-energy heliosheath ions and the reappearance of low-energy heliosheath electrons. Results from Voyager 1 not previously published include the energy dependence of ion intensity decreases prior to the heliopause crossing and a quantitative measure of the evolution of low-energy heliosheath proton pitch angle distributions that extend across the heliopause. For Voyager 2 we describe the evolution in time and with ion energy of the tangential streaming of ions directed from the nose toward the tail of the heliosheath, summarize the recovery of low-energy heliosheath ion intensities since their decline to minimum levels during 2013.0-3013.3, and discuss the effects of this intensity minimum and subsequent recovery on the ion partial pressure in the heliosheath and on its magnitude relative to that of the thermal plasma and the magnetic field.

1. Introduction

Voyager 1 (V1) and 2 (V2) are receding from the Sun at 3.58 AU yr^{-1} (17.0 km s^{-1}) and 3.24 AU yr^{-1} (15.4 km s^{-1}) respectively, and are now at respective heliocentric radii 127.8 AU and 104.8 AU (23.0 AU separation), heliographic latitudes N34.6° and S30.7° (65.3° separation), heliographic inertial longitudes 174.4° and 217.6° (43.2° separation; the latitude and longitude of inflow from the local interstellar medium (LISM) are ≈0° and 175°, respectively; so V1 is near the heliospheric nose). Their rectilinear separation is 145.5 AU. V1 crossed the termination shock (TS) of the solar wind on 16 Dec. 2004 at 94.0 AU [1-4], and explored the shocked solar wind in the heliosheath (HS; ≈30 AU wide along the V1 path) for ≈8.2 years. On or about 25 Aug. 2012, V1 at 121.6 AU entered a new region characterized by an increase in the magnitude of the magnetic field and of the intensity of galactic cosmic rays (GCR), and by large reductions in the intensity of TS ions and electrons and of anomalous cosmic rays (ACR) [5-8]. It is believed that around that time V1 crossed the heliopause (HP), the boundary separating plasma of solar origin from that of the LISM, and is now in the LISM [5,9]. V2 crossed the TS several times on 29-31 Aug. 2007 at 83.65 AU [10-13] and is now in the HS, ≈21 AU beyond the TS, as it heads toward the HP. Each Voyager carries a nearly identical set of instruments that continue to monitor plasma, field, and energetic particle activity in the HS and LISM (the exception is that only V2 has a working plasma instrument). This paper is a survey of recent measurements made during ≈2012-2014.4 by the Low Energy Charged Particle (LECP) instruments [14] on V1 and V2.

[1] To whom any correspondence should be addressed.

2. Voyager 1 near the heliopause crossing

Figure 1 shows particle intensities measured at V1 during 2012.5-2012.7 in seven low-energy ion (single-parameter: total energy) channels covering 0.05-4.0 MeV, two proton (two-parameter: total energy, species) channels covering 3-31 MeV, a galactic cosmic ray (GCR) proton channel >211 MeV, and a low-energy electron channel 0.04-0.07 MeV. All traces are daily averages and were corrected for background as necessary. The two grey-shaded intervals during days 210-215 and days 226-234 are periods when intensities of low-energy ions and protons 0.05-31 MeV and low-energy electrons decreased while those of GCR protons and electrons increased and the magnitude of magnetic field increased [5-8]. The grey-shaded period from day 238 onward denotes V1 beyond its HP crossing. In the LISM the intensities of the ion channels and the electron channel decrease to background levels, the 3-17 and 22-31 MeV proton intensities, which are mainly ACRs, extend into the LISM with decreasing intensities, and the GCR protons increase to their highest levels measured prior to day 238 [5-7]. Aside from the GCRs, all channels displayed in figure 1 are evidently of heliospheric origin.

2.1. Energy dependence of intensity decreases in the first depletion region

We refer to days 210-215 and 226-234 of 2012 as the first and second depletion regions respectively. In the first depletion region intensities decrease in all channels except that of >211 GCR protons, which increases, and all remain above background except the electron channel. In the second depletion region the ion and proton intensity decreases are larger, and the lower-energy ion channels and the

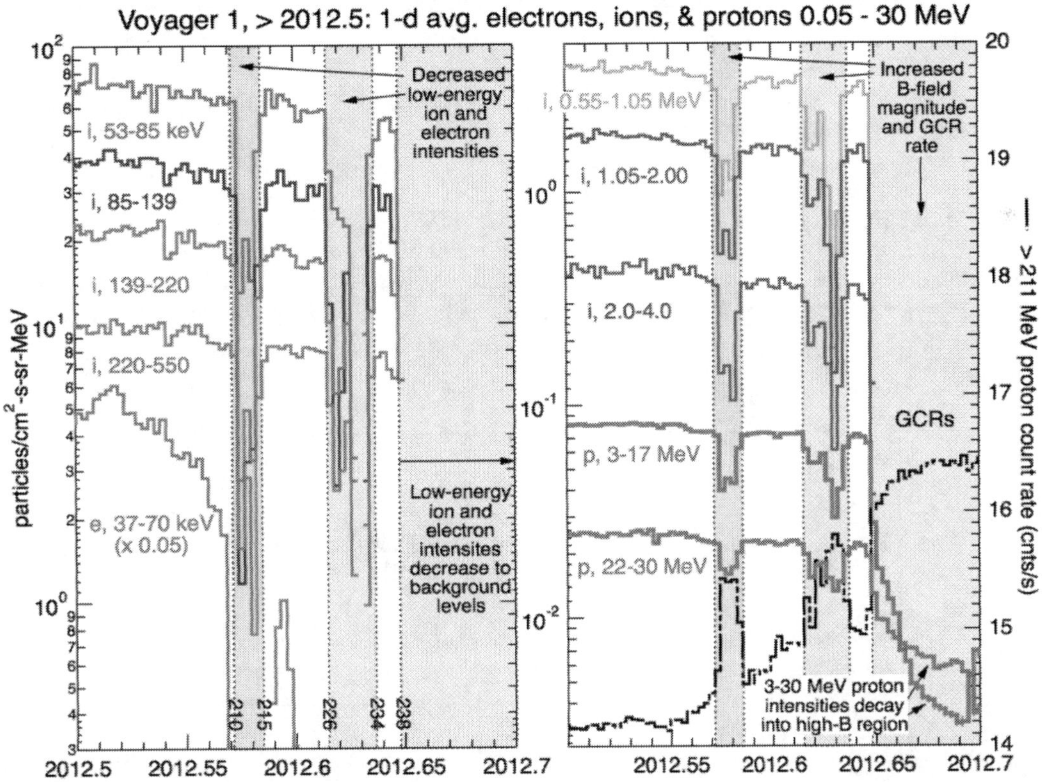

Figure 1. Daily averaged intensities of low-energy ions (0.05-4.0 MeV), protons (3-31 MeV), electrons (0.04-0.07 MeV), and >211 MeV GCR proton count rates measured at V1 during 2012.5-2012.7. Grey-shaded periods during day 210-215, 226-234, and after day 238 indicate periods when V1 was in the first and second depletions regions and then evidently in the LISM.

electrons drop to background levels. A noted in [6], the electron intensity begins a gradual exponential decrease starting ≈ 2012.52 that continues until the first depletion region, then recovers slightly for 3-4 days between the two depletion regions, and then drops to background levels. This trend is markedly different from those of the low-energy ions and protons, which decrease only gradually, if at all, before, between and after the depletion regions and up until the HP crossing into the LISM. The nature of the HS and the HS depletion regions and variations of the particle populations and magnetic field [7] have been interpreted differently by various authors [6,9,16-18].

We examine in figure 2 the energy dependence of the 0.05-30 ion and proton intensity decreases in the first depletion region. For each channel we calculated the average intensity during the three-day depletion period 211-213, which excludes intensities within one day of the "edges" on days 210 and 215, and during the three-day pre-depletion period 207-209 just ahead of the first depletion. The ratio of intensity before to that during the first depletion, denoted as R_D in figure 2, is plotted versus the particle's geometric mean speed, $v = [2(E_L E_H)^{1/2} / m_p]^{1/2}$, where E_L and E_H are the lower and upper energy passband of a channel, m_p is the proton mass, and we have assumed that protons dominate the intensities in the seven ion channels. The large error bars in the lowest three channels are due to reduced count rates when the large background contributions are removed from these channels. The nine error-weighted points are well fit by a power-law in v with exponent -0.44 ± 0.03 (-0.45 ± 0.03 if we neglect the lowest three channels), indicating a rather weak dependence, $R_D \approx r_g^{-1/2}$, on the proton gyroradius r_g. This relation may be useful in future modelling to determine the energy dependence of, e.g., the escape of ions from HS flux tubes reconnected to LISM fluxes tubes [15,17], or the filling of LISM flux tubes that intrude into the HS [6]. For example, in the interchange scenario advocated by Krimigis et al. [6], we would expect higher energy ions with larger gyroradii to fill empty LISM tubes more rapidly, as observed in the two depletion regions. However, details remain to be worked out.

2.2. Angular distributions of 3-31 MeV protons after the HP crossing

Krimigis et al. [6] discuss the evolution of the 3-17 MeV proton pitch-angle distributions (PADs) before and after the HP crossing (red trace in the right panel of figure 1). Florinski et al. [19] published model calculations that reproduce these data. For completeness, we show in figure 3 the PADs of higher-energy, 22-31 MeV ACR protons. The time evolution and interpretation of these PADs are

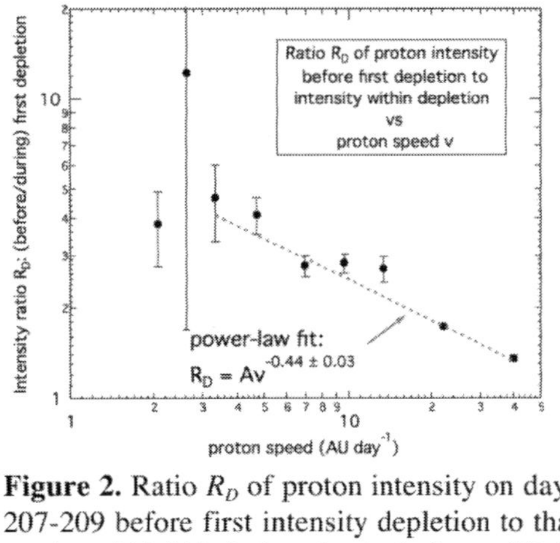

Figure 2. Ratio R_D of proton intensity on days 207-209 before first intensity depletion to that on days 211-213 during the depletion at V1 in 2012 (see figure 1) versus geometric mean speed v of the energy channel. Fit is an error-weighted power law $R_D = Av^{-0.44 \pm 0.03}$.

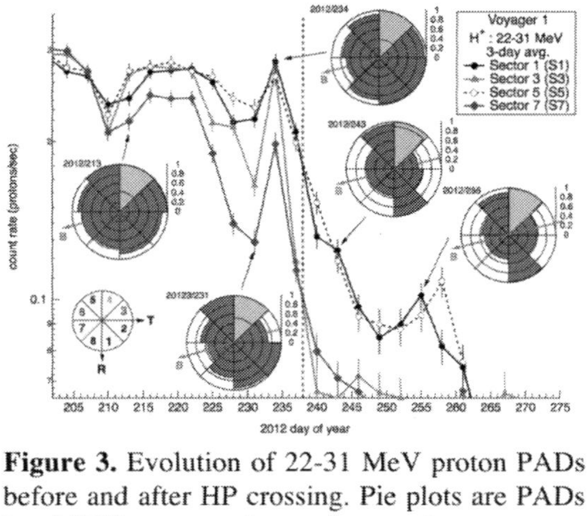

Figure 3. Evolution of 22-31 MeV proton PADs before and after HP crossing. Pie plots are PADs in LECP scan plane (inset shows sector numbers). Light-shaded sector 4 is blocked. Red vector in pie plots is the direction of **B**, measured by V1 MAG, that lies nearly in LECP scan plane.

very similar to those of the 3-17 MeV protons, and we refer to Krimigis et al. [6] for details. We simply note that the first three PADs in figure 3 indicate that before the HP crossing, magnetic conditions along field lines were producing asymmetric loss-cone distributions along the magnetic field **B**, while after the HP crossing, the PADs developed pancake distributions, i.e., intensities are peaked transverse to **B**, and in the LISM intensities in all sectors rapidly decrease.

In figure 4 we examine the evolution of the 3-17 MeV proton anisotropy after the HP crossing. The lower panel shows intensities in four of the seven active sectors, color-coded as indicated in the legend, with sector view direction in the LECP scan plane indicated in the inset pie plot. As a measure of the anisotropy, we plot in the upper panel the ratio (R) of the sum of sector 1 (S1) and S5, which measure protons spiralling nearly perpendicular to the **B**, to the sum of S3 and S7, which measure protons moving nearly parallel and anti-parallel to the **B**, respectively. The number of counts in S3 and S7 are sufficient to give a statistically significant ratio R for ≈10 days after the HP crossing (grey-shaded period). The red dashed line is an exponential fit to the points during this period. The anisotropy increases away from the HP with an e-folding time of 4.5 days, which translates to an e-folding distance of 0.045 AU or ≈7 gyroradii of a 3-17 MeV proton in a 0.4 nT magnetic field [7] when we convert time to distance using the radial speed of the spacecraft (≈0.01 AU day^{-1}).

3. Voyager 2 observations in the heliosheath

3.1. Tangential anisotropy of 0.03-30 MeV protons in 2013

We turn now to measurements made by the V2 LECP in the HS. Figure 5(a) shows five-day running averages of daily intensities, from the TS crossing on 2007.65 to 2014.4, of 28-43 keV ions moving in

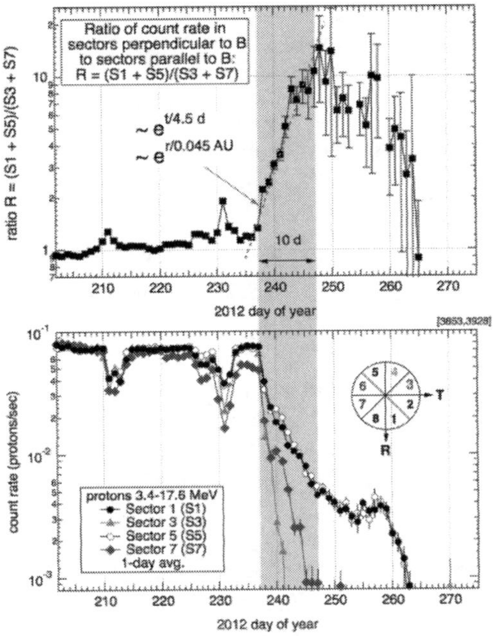

Figure 4. Bottom: Count rates in four of seven active sectors (S4 is blocked) of 3-17 MeV protons at V1. Top: Ratio of S1+S5 that detect protons gyrating nearly transverse to magnetic field vector **B**, to sectors S3+S7 that detect protons moving along **B**.

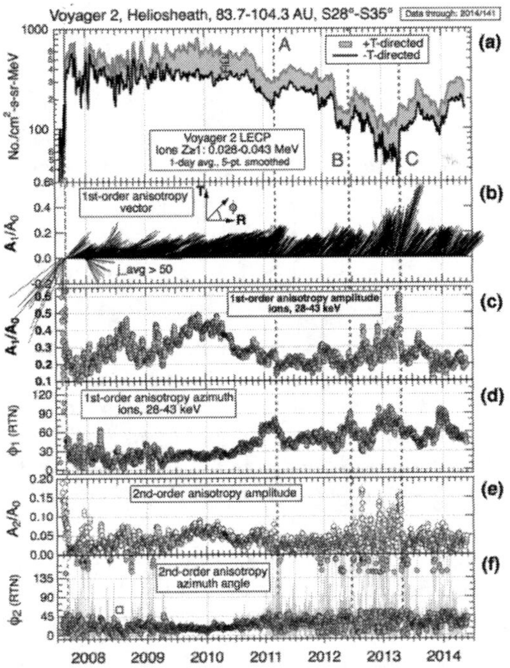

Figure 5. (a) Intensity of 28-43 keV ions moving in $+T$ (red) and $-T$ (black) direction. (b) 1st-order anisotropy vector, (b) associated relative amplitude A_1 / A_0 and (c) azimuth angle ϕ_1. (d) Relative 2st-order anisotropy amplitude and (e) associated azimuth angle ϕ_2.

the $+T$ direction (red trace and shading) and the $-T$ direction (black trace) (the $\pm T$ direction conforms to the usual heliospheric *RTN* coordinate system). This is the lowest-energy ion channel on either V1 or V2, with the mean speed of measured protons being ≈1.5 AU day^{-1}. The larger intensity in the $+T$ direction is generally consistent with a convective anisotropy due to the HS plasma velocity having components in the $+R$ and $+T$ directions that are of comparable magnitudes at V2 [20]. However, as described below, there are period of interest that indicate contributions to the low-energy ion channels from heavy ions and from *bona fide* field-aligned anisotropies in the HS. Figures 5(b)-5(f) show the results of a least squares Fourier fit to the angular data taken in eight 45° sectors in the V2 LECP scan plane, which is nearly parallel to the *R-T* plane [14,21]. These data are fit to the function

$$j(\phi) = a_0 + a_1 \cos(\phi) + b_1 \sin(\phi) + a_2 \cos(2\phi) + b_2 \sin(2\phi), \quad (1)$$

where j is intensity, the fit coefficients are $(a_0, a_1, b_1, a_2, b_2)$, and azimuth angle ϕ is measured in a right-handed sense from the R- to the T-axis. Equation (1) can be written in the more convenient form

$$j(\phi) = A_0 + A_1 \cos(\phi - \phi_1) + A_2 \cos[2(\phi - \phi_2)]. \quad (2)$$

Here, $A_0 = a_0$, $A_k = (a_k^2 + b_k^2)^{1/2}$ and $\tan\phi_k = b_k / a_k$ ($k = 1,2$) are calculable from $(a_0, a_1, b_1, a_2, b_2)$.

Figure 5(c) is the relative first-order anisotropy amplitude A_1 / A_0, where A_0 is the sector-averaged intensity, and panel (d) the associated direction ϕ_1. Figure 5(b), constructed from the best-fit coefficients in panels (c) and (d), is a vector pointed in the direction that particles are moving (see inset showing R, T axes), with magnitude given in panel (c). Figure 5(e) is the relative second-order anisotropy amplitude A_2 / A_0 and panel (f) the associated direction azimuth ϕ_2 (see [21] for details).

We will focus on the period in 2012-13 leading up to the large increase of A_1 / A_0. During this period, which corresponds roughly to the interval 2012.45-2013.30 between the vertical dashed lines labelled B and C in figure 5, the intensity drops while the angle ϕ_1 rotates from ≈45°to ≈80° (not quite 90° because intensities in the spacecraft frame include radial convection). Around 2013.3 the intensity reaches its lowest value while A_1 / A_0 reaches its highest value yet observed at V2 in the HS. This period is also marked by large and sporadic fluctuations in A_2 / A_0 due to the sporadic increases in the anisotropy amplitude. We attribute the enhanced A_1 / A_0 during 2009.3-2010.5 to suprathermal pickup He^+ or O^+ ions, or both, that were boosted in energy at the TS, convected to V2 by the ~200 km s^{-1} HS flow, and measured by the LECP instrument, which responds to He and O ions with energies 9-40 keV nuc^{-1} and 4-15 keV nuc^{-1} respectively. Although it is unclear why these heavy ions are prevalent at V2 during 2009.3-2010.5 (solar minimum), we note that pick-up O^+ ions were measured by the V1 and V2 LECP instruments in high-speed solar wind flows during the 1995-1998 solar minimum [22].

Figure 6(a) shows daily intensities of $+T$-directed (blue) and $-T$-directed (black) 43-80 keV ions at V2 during 2010-2014.4. Also shown for reference is the sector-averaged intensity of a 43-80 keV ion channel from V1 (this channel is synthesized using two V1 energy channels 40-53 and 53-85 keV). Figure 6(b) shows the T-component of the relative first-order anisotropy vector $\xi_{1T} = (A_1 / A_0)\sin\phi_1$. Panel (b) illustrates the gradual increase in ξ_{1T} and then a spike-like jump as the 43-80 keV ion intensity approaches a minimum, followed by an abrupt drop of ξ_{1T}, coincident with an increase in the intensity on 2013.30 (line C). The essential point is that although both $+T$- and $-T$-directed ions were decreasing in a quasi-exponential manner during ≈2012.5-2013.30, the $-T$-directed intensity decreased more rapidly during ≈2013.0-2013.3. It is this relative reduction in the ions reaching V2 from the tailward direction relative to those reaching V2 from the noseward direction of the HS that produced the bulk of the growing $+T$-directed anisotropy. In addition, close examination of the V2 traces in figure 6(a) just prior to the peak of ξ_{1T} shows that the $+T$-directed ions actually began increasing at least ten days before the -*T*-directed ions, and it is this early rise in tailward-directed ions that gave rise to the spike.

In summary, the gradual increase in ξ_{1T} during the intensity drop results from fewer ions coming from the HS tailward of V2 compared to those coming from the HS noseward of V2, and the large spike in ξ_{1T} results from a +*T*-directed (or tailward-directed) increase in ion intensity from a probable "source" noseward of V2 arriving several days before the corresponding increase in the –*T*-directed (or noseward-directed) intensity increase. It is plausible that the ≈10-day delay in the rise time of these two oppositely directly populations represents the scale time required for the tailward moving ion population to scatter or mirror and return to V2 as part of the noseward moving component (e.g., the average proton speed ≈1.9 AU/day at a 30° pitch angle implies a transit distance ≈ 16 AU).

There was a similar, but smaller, increase in ξ_{1T} as the 43-80 ion intensity decreased during 2010.7-2011.2, followed by an abrupt decrease in ξ_{1T}, again coincident with an intensity increase on 2011.2 (line A in figure 6(b)). The anisotropy evolution during these two intensity decreases shows that as the ion intensity drops, there is a net streaming of ions measured at V2 that is toward the flanks of the HS, i.e., again away from the nose (from the direction of the V1 azimuth) and toward the tail. Figure 7 shows that this net streaming is present in ion and proton channels that extend from ~0.03 to 20 MeV (the same streaming is seen in the 22-30 MeV proton channel, not shown here, extending this range to 30 MeV). Periods with large +*T*-directed streaming are the only periods when the *T*-component of plasma flow measured by the V2 plasma instrument and also estimated from Compton-Getting analysis of the LECP low-energy ion angular distributions show significant differences [20].

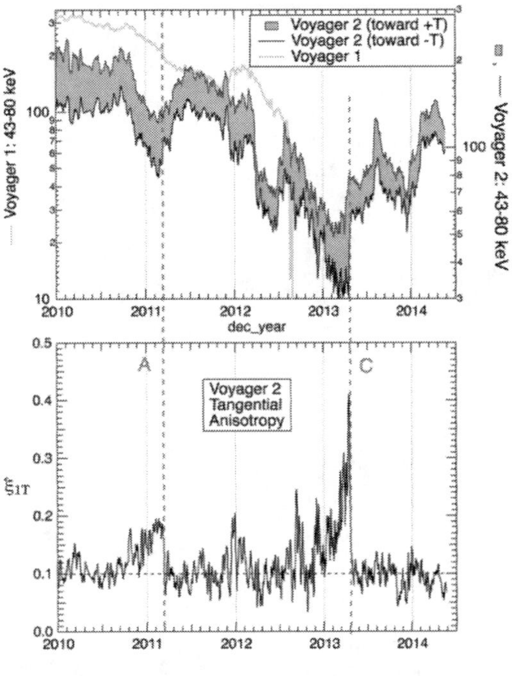

Figure 6. (a) Intensity of ions 43-80 keV moving in the *+T* (blue) and *–T* direction (black) at V2, and sector-averaged ions 43-80 keV at V1 (green). (b) *T*-component of 1st-order anisotropy vector $\xi_{1T} = (A_1 / A_0)\sin\phi_1$ (in *RTN* system).

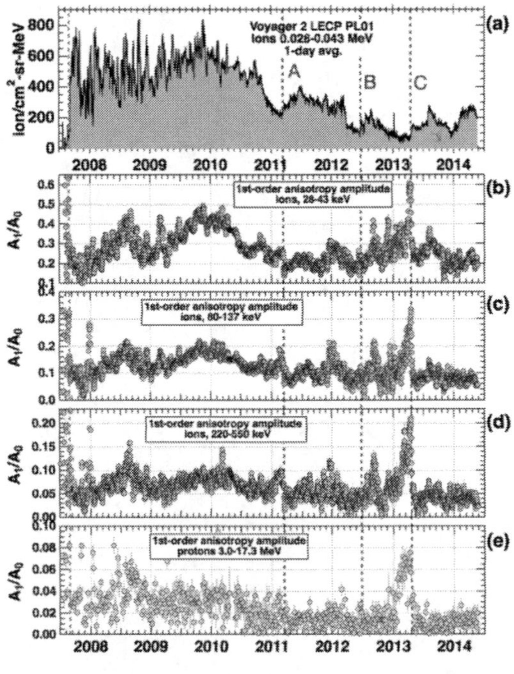

Figure 7. (a) Sector-averaged intensity of ions 28-43 keV at V2. (b) 1st order anisotropy of ions 28-43 keV, (c) ions 80-137 keV, (d) ions 220-550 keV, and (d) protons 3-17 MeV. Vertical dashed lines labelled A-C are defined in the text.

Journal of Physics: Conference Series **577** (2015) 012006 doi:10.1088/1742-6596/577/1/012006

3.2. Partial recovery heliosheath particle intensities at V2

Figure 8 is an overview of energetic particle activity at V2 during the 1.8-year period 2012.5-2014.3. Plotted are daily intensities in eight ion channels covering 0.03-3.5 MeV, 2 proton channels covering 3-30 MeV, and an electron channel 0.04-0.06 MeV. There is a decline of all ion intensities to minima at slightly different times that depend on energy. Ions with energies >0.5 MeV reach sharp minima around 2013/008 (year/doy; first vertical dashed line), while ions 0.03-0.5 MeV reach broad minima in the ≈100 day period between the first dashed vertical line at 2013/008 and line C at 2013/110, where the lower energy ion channels increase abruptly, while higher energy ions show local peaks. Line C also marks the return at V2 of low-energy heliospheric electrons with energies from a few tens of keV to one MeV (only the 0.04-0.06 MeV electrons are shown here). Note that the abrupt increase of the intensities of lower-energy ions, generally consistent with being the suprathermal tail created when pickup protons are heated and mildly accelerated at the TS [23], and the electrons and the abrupt cessation of the $+T$-directed ion streaming anisotropies coincide. At 2013/273 (line D) there are increases in all the ion channels that are qualitatively similar to those that occurred on 2013/110. Although another decrease in ion intensities occurred late ≈2013.9, the intensities of the lower energy ions and the electrons have continued to increase in the most recent data examined.

3.3. Variations of suprathermal ion partial pressures

Figure 9(a) shows intensities of ≈80-140 keV ions at V1 and V2 during 2007.5-2014.4, the roughly 7-year period V2 has been in the HS. Figure 9(b) compares the partial pressures of suprathermal ions

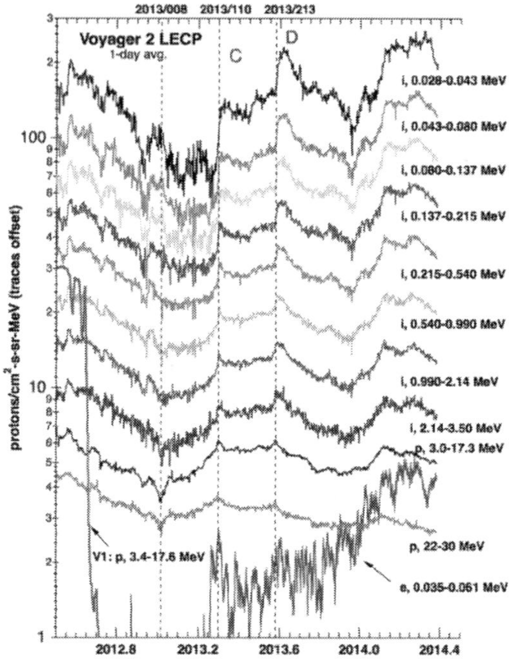

Figure 8. Stack plot of daily-averaged intensities in 8 ion and 2 proton channels covering 0.03-30 MeV and one electron channel 35-61 keV from the V2 LECP in 2012.5-2014.4. Traces have been offset vertically for clarity. Plotted for reference is the V1 LECP 3.4-17.6 MeV proton channel showing the abrupt drop on ≈2012/238 when V1 crossed the HP.

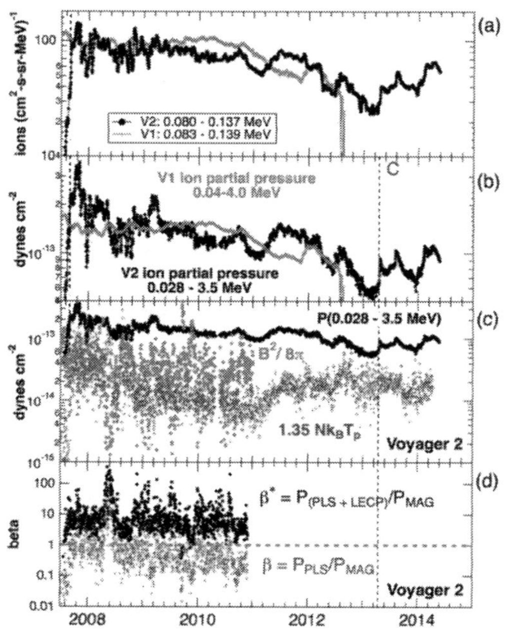

Figure 9. (a) Daily intensities of ions ≈80-140 keV at V1 and V2, 2007.5-2014.4. (b) Partial pressure of protons at V1 (0.04-4.0 MeV) and V2 (0.03-3.5 MeV). (c) V2 ion pressure from panel (b) compared to plasma and magnetic pressures. (d) Plasma beta with (β^*) and without (β) low-energy ion pressure included.

(assumed to be protons) at V1 and V2. Line C indicates where +*T*-directed ion streaming at V2 abruptly stopped and ion intensities increased. Figure 9(c) compares the partial ion pressure at V2, reproduced from panel (b), with the magnetic field pressure $P_{MAG} = B^2/8\pi$, where B is the field magnitude measured by the V2 Magnetometer (MAG), and plasma pressure $P_{PLS} = (1+0.35)Nk_BT_p$, where N is the plasma density and T_p the proton temperature measured by the V2 Plasma Science (PLS) instrument [10], and k_B is the Boltzmann constant (corrected data from the V2 MAG beyond 2010 were unavailable at this writing). We have set the electron temperature to $0.35\,T_p$, consistent with measured post-TS temperatures $T_p \approx 10^5$K and $T_e \approx 3\text{-}4\text{x}10^4$K [10]. Figure 9(d) shows the plasma beta (ratio of particle pressure to magnetic pressure), with $\beta = P_{PLS}/P_{MAG}$ calculated using the plasma pressure only, and $\beta^* = P_{PLS+LECP}/P_{MAG}$ calculated using the combined plasma and low-energy ion pressures. Quantity β varied about unity from the TS crossing until mid-2008 and then decreased by a factor ≈2-3 through 2010. Quantity β^* remained well above unity and showed large fluctuations about an average ≈5 that are due mainly to variations in B from the TS crossing through 2010. The large value of β^* reflects the dominant role played by pickup protons in maintaining pressure balance in the HS [10,11,24]. However, since the bulk of the HS pickup proton population with energies above a few hundred eV are not measured by either the V2 PLS or LECP instruments, β^* greatly underestimates the actual HS beta that includes the full energy spectrum of pickup protons heated and accelerated at the TS and possibly in the HS [10,11,16,24].

The evolution of the low-energy ion intensities in the 80-140 keV and other channels from the V1 and V2 LECP were remarkably similar, on average, until the V1 HP crossing. The partial pressures in panel (b) also evolved similarly, showing about the same magnitude and decay rate in ≈2012.0-2013.0. The lowest pressure at V2 of ≈$5.5\text{x}10^{-14}$ dyne cm^{-2} occurred in ≈2013.0-2013.15, representing a 7-fold drop from the post-TS peak of $3.5\text{x}10^{-13}$ dyne cm^{-2} in 2007.8, and a 3-fold drop from the peak pressure in 2011. Since the suprathermal ion partial pressure is only ≈15% of the total pressure in the heated pickup proton population in the HS [24], it is clear from figure 9(c) that the HS remains a high beta plasma, unless of course there are large increases in the magnetic field amplitude that have not yet been reported.

4. Summary

We have briefly summarized measurements made by the LECP instrument on V1 near its crossing the HP and entering the LISM on or about day 238 of 2012 at 121.6 AU, and on Voyager 2 mainly during the period 2012-2014.4 characterized by large variations in the intensities and angular distributions of low-energy HS ions and the reappearance of low-energy HS electrons. Results from V1 not previously published included the energy dependence of ion intensity decreases prior to the HP crossing and a quantitative measure of low-energy HS proton pitch angle distributions extending across the HP into the LISM. For V2 we described the evolution in time and with ion energy of the tangential streaming of ions that is directed from the nose toward the tail of the HS, summarized the recovery of low-energy HS ion intensities since their decline to minimum levels during 2013.0-3013.3, and discussed the implications of this intensity minimum and its subsequent increase on the ion partial pressure in the HS and its magnitude compared to those of the thermal protons and the magnetic field.

5. References

[1] Stone E C, Cummings A C, McDonald F B, Heikkila B C, Lal N and Webber W R 2005 *Science* **309** 2017
[2] Decker R B, Krimigis S M, Roelof E C, Hill M E, Armstrong T P, Gloeckler G, Hamilton D C and Lanzerotti L J 2005 *Science* **309** 2020

[3] Burlaga L F, Ness N F, Acuña M H, Lepping R P, Connerney J E P, Stone E C 2005 *Science* **309** 2027
[4] Gurnett D A and Kurth W S 2005 *Science* **309** 2025
[5] Webber W R and McDonald F B 2013 *Geophys. Res. Lett.* **40** 1665
[6] Krimigis S M, Decker R B, Roelof E C, Hill M E, Armstrong T P, Gloeckler G, Hamilton D C and Lanzerotti L J 2013 *Science* **341** 144
[7] Burlaga L F, Ness N F and Stone E C 2013 *Science* **341** 147
[8] Stone E C, Cummings A C, McDonald F B, Heikkila B C, Lal N and Webber W R 2013 *Science* **341** 150
[9] Gurnett D A, Kurth W S, Burlaga L F and Ness N F 2013 *Science* **341** 27
[10] Richardson J D, Kasper J C, Wang C, Belcher J W and Lazarus A 2008 *Nature* **454** 63
[11] Decker R B, Krimigis S M, Roelof E C, Hill M E, Armstrong T P, Gloeckler G, Hamilton D C and Lanzerotti L J 2008 *Nature* **454** 67
[12] Stone E C, Cummings A C, McDonald F B, Heikkila B C, Lal N and Webber W R 2008 *Nature* **454** 71
[13] Burlaga L F, Ness N F, Acuña M H, Lepping R P, Connerney J E P, Richardson J D 2008 *Nature* **454** 75
[14] Krimigis S M, Armstrong T P, Axford W I, Bostrom C O, Fan C Y, Gloeckler G and Lanzerotti L J 1977 *Space Sci. Rev.* **21** 329
[15] Swisdak M, Drake J F and Opher M 2013 *Astrophys. J.* **774** L8
[16] Fisk L A and Gloeckler G 2013 *Astrophys. J.* **776** 1
[17] Schwadron N A and McComas D J 2013 *Astrophys. J.* **778** L33
[18] Borovikov S N and Pogorelov N V 2014 *Astrophys. J.* **783** L16
[19] Florinski V, Jokipii J R, Alouani-Bibi F and LE Roux J A 2013 *Astrophys. J.* **776** L37
[20] Richardson J D and Decker R B 2014 *These Proceedings*
[21] Decker R B, Krimigis S M, Roelof E C, Hill M E 2012 *Nature* **489** 124
[22] Decker R B, Krimigis S M, Armstrong T P, Mosley C J, Gloeckler G and Hamilton D C 2002 *Ad. Space Res.* **32** 591
[23] Giacalone J and Decker R B 2010 *Astrophys. J.* **710** 91
[24] Roelof E C, Krimigis S M, Mitchell D G, Decker R B, Richardson J D, Gruntman M and Funsten 2010 *AIP Conf. Proc.* **1302** 133

Acknowledgments
This work was supported by the Voyager Interstellar Mission under NASA Contact NNX07AB02G. We thank the Voyager 2 PLS and MAG teams for making their data available at NSSDC and on their instrument web sites.

Energetic Neutral Atom (ENA) intensity gradients in the heliotail during year 2003, using Cassini/INCA measurements

K. Dialynas[1], **S. M. Krimigis**[1,2], **D. G. Mitchell**[2], **and E. C. Roelof**[2]

[1]Office of Space Research and Technology, Academy of Athens, 106 79 Athens, Greece.

[2]Applied Physics Laboratory, Johns Hopkins University, Laurel, MD, USA.

Email: kdialynas@phys.uoa.gr

Abstract. In the present study we use all-sky energy-resolved (5-55 keV) energetic neutral atom (ENA) maps obtained by the Ion and Neutral CAmera (INCA) on board Cassini during the time period DOY 265/2003 to 268/2003, to investigate the properties of the peak-to-basin ENA emissions in the direction of the heliotail. Our conclusions can be summarized as follows: (1) a relatively smooth boundary (called "transition region") between the very low (basin) and high (tail) ENA emissions from the heliosheath, with a spatial width of ~30° deg in ecl. longitude, that no theory had predicted to date, is identified in the energy range of 5-55 keV; (2) the ENA intensity gradient in this transition region is almost invariant as a function of both ecl. Latitude and energy, with an average value of ~2.4% per degree; (3) the deduced partial plasma pressure distributions in the 5-55 keV energy range are consistent with the ENA intensity distributions in the same energy range, while the ENA intensity gradient translates to a corresponding partial pressure gradient that occurs in the transition region; and (4) this partial pressure gradient is possibly not consistent with a tail magnetic field configuration that is similar to the measured magnetic fields by the Voyagers in the nose hemisphere.

1. Introduction

New imaging using Energetic Neutral Atom (ENA) imagers (e.g. Ion and Neutral Camera-INCA and Interstellar Boundary Explorer-IBEX) have led to new aspects of heliospheric physics and insights about heliosphere interactions with the local interstellar medium (LISM) [1],[2]. ENAs created by charge-exchange (CE) interactions in the heliosphere, can be thought of as long-distance communicators between different locations in space, i. e. their creation and detection regions. With that in mind and the knowledge of the powerful synergism between ENA measurements and in-situ energetic ion measurements by the Voyagers (e.g. [3]) a number of previously unknown parameters in the heliosphere, that were substantially at variance with theoretical models have been recently analyzed and interpreted.

The INCA detector onboard Cassini [4], capable of producing energy-resolved ENA maps of the heliosphere in the energy range of 5-55 keV, has significantly contributed to advancing our understanding on the heliosphere's interaction with the LISM. An unexpected heliospheric signature, known as "Belt" [1], a high ENA intensity region that loops through a rough circle in the sky sphere presenting a remarkable degree of symmetry in ecliptic coordinates, has been identified in all INCA channels. The Belt was found to be energy independent in the 5-55 keV range and possibly corresponds to a reservoir of particles that exist within the heliosheath, moving in a great circle along the nose to tail direction, passing through the ecliptic poles, constantly replenished by new particles from the solar wind, that is presumably the original source of the observed ENAs in the eV to keV energy range [5]. The high flux ENA emissions are moderately well organized in Galactic coordinates (with a ~30° tilt with respect to the galactic equator) as the ENA minima appear in the vicinity of the north and south galactic poles, but these ENA emissions are minimized in a rotated frame where its north pole points toward 190° in ecliptic longitude and 15° in ecliptic latitude.

Although a lot of effort has been expended over the past decades to estimate the overall size of the heliosphere (e.g. [6]), new analysis of the INCA ENA data [7] have shown that the heliosheath appears twice as thick along the Voyager 2 (VGR-2) line of sight (LOS; L_{VGR2}~71(+30, -15) AU) compared to the Voyager 1 (VGR-1) LOS (L_{VGR1}~31(+31, -18)AU). We know now from measurements of VGR-1 and VGR-2 that the termination shock (TS) was crossed at distances of ~94 AU and ~84 AU in 2004 and 2007, respectively [8],[9],[10],[11], while VGR-1 has probably reached the "depletion region", i.e the interface between the solar wind plasma and the interstellar plasma, at ~121.6 AU [12]. These observations verify the calculations [13] (at least for the VGR-1 direction) using Cassini/INCA ENA measurements. Subsequent measurements of the plasma density in this region [14] strongly suggest that the boundary crossed is indeed the heliopause.

In the present study we re-visit the 5-55 keV INCA ENA images during the year 2003, to determine the properties of the peak to basin ENA emissions in the direction of the heliotail.

2. Instrumentation and data set information

As part of the Magnetospheric Imaging Instrument (MIMI), the Ion and Neutral Camera (INCA) onboard Cassini spacecraft is designed to perform ENA measurements originating from the vicinity of Saturn (e.g [15], [16]). INCA is a large geometry factor detector (G~0.6 cm^2 sr for H) that possesses high sensitivity to remotely detect even the lowest intensity events from large distances [4]. As shown in previous studies [1],[7], imaging of the heliosphere lies well within the instrument capabilities. The detection of the ENAs is based on the time-of-flight (TOF) technique, and several discrete energy passbands are defined within the energy range of ~5.2 to >220 keV. INCA utilizes a broad field of view (FOV) of 90° in the nominal Cassini roll direction (azimuthal direction) and 120° in the direction perpendicular to the spacecraft roll plane (elevational direction). The detector's nominal angular resolution is ~5° for hydrogen (although it is energy dependent and is more than 5° for low energies).

The time period just prior to Saturn Orbit Insertion (SOI, July of 2004), provided a unique opportunity for the INCA imager to obtain neutral H images that correspond to extensive regions in the heliosphere, away from possible contaminating sources, such as Saturn's magnetosphere. Figure 1 illustrates four all-sky (320×160 pixels), energy resolved, color-coded intensity images of heliospheric ENAs (H) in the energy range of 5–55 keV, for the time period DOY 265/2003 to 268/2003, organized in ecliptic coordinates. The data shown in this figure are the accumulation of suitable images that result from a "stare" mode of the Cassini spacecraft (i.e. the spacecraft was not rotating about its spin axis that can result in a 360° scan across the sky), where INCA FOV was pointing in the direction of the heliotail, away from the Sun, that was located towards the "nose' during that time. Blackened pixels centered at ~0° in ecliptic latitude and ~30° in ecliptic longitude correspond to the direction of Saturn and as a consequence, no observations where available for this particular region (from a distance of ~0.88 AU away from Saturn, the planet's magnetosphere occupies only a few pixels in the sky, which were excluded from the analysis).

The high ENA intensities shown in figure 1 that occur in the heliotail (~60° to 110° in ecliptic longitude and ~-40° to 30° in ecliptic latitude), identified in all energy ranges, are part of the previously defined "Belt" [1] and are associated with the descending phase of solar cycle 23 (SC23) (approximately 2 yrs after the solar maximum of SC23). The general agreement between the spectra of the proton intensities inferred from the ENA measurements [17] and the Voyager ion measurements [18] suggest that the source of these ENA emissions is most likely in the heliosheath. The minima of the ENA emissions (Basin) that reside in the center of the image, at ~-30° to 30° in ecliptic longitude and -60° to 30° in ecliptic latitude, roughly coincide with the galactic South pole, with a ~30° tilt with respect to the galactic equator [6].

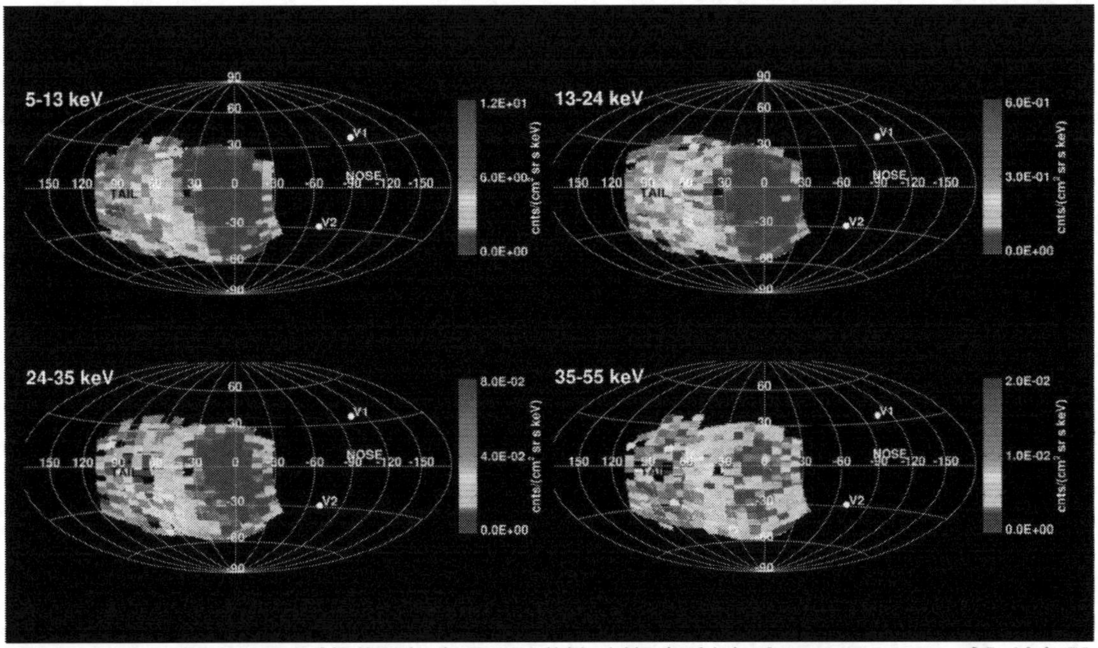

Figure 1. Intensity images of heliospheric ENAs (320x160 pixels) in the energy ranges of 5–13 keV, 13–24 keV, 24–35 keV, and 35–55 keV, organized in ecliptic coordinates (Hammer projection).

The role of all possible background sources, namely X-ray, ultraviolet, and extreme ultraviolet radiation, together with possible background from galactic Cosmic Rays (CR), have been evaluated elsewhere [1],[5] and proved to be uncorrelated with the ENA emissions that INCA detects from the heliosphere. In addition, while the Sun is located opposite of the imaging direction (i.e. no pixels are contaminated by sunlight), and the ENA emissions from Saturn's magnetosphere are treated with exclusion criteria, we assume that our ENAs are consistent with Poisson counting statistics, i.e. all measured ENA counts C are bounded to an uncertainty of $\sqrt{C}$. The measured H ENA differential intensities as a function of energy ($j_{ENA} = C/(cm^2 \cdot sr \cdot s \cdot keV)$) are calculated as the LOS integral of the emission region (i.e. the heliosheath, that contains the bulk of the energetic ion population) over the intensity of the energetic ions $j_{ion}(E)$ and the interstellar flow for H-Atoms $n_H(S)$, multiplied by the energy depended charge-exchange cross section for the interaction between ion and neutral species ($\sigma^{10}(E)$).

$$j_{ENA}(E) = \sigma^{10}(E) \cdot \int_L j_{ion}(E) \cdot n_H(L) dL \tag{1}$$

As explained previously [5], the identified minima and maxima in the ENA intensities shown in figure 1, do not correspond to low and high total accumulated count regions respectively, i.e. the belt is not necessarily a high count region and becomes a high intensity region after the ENA count maps are normalized over the accumulation time per each pixel. To ensure the credibility of our results, we have averaged our original intensity maps shown in figure 1 over 11.25° in ecl. latitude and 9° in ecl.

longitude. With this averaging in intensities, the total accumulated counts found at each 11.25°x9° bin in the tail for the time period examined (e.g. for the 5-13 keV channel) is ~10-40 counts (on average), while in the basin, the total counts per each bin is >100 counts.

3. Intensity gradients in the heliotail

One of the most interesting features in the images shown in figure 1, concerns the transition of the high ENA intensities in the tail, to the low ENA intensities in the Basin, that occurs at ~60° to 30° in ecl. longitude, henceforth called "transition region". This region is identified in all energy ranges and serves as a relatively smooth boundary, with a spatial width of ~30° deg in ecl. longitude, between the very low and high ENA emissions in the heliosheath, that no theory had predicted to date.

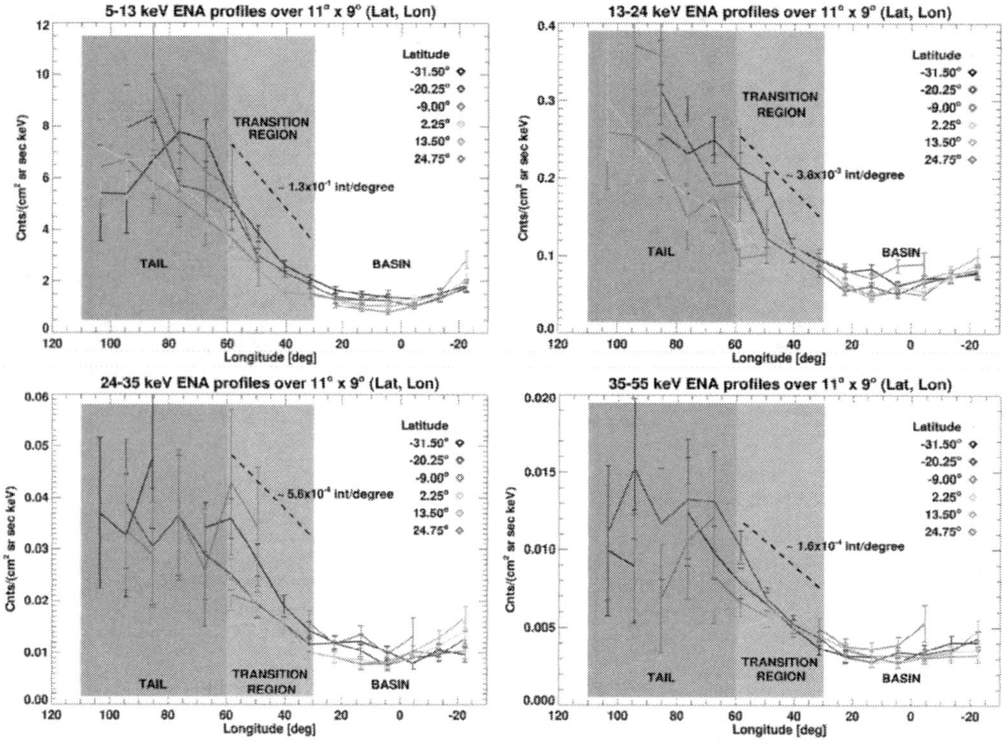

Figure 2. ENA longitudinal profiles for various latitudes in the energy range of 5-55 keV (ecliptic coordinates). Shaded grey areas are used as a visual aid to distinguish between the heliotail, the transition region and the basin. Dashed lines are average linear ENA intensity drop rates as a function of longitude, as explained in the text. Mean χ^2 values for these dashed lines are ~0.9, ~1.8, ~0.6 and ~0.4 for 5-13 keV, 13-24 keV, 24-35 keV and 35-55 keV channels respectively.

Although the transition region is apparent in the INCA data, it is shown fairly quiet in the low energy IBEX maps [19] and despite the fact that the 4.29 keV IBEX map [20] is very consistent with the > 5keV INCA maps, the ENA emissions are still minimal in the transition region in that particular IBEX channel. The identified INCA "basin" is somewhat smaller in the longitude direction, while the ENA intensities in the 4.29 keV IBEX map start to build up ~15-20° closer in the tail direction compared to the > 5keV INCA channels. Nevertheless, a similar, slightly displaced transition region between the high ENA intensities in the tail and the low ENA intensities in the so-called "offset heliotail" (part of which is the INCA basin) is still visible in the 4.29 keV IBEX data [21]. We must keep in mind, however, that our data were taken some six years prior to the launch of IBEX, so some time variability cannot be excluded. Our primary purpose is to examine the basic properties of this unexpected, INCA defined, transition region and its possible energy dependence.

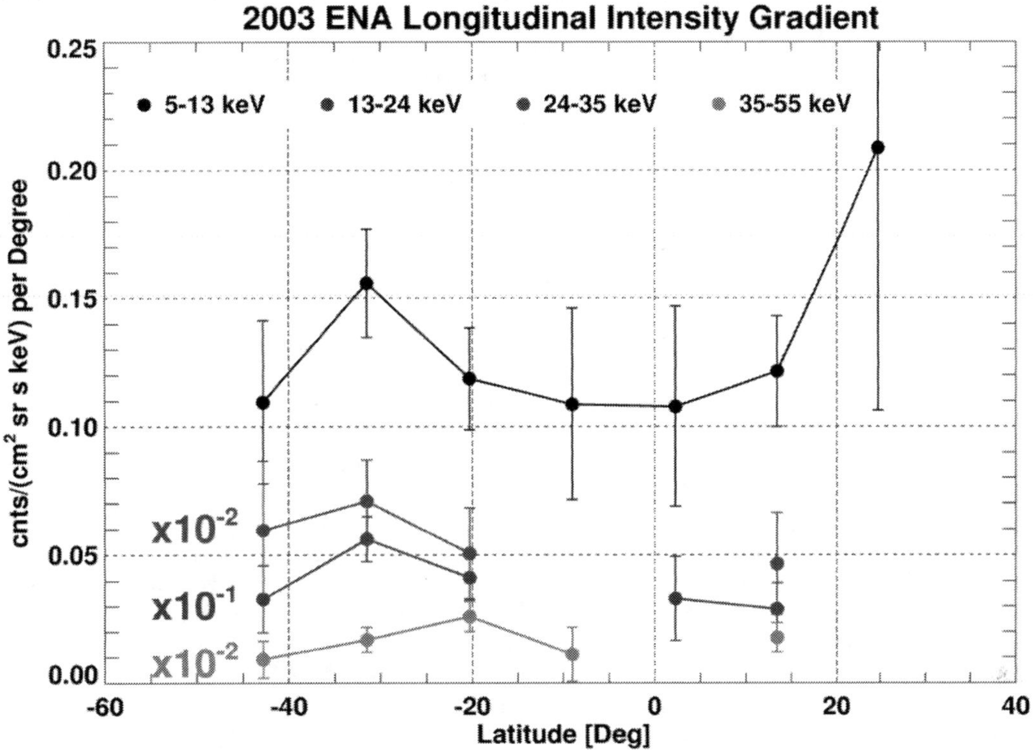

Figure 3. Longitudinal ENA intensity decrease per degree profiles as a function of Latitude that corresponds to the transition region, for all TOF channels, that resulted after fitting the ~60° to 30° in ecl. longitude ENA distributions seen in figure 2 with a linear function.

As can be seen in figure 2, the ENA intensities in the transition region drop almost linearly as a function of longitude and the basic question that needs to be answered is: How sharp is this transition region? Figure 3 shows the longitudinal ENA intensity drop rates as a function of latitude that resulted after performing normal weighted linear fits to the ENA distributions seen in figure 2, for all INCA TOF channels. Strictly speaking, these intensity drop rates represent the intensity decrease per degree that corresponds roughly to the transition region and not the "Peak to basin" intensity decrease. However, if one "extrapolates" the fitting lines (e.g. dashed lines in figure 2) to reach the longitudes where the peaks of the ENA longitudinal distributions occur, it becomes clear that these calculated rates do in fact reflect what would be the "peak to basin" intensity drop rate per each INCA TOF channel.

Given the calculated error bars, the rates shown in figure 3 are undoubtedly constant per each TOF channel as a function of ecl. latitude and an average of these rates per each channel can be adopted, corresponding to a characteristic intensity gradient rate. These characteristic intensity rates can be found in table 1 (3rd column), together with a calculation of the percentage of the intensity decrease over this 30° degree wide transition region per each TOF channel (5th column). By taking an average of the ENA intensities at ~60° ecl. long., that corresponds to the edge of the transition region towards the tail, we conclude that the intensity gradient over this region is almost constant as a function of energy (within the calculated error bars shown in table 1) and takes an average value of R~2.4% per degree.

Table 1. Transition region intensity gradient characteristic numbers.

INCA TOF channel [keV]	Average ENA intensity at ~60° ecl. Longitude [cnts/(cm^2 sr s keV)]	Mean calculated rate R [cnts/(cm^2 sr s keV) per degree]	Mean rate relative percentage error (dR/R %)	Percentage of Intensity decrease per degree (R%)
5-13	4.8×10^{0}	$(1.3\pm0.2)10^{-1}$	15.4	2.8±0.4
13-24	1.5×10^{-1}	$(3.8\pm0.5)10^{-3}$	13.1	2.6±0.3
24-35	2.7×10^{-2}	$(5.6\pm1.0)10^{-4}$	17.8	2.1±0.4
35-55	7.5×10^{-3}	$(1.6\pm0.3)10^{-4}$	18.7	2.1±0.4

The ratio of the peak ribbon flux to the underlying Globally Distributed Flux (GDF), as a function of the IBEX energy channels (e.g. in every ~5° swath of ecliptic latitude [20]) yields a number of <7% per degree at ~1 keV and <3% per degree at ~4 keV. As shown in the literature (e.g. [20]), the ribbon is most distinctly apparent between 1 keV and 2 keV (average width of ~20° in the 0.7 to 2.7 keV energy range), but broadens and changes its shape with increasing energy (width of ~60° at 4.29 keV) so as to match the belt as identified by INCA in the > 5 keV energy range [17],[5].

The width of the belt, as shown in the ENA emissions in the > 5 keV energy range, is almost energy independent [5]. Accordingly, the ENA intensity decrease per degree rates are almost energy independent, consistent with the ratio of the peak ribbon flux to the GDF at the 4.29 keV IBEX channel. While the < 5 keV ENA emissions, as detected by IBEX, evolve with increasing energy, the consistency between the intensity decrease rates in both the 4.29 keV IBEX channel and the > 5 keV INCA channels, provides further evidence on the concept put foreword in the literature [5] concerning the interpretation of the observed global ENA heliosphere from eV to keV energy range.

4. Partial 5-55 keV INCA pressure

By taking a simpler form of equation (1), i.e. $j_{ENA} = \sigma^{10} n_H j_{ion} L_{ion}$, one can easily convert the measured ENA intensities to their parent H^+ intensities.

$$j_{H^+} = j_{ENA} / (\sigma^{10} n_H L_{ion}) \quad (2)$$

For a given H^+ momentum (p) and measured energy ΔE (INCA energy passbands), it follows [9] that we are able to calculate the partial H^+ pressure ($\Delta P = (4\pi p/3) j_{ion} \Delta E$).

$$\Delta P(pPa) = (7.0\times10^{-8}) \cdot v(km/s) \cdot j_{ENA} \cdot \Delta E(cm^2 \cdot sr \cdot s)^{-1} / \sigma^{10} n_H L_{ion} \quad (3)$$

Figure 4 illustrates the partial H^+ pressure for energies > 5 keV, that resulted by the use of equation (3) that is solved separately for each 11.25° x 9.0° (ecl long. x ecl lat.) pixel of the measured ENA intensities as a function of energy, as explained in section 3. The plasma pressure distributions shown in figure 4 are very consistent with the ENA intensity distributions shown in figure 1. Namely, the high ENA intensities in the tail are associated with high partial pressures, while the partial pressures drop to minimum values in the basin. The 5-55 keV partial pressure can be calculated simply by adding the individual pressure distributions shown in figure 4. We note that the plasma pressures shown in figure 4 are consistent with the corresponding plasma pressure calculations performed in the past [1].

Interestingly, the ENA intensity gradient shown in the transition region, translates to a partial pressure gradient in the same region for all INCA TOF channels, as shown in figure 4. Krimigis and his colleagues [1], provided a modified "diamagnetic bubble" concept concerning the shape of the heliosphere, explained earlier in the literature [22], in which the ENAs that INCA detects form a region of enhanced particle pressure inside the heliopause and contribute in balancing the pressure of the ISMF, moving the interaction with the ISMF from the TS to the HP. In situ ion measurements in the > 28 keV energy range obtained by the two Voyagers during their survey through the heliosheath, showed that the PUI pressure in the heliosheath is almost persistently higher than the magnetic field pressure [17] (i.e. plasma $\beta > 1$), which is a basic premise for the diamagnetic behavior.

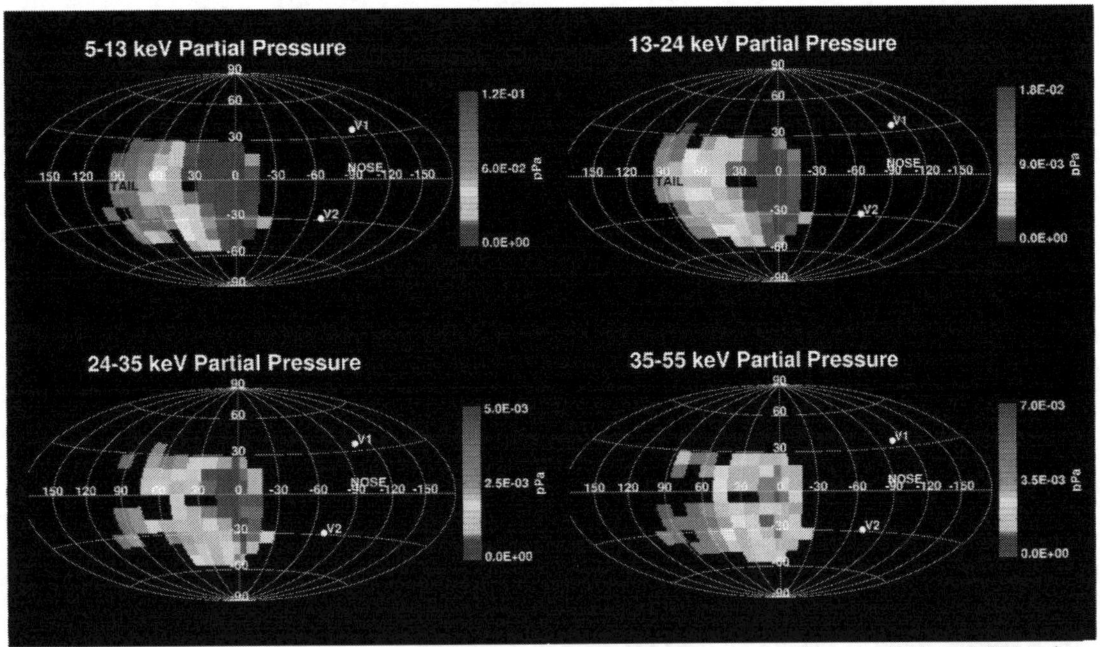

Figure 4. H^+ particle partial pressure in the 5-55 keV energy range exerted by the INCA H ENA data, assuming a heliosheath thickness of ~50 AU, organized in ecliptic coordinates (Hammer projection) (1 pPa $=10^{-11}$ dynes/cm^2). The mean relative percentage errors for these images are ~14% for the 5-13 keV, 13-24 keV channels and ~17% for the 24-35 keV, 35-55 keV channels.

In such case, the partial pressure gradient shown in figure 4 is expected to produce a "shielding current" (J) in the heliopause that, with the assumption of isotropic pressure balance and in the absence of convection, can be expressed as follows.

$$\nabla P = \vec{J} \times \vec{B} \tag{4}$$

The 2003 INCA images indicate that the gradients of the LOS-integrated partial pressure (5<E<55 keV) in the general direction of the anti-nose (or tail) region of the heliosheath (HS) is $\partial\Delta P/\partial\varphi > 0$ (finite) and $\partial\Delta P/\partial\theta \approx 0$ (much smaller than $\partial P/\partial\varphi$). Assuming that these pressures reside in the HS, we can test against these data the implications for (or against) the existence of the simplest relation for total plasma pressure balance with the total electrical forces on the plasma, to wit, $\nabla P = \vec{J} \times \vec{B}$, which written out in the usual spherical polar coordinates becomes:

$$\left\{\begin{aligned} J_\theta B_\varphi - J_\varphi B_\theta &= \frac{\partial P}{\partial r} \\ J_\varphi B_r - J_r B_\varphi &= \frac{1}{r}\frac{\partial P}{\partial \theta} \\ J_r B_\theta - J_\theta B_r &= \frac{1}{r\sin\theta}\frac{\partial P}{\partial \varphi} \end{aligned}\right\} \tag{5}$$

If these relations hold, they immediately imply that the total pressure gradient along the magnetic field vanishes identically, i.e. $\vec{B}\cdot\vec{\nabla}P = 0$. Therefore,

$$B_r\frac{\partial P}{\partial r} + \frac{B_\theta}{r}\frac{\partial P}{\partial \theta} + \frac{B_\varphi}{r\sin\theta}\frac{\partial P}{\partial \varphi} = 0 \tag{6}$$

Assuming that the relationships for the INCA partial pressure (ΔP) also hold (qualitatively) for the total pressure (P), then the relation among the pressure gradient components becomes:

$$\frac{\partial P}{\partial \varphi} \approx -r\sin\theta\frac{B_r}{B_\varphi}\frac{\partial P}{\partial r} \tag{7}$$

which cannot be influenced by B_θ because $\partial P/\partial\theta \approx 0$. Of course, INCA gives us no direct information on the radial pressure gradient ($\partial P/\partial r$), but nonetheless, equation (7) can be useful if we have information about the configuration of the magnetic field in the HS.

From Voyagers 1/2 (in the nose hemisphere of the HS), we know that the magnetic field is nearly Parker-like throughout the HS [23], so that $-B_r/B_\varphi \approx |B_r/B_\varphi| < 1$ (the ratio of field components is, ideally, independent of polarity). The LECP energetic ion measurements (~30 keV<E<1 MeV/nuc) indicated that the partial ion pressure at both Voyagers (ΔP) was relatively constant [24] over radial distances ~30AU. Thus, $r\partial\Delta P/\partial r = \Delta P(\partial \ln \Delta P/\partial r) << \Delta P$. On the other hand, our INCA measurements (in the anti-nose hemisphere) gave $\partial\Delta P/\partial\varphi \sim \Delta P$ over a longitude range of $\Delta\varphi \approx 30^0 \approx 0.5\ rad$. Consequently, we can see that (given all the foregoing assumptions) if INCA is indeed measuring ENAs generated in the anti-nose hemisphere of the HS, that region cannot have a pressure distribution and a magnetic field configuration that are similar to those measured by the Voyagers in the nose hemisphere.

Equation (7) also allows us to speculate on the possibility of a non-Parker field in the anti-nose hemisphere of the HS. If the field is more tail-like, then we would have $|B_r/B_\varphi| > 1$, and the INCA data could be consistent with a substantial radial gradient ($\partial P/\partial r$). However, if the field had the same polarity properties as a Parker field ($-B_r/B_\varphi > 0$), the radial gradient would be positive (because the INCA measurement yields $\partial\Delta P/\partial\varphi > 0$), and the pressure would be increasing down the tail. In any case, equation (7), would still imply a finite radial gradient, but it would have the uncomfortable property that the sign of the radial gradient depends on the signs of the components of the (non-Parker) field in the anti-nose sector viewed by INCA.

The discussion presented here is based on several assumptions, as noted earlier, and that all contributions in our 2003 ENA image came from ENAs generated by energetic protons in the heliosheath (which, as explained earlier, is most likely). We should note that an attempt to quantitatively derive electric currents from equation (4) would be risky, because it would depend on the assumed behavior of the magnetic field in the anti-nose hemisphere (where our ENA view is in 2003) and (even if we appeal to simulations) we have no means of independent confirmation by direct *in situ* measurements (as we have from the Voyagers in the noseward hemisphere).

5. Summary and Conclusions

By utilizing all available Cassini/INCA remote observations over the time period 265/2003 to 268/2003, we have produced all-sky, energy resolved (5–55 keV energy range) H ENA maps of the heliosphere in ecliptic coordinates. These H ENA maps were integrated over 11.25° x 9.0° ecl. long x ecl. lat to investigate the peak-to-basin ENA intensity gradient in the heliosheath, that translates to a corresponding tail–to-basin partial plasma pressure gradient. Primary results of this study are summarized as follows:

- A relatively smooth boundary (called "transition region") between the very low (basin) and high (tail) ENA emissions from the heliosheath, with a spatial width of ~30° deg in ecl. longitude, that no theory had predicted to date, is identified in the energy range of 5-55 keV.
- The ENA intensity gradient in this transition region is almost invariant as a function of both ecl. latitude and energy (for >5 keV), with an average value of ~2.4% per degree.
- The deduced partial plasma pressure distributions in the 5-55 keV energy range are consistent with the ENA intensity distributions in the same energy range, while the ENA intensity gradient translates to a corresponding partial pressure gradient that occurs in the transition region.
- This partial pressure gradient is possibly not consistent with a tail magnetic field configuration that is similar to the measured magnetic fields by the Voyagers in the nose hemisphere.

Acknowledgments
The authors thank M. Kusterer (Johns Hopkins University Applied Physics Laboratory) for assistance with the INCA data processing. We are grateful to all colleagues on the MIMI team, who provided valuable comments that have improved the presentation. Work at JHU/APL was supported by NASA under contract NAS5 97271 and NNX07AJ69G and by subcontracts at the University of Maryland and the Office for Space Research and Applications of the Academy of Athens.

References

[1] Krimigis, S. M., Mitchell, D. G., Roelof, E. C., Hsieh, K. C., & McComas, D. J. 2009, *Science*, 326, 971.
[2] McComas, D. J., Allegrini, F., Bochsler, P., et al. 2009, *Science*, 326, 959.
[3] Decker, R. B., Krimigis, S. M., Roelof, E. C. and Hill, M. E. 2012, *Nature*, 489, 124.
[4] Krimigis, S. M., et al. 2004, *SSRv*, 114, 233.
[5] Dialynas K., Krimigis S. M., Mitchell D. G., Roelof E. C., and Decker R. B. 2013, *ApJ*, 778:40 (13pp), doi: 10.1088/0004-637X/778/1/40.
[6] Stone, E. C. 2001, *Science*, 293, 5527.
[7] Roelof, E. C., Krimigis, S. M., Mitchell, D. G., Decker, R. B., & Dialynas, K. 2012, *AIP Conf. Proc.*, 1436, 239.
[8] Decker, R. B., Krimigis, S. M., Roelof, E. C., et al. 2005, *Science*, 309, 2020.
[9] Decker, R. B., Krimigis, S. M., Roelof, E. C., et al. 2008, *Nature*, 454, 67.
[10] Stone, E. C., Cummings, A. C., McDonald, F. B., et al. 2005, *Science*, 309, 2017.
[11] Stone, E. C., Cummings, A. C., McDonald, F. B., et al. 2008, *Nature*, 454, 71.
[12] Krimigis S. M. et al. 2013, *Sience*, Vol 341, pp144-147.
[13] Krimigis, S. M., E. C. Roelof, R. B. Decker, and M. E. Hill, 2011, *Nature, 474,* pp. 359-361, doi: 10.1038/nature10115, 2011.
[14] Gurnett, D. A. et al., 2013 *Science 341*, 1489, DOI: 10.1126/science.1241681, 2013.
[15] Mitchell, D. G., Brandt, P. C., Roelof, E. C., et al. 2005, *GeoRL*, 32, 20501.

[16] Dialynas, K. et al. 2013, *J. Geophys. Res. Space Physics*, 118, 3027–3041, doi:10.1002/jgra.50295.
[17] Krimigis, S. M., Mitchell, D. G., Roelof, E. C., & Decker, R. B. 2010, in *AIP Conf. Proc.*, 1302, Pickup Ions Throughout the Heliosphere and Beyond, ed. J. Roux, G. Zank, A. J. Coates, & V. Florinski (College Park, MD: AIP), 79.
[18] Decker, R. B., et al. 2009, in *Proc. of Twelfth International Solar Wind Conference*, Solar Wind 12, ed. M. Macksimovic, N. Meyer-Vernet, M. Moncuquet, & F. Pantellini (Sant-Malo: AIP).
[19] Fuselier S. A. et al. 2009, *Science*, Vol 326, pp962-964.
[20] Schwadron, N. A., Allegrini, F., Bzowski, M., et al. 2011, *ApJ*, 731, 56.
[21] McComas, D.J. et al. 2013, *ApJ*, 771:77 (9pp), doi: 10.1088/0004-637X/771/2/77.
[22] Parker, E. N. 1961, *ApJ*, 134, 20.
[23] Burlaga, L. F. and N. F. Ness, 2014, *The Astrophysical Journal*, Volume 784, Issue 2, article id. 146, 14 pp. (2014).
[24] Decker, R. B., S. M. Krimigis, E. C. Roelof and M. E. Hill, Proceedings of *This Conference*, 2014.

Interstellar magnetic fields: from Galactic scales to the edge of the heliosphere

Katia Ferrière

IRAP, Université de Toulouse, CNRS, 9 avenue du Colonel Roche, BP 44346, F-31028 Toulouse Cedex 4, France

Abstract.
I review the observational properties of interstellar magnetic fields in our Galaxy, starting from the large Galactic scales and gradually zooming in toward the heliosphere. It emerges that the large-scale magnetic field is predominantly azimuthal in the Galactic disk, but that this global field orientation is severely perturbed in the small-scale surroundings of the heliosphere.

1. Introduction

It was Alfvén (1937) who opened the era of interstellar magnetic fields (ISMFs) when he correctly pointed out that a magnetic field threading interstellar space could naturally explain the confinement of cosmic rays in our Galaxy. Pushing this notion one step further, Fermi (1949) wrote in a seminal paper on the origin of cosmic radiation [1] that "cosmic rays are originated and accelerated primarily in the interstellar space of the Galaxy by collisions against moving magnetic fields." He went on to add, with surprising accuracy for the time, that "the magnetic field in the dilute matter is of the order of magnitude of 5×10^{-6} Gauss, while its intensity is probably greater in the heavier clouds."

At about the same time, two major advances occurred in observational astronomy. First, Hall (1949) and Hiltner (1949a,b) independently discovered that the optical light from nearby stars is linearly polarized [2–4] – a phenomenon that Davis & Greenstein (1951) attributed to dichroic extinction by elongated dust grains which are aligned by a coherent ISMF [5]. Second, Kiepenheuer (1950) realized that the general radio continuum emission from our Galaxy is synchrotron emission [6] – a process that implies the presence of relativistic electrons spiraling about magnetic field lines.

Thus, more than 60 years ago, two pioneering theoretical ideas as well as two ground-breaking observational discoveries pointed to the existence of ISMFs in our Galaxy. Since then, a multitude of observations have provided convincing evidence that strong ISMFs do indeed pervade our Galaxy. It has also become clear, from a theoretical point of view, that these magnetic fields play a crucial role in the interstellar medium (ISM): not only are they responsible for the acceleration and confinement of cosmic rays, but they also have a profound impact on the spatial distribution, dynamics, and energetics of the interstellar matter, and they influence – possibly even regulate – the vital process of star formation.

In this paper, I review the main observational methods employed to detect and measure ISMFs, and I summarize what each of these methods has taught us about their typical strength and direction, their overall configuration, and their spatial distribution. In Section 2, I consider the large-scale ISM, and in Section 3, I focus on an increasingly close vicinity of the heliosphere. Throughout this paper, $\vec{B}$ denotes the magnetic field vector, $\vec{B}_{\perp}$ the projection of $\vec{B}$ onto the plane of the sky, and $B_{\parallel}$ the component of $\vec{B}$ along the line of sight.

2. Large-scale interstellar medium

2.1. Polarization of starlight and dust thermal emission

Interstellar dust grains generally have irregular, elongated shapes, which, in a directional stellar radiation field, cause them to feel radiative torques. These torques have two important effects [see 7–9]. First, they spin up the grains to suprathermal rotation about their short axes. Second, they gradually bring the grain spin axes into alignment with the local ISMF. As a result, dust grains tend to line up with their long axes perpendicular to the local ISMF.

The magnetically aligned dust grains collectively act like a polarizing filter for starlight. Since they preferentially block the starlight component with polarization vector parallel to their long axes in the sky, the starlight that passes through is linearly polarized in the direction parallel to the sky-projected ISMF, $\vec{B}_{\perp}$. In consequence, measuring the direction of starlight polarization directly reveals the orientation of $\vec{B}_{\perp}$. This technique applied to a large number of stars indicates that the ISMF in the Galactic disk within a few kpc of the Sun is horizontal, i.e., parallel to the Galactic plane, and nearly azimuthal [10], making a small angle $\simeq -7°$ to the azimuthal direction [11]. See Figure 1.

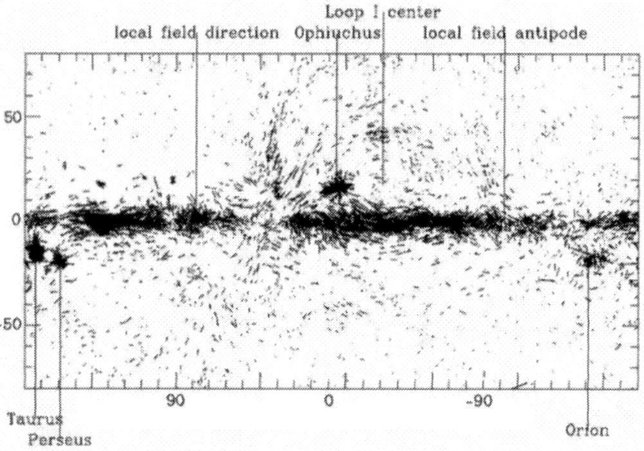

Figure 1. All-sky map in Galactic coordinates (with the Galactic center in the middle, longitude increasing to the left, and latitude increasing upward) of the polarization vectors of 8 662 stars from the compilation by [12]. The polarization vectors trace out magnetic field lines in the plane of the sky.

In addition to polarizing starlight, the magnetically aligned dust grains also emit infrared/submm thermal radiation in a dichroic manner. Since the emissivity is strongest for the radiation component with polarization vector parallel to the grains' long axes in the sky [13], dust thermal emission is linearly polarized in the direction perpendicular to $\vec{B}_{\perp}$. The 353 GHz observations of the *Archeops* balloon experiment [14] and, much more recently, the 353 GHz all-sky map from the *Planck* satellite [15] nicely bear out the ISMF orientation inferred from starlight polarimetry. See Figure 2.

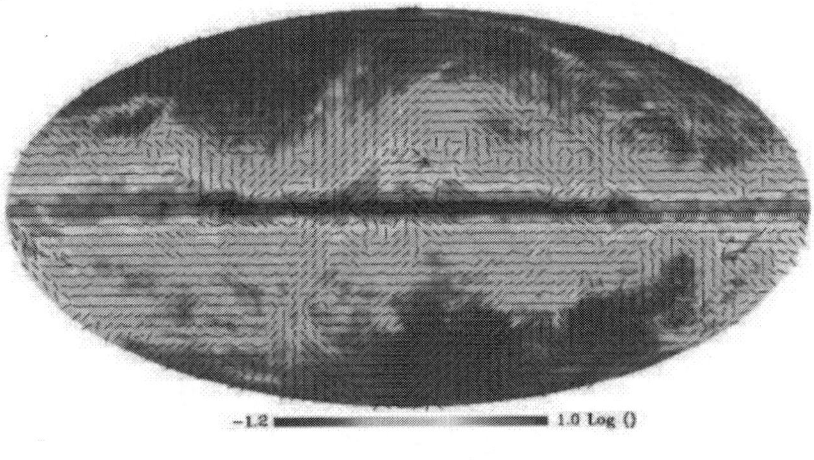

Figure 2. All-sky map in Galactic coordinates (with the Galactic center in the middle) of the 353 GHz dust thermal emission measured by *Planck*. The polarization vectors were rotated by 90° to trace the orientation of the sky-projected ISMF. Figure credit: Jean-Philippe Bernard, *Planck* collaboration (ESLAB 2013).

2.2. Synchrotron emission

A global method to map out the ISMF in the general ISM rests on the observed Galactic synchrotron emission.

Synchrotron emission is produced by relativistic electrons gyrating about magnetic field lines. The synchrotron emissivity at frequency ν due to a power-law energy spectrum of relativistic electrons, $f(E) = K_e\, E^{-\gamma}$, is given by

$$\mathcal{E}_\nu = \mathcal{F}(\gamma)\, K_e\, B_\perp^{\frac{\gamma+1}{2}}\, \nu^{-\frac{\gamma-1}{2}} , \quad (1)$$

where $\mathcal{F}(\gamma)$ is a known function of the electron spectral index and $B_\perp$ it the strength of the sky-projected magnetic field.

The spatial distribution of the Galactic synchrotron emissivity was modeled by [16], based on the all-sky 408 MHz radio continuum map of [17]. Several authors then used this synchrotron distribution model to derive the ISMF distribution in our Galaxy. To do so, the vast majority of them resorted to the standard double assumption that (1) relativistic electrons represent a fixed fraction (generally taken to be 1/100) of the cosmic-ray population and (2) cosmic rays and magnetic fields are in (energy or pressure) equipartition.[1] Relying on the cosmic-ray ion and electron spectra directly measured by the *Voyager* spacecraft, [19] verified that, near the Sun, cosmic rays and magnetic fields are indeed close to (pressure) equipartition, with a total magnetic field strength $B \simeq 5\ \mu$G. She also found that the total field strength has a radial scale length $\simeq 12$ kpc and a vertical scale height near the horizontal position of the Sun $\simeq 4.5$ kpc. This scale height is significantly greater than that of the Warm Ionized Medium ($H_{\rm WIM} \simeq 1.8$ kpc [20]), which in turn is significantly greater than the scale heights of the neutral phases [19], such that the plasma β changes from $\beta \sim 1$ in the Galactic disk to $\beta \ll 1$ in the Galactic halo.

Because synchrotron emission is linearly polarized perpendicular to $\vec{\boldsymbol{B}}_\perp$, information can also be gained on the orientation of $\vec{\boldsymbol{B}}_\perp$. Evidently, if the observing frequency is too low to avoid Faraday rotation (see Section 2.3), the received polarized signal must somehow be "de-rotated" in order to recover the true field orientation. In addition, if the ISMF has a randomly fluctuating component, the contributions from isotropic magnetic fluctuations to the polarized

[1] The argument usually put forward to justify the equipartition assumption is that energy equipartition corresponds almost exactly to the condition that the total (cosmic-ray + magnetic) energy is minimum [18].

Journal of Physics: Conference Series **577** (2015) 012008 doi:10.1088/1742-6596/577/1/012008

emission cancel out, leaving only the contribution from the ordered (i.e., regular + anisotropic random) ISMF, $\vec{B}_{\rm ord}$. Thus, while the *total* synchrotron intensity yields the strength of the *total* sky-projected ISMF, the *polarized* synchrotron intensity yields the strength and the orientation of the *ordered* sky-projected ISMF. For illustration, all-sky maps of the total and polarized synchrotron intensities are displayed in Figure 3.

In the large-scale vicinity of the Sun, the ratio of ordered to total ISMF strengths turns out to be $\simeq 0.6$ [21]. Together with $B \simeq 5\ \mu$G, this ratio implies an ordered ISMF strength $B_{\rm ord} \simeq 3\ \mu$G. Besides, the radio polarization vectors rotated by 90° (plotted in the right panel of Figure 3) are fully consistent with their optical and submm counterparts (plotted in Figures 1 and 2, respectively), thereby confirming that the large-scale ordered ISMF in the Galactic disk is horizontal.

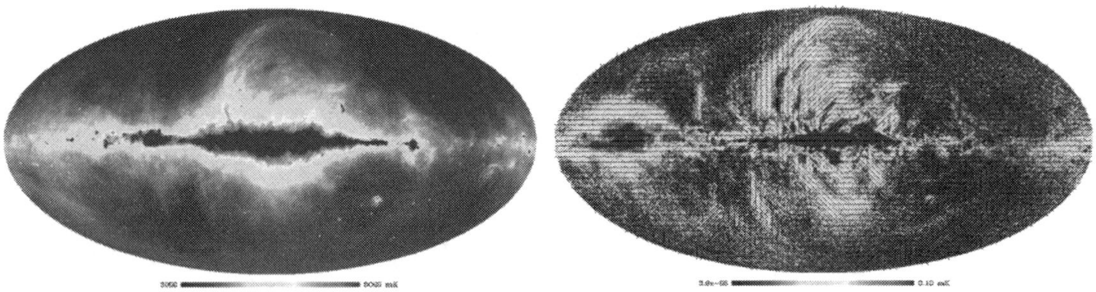

Figure 3. All-sky maps in Galactic coordinates (with the Galactic center in the middle) of the 1.4 GHz total intensity measured by the 25 m Stockert and 30 m Villa Elisa radio telescopes (*left*) and the 23 GHz polarized intensity measured by WMAP, with the polarization vectors rotated by 90° to trace the orientation of the sky-projected ISMF (*right*). Figure credit: Tess Jaffe.

2.3. Faraday rotation

In ionized regions of the ISM, the ISMF can be probed with Faraday rotation measures of Galactic pulsars and extragalactic sources of linearly polarized radio waves.

Faraday rotation of a linearly polarized radio wave occurs when the wave propagates along the magnetic field of an ionized region, as a result of its interaction with the free electrons gyrating about magnetic field lines. The angle by which the direction of polarization rotates is equal to the observing wavelength squared times the so-called rotation measure,

$$\mathrm{RM} = \mathcal{C} \int_0^L n_e\, B_\parallel\, ds\ , \tag{2}$$

where $\mathcal{C}$ is a numerical constant, n_e the free-electron density, $B_\parallel$ the line-of-sight component of the magnetic field, and L the path length from the source to the observer. In practice, the rotation measure of a given radio source can be determined by measuring the direction of polarization of the incoming radiation at at least two different wavelengths.

Rotation measures have now been obtained for about 1 200 Galactic pulsars and 42 000 extragalactic radio point sources (see Figure 4). Used in conjunction either with a model for the spatial distribution of free electrons or, in the case of Galactic pulsars, with their distances and dispersion measures, these numerous rotation measures have made it possible to gather a wealth of information on the strength, direction, and spatial configuration of the ISMF in ionized regions. Here are the key points that have emerged from Faraday rotation studies.

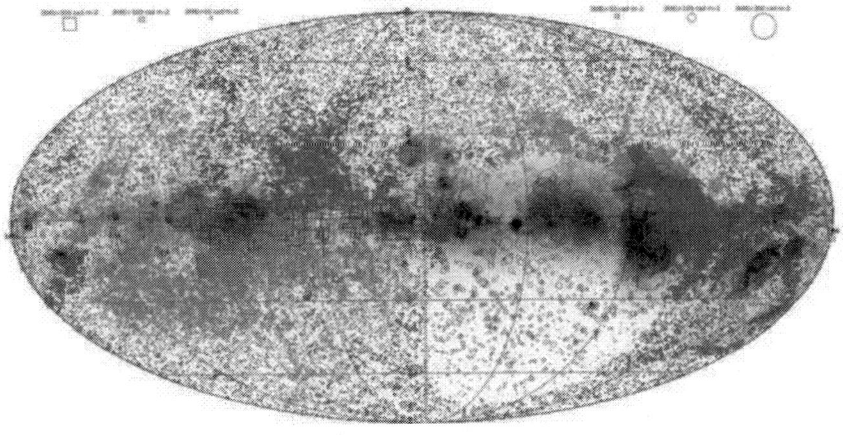

Figure 4. All-sky map in Galactic coordinates (with the Galactic center in the middle) of the rotation measures of $\simeq 42\,000$ extragalactic radio point sources from the NVSS ($\delta > -40°$) and S-PASS ($\delta < 0°$) surveys. Positive (negative) rotation measures, wich correspond to a magnetic field pointing on average toward (away from) the observer, are plotted in blue (red). Figure credit: Dominic Schnitzeler.

1) The ISMF has a regular component, $\vec{B}_{\rm reg}$, and a randomly fluctuating component, $\vec{B}_{\rm fluct}$. In the large-scale vicinity of the Sun, $B_{\rm reg} \simeq 1.5\ \mu$G and $B_{\rm fluct} \sim 5\ \mu$G [22]. $B_{\rm reg}$ increases toward the Galactic center, to $\gtrsim 3\ \mu$G at Galactocentric radius $R = 3$ kpc [23], i.e., with an exponential scale length $\lesssim 7.2$ kpc. Moreover, $B_{\rm reg}$ decreases away from the Galactic plane, albeit at a very uncertain rate. For reference, from the rotation measures of extragalactic radio point sources, [24] derived an exponential scale height ≈ 1.4 kpc; however, repeating their derivation with the WIM parameter values updated by [20] raises this scale height to ≈ 3.8 kpc, much closer to the scale height deduced from synchrotron data (see Section 2.2).

2) In the Galactic disk, $\vec{B}_{\rm reg}$ is nearly horizontal and generally dominated by its azimuthal component. In the large-scale vicinity of the Sun, $\vec{B}_{\rm reg}$ runs clockwise at an angle $\simeq -8°$ to the azimuthal direction [25], which is very close to the pitch angle $\simeq -7°$ inferred from starlight polarization (see Section 2.1). $\vec{B}_{\rm reg}$ reverses direction at least a couple of times with decreasing Galactocentric radius, but the exact number and radial locations of the field reversals are still highly controversial [23; 25–29]. These field reversals have often been interpreted as evidence that $\vec{B}_{\rm reg}$ is bisymmetric (azimuthal wavenumber $m = 1$), while an axisymmetric ($m = 0$) field would be expected from dynamo theory. In reality, [30] showed that neither the axisymmetric nor the bisymmetric picture is consistent with the existing pulsar rotation measures, and they concluded that $\vec{B}_{\rm reg}$ must have a more complex pattern.

3) In the Galactic halo, $\vec{B}_{\rm reg}$ could have a significant vertical component. Near the horizontal position of the Sun, [31] obtained $(B_{\rm reg})_z \simeq -0.14\ \mu$G above the Galactic midplane ($z > 0$) and $(B_{\rm reg})_z \simeq +0.30\ \mu$G below the midplane ($z < 0$), whereas [32] obtained $(B_{\rm reg})_z \simeq 0.00\ \mu$G toward the north Galactic pole and $(B_{\rm reg})_z \simeq +0.31\ \mu$G toward the south Galactic pole. In contrast to the situation prevailing in the Galactic disk, the horizontal component of $\vec{B}_{\rm reg}$ shows no sign of reversal with decreasing Galactocentric radius.

4) $\vec{B}_{\rm reg}$ exhibits some symmetry properties with respect to the Galactic midplane. At low latitudes (basically, in the disk), $\vec{B}_{\rm reg}$ appears to be roughly symmetric in z [26; 33], while at high latitudes (in the halo), the rotation-measure sky features a rather striking antisymmetry/symmetry in z in the inner/outer Galactic quadrants [25; 34; 35], which suggests

that $\vec{B}_{\rm reg}$ is roughly antisymmetric in z inside the solar circle [25; 35, but see also [33]]. Finding $\vec{B}_{\rm reg}$ to be symmetric in the disk and antisymmetric in the inner halo is consistent with the predictions of dynamo theory and with the results of galactic dynamo calculations [e.g., 36–38].

2.4. Zeeman splitting

In neutral (atomic or molecular) regions of the ISM that are sufficiently cold and dense, ISMF strengths can be inferred from Zeeman splitting measurements.

Zeeman splitting of a given atomic or molecular spectral line results from the interaction between the magnetic moment of the valence electrons and an external magnetic field. The amount of Zeeman splitting, $\Delta\nu$, is directly proportional to the magnetic field strength, B, so that, in principle, it suffices to measure $\Delta\nu$ in order to obtain B in the region of interest. In practice, however, $\Delta\nu$ is usually too small compared to the line width to be measurable. Under these conditions, one measures instead the Stokes parameter V, which is directly proportional to the line-of-sight component of the magnetic field, $B_\parallel$.

A vast body of Zeeman splitting data has now built up, both for the 21 cm line of HI (in atomic clouds) and for several centimeter lines of OH and other molecules (in molecular clouds). [39], who conducted a series of Zeeman measurements of the HI 21 cm absorption line, found that in atomic clouds B is typically a few μG, with a median value $\simeq 6\ \mu$G, and they confirmed earlier findings that B shows no tendency to increase with increasing gas density, n [e.g., 40; 41]. More recently, [42] performed a Bayesian analysis of Zeeman data from HI, OH, and CN surveys, on the basis of which they identified two regimes of B as a function of n: for $n < 300\ {\rm cm}^{-3}$ (atomic clouds), B appears to be independent of n, while for $n > 300\ {\rm cm}^{-3}$ (molecular clouds), B is randomly distributed between very small values and a maximum value $\propto n^{0.65}$. These results suggest that the ISMF in the diffuse ISM is generally strong enough (compared to turbulence) to inhibit the formation of atomic clouds other than through gas flows along field lines, whereas the ISMF in molecular clouds is often too weak (compared to self-gravity) to cause large departures from isotropic contraction.

3. Zooming in toward the Sun

One of the important conclusions of Section 2 is that the large-scale ISMF in the Galactic disk is nearly horizontal, with a dominating azimuthal component. However, in the presence of small-scale magnetic fluctuations, we do not expect the large-scale ISMF orientation to persist all the way in to the surface of the heliosphere. So the question that naturally comes to mind is how the ISMF orientation changes as one gradually approaches the Sun.

The two tools that can be used to probe the ISMF closer and closer to the Sun are the rotation measures of nearby pulsars and the polarization vectors of nearby stars. Dust thermal emission and synchrotron emission are of little help here, because their measurement in any given direction only provides the total emission produced along the entire line of sight through the Galaxy, with no possibility of separating out the contribution from the local ISM.

3.1. Faraday rotation of nearby pulsars

[43] collected all the available rotation measures, dispersion measures, and distances of pulsars within 2 kpc of the Sun, and he binned the pulsars into three distance ranges. For each distance range, he assumed a uniform ISMF and calculated the magnetic field vector, $\vec{B}$, that yielded the best fit to the pulsar data. The strength and the direction of each best-fit $\vec{B}$ are listed in Table 1, and the corresponding normalized $\vec{B}$ vector is plotted in Figure 5.

Because of the small number of pulsars, especially in the two closest distance ranges, the results of [43] are actually subject to very large uncertainties. The author did not attempt to compute error bars, but from his Table 2, the uncertainties in the ISMF strength and direction can be roughly estimated at $\sim 1\ \mu$G and $\sim 20°$, respectively.

Journal of Physics: Conference Series **577** (2015) 012008 doi:10.1088/1742-6596/577/1/012008

Despite the very large uncertainties, the work of [43] provides clear evidence that the ISMF fluctuates widely as one approaches the Sun and completely loses the nearly azimuthal direction that prevails in the large-scale ISM in to ~ 500 pc of the Sun. These wide magnetic fluctuations can most likely be attributed to the presence of nearby discrete structures, such as the Local Bubble and the Loop I superbubble.

Table 1. Strength, B, Galactic longitude, ℓ_B, and Galactic latitude, b_B, of the ISMF vector that best fits the pulsar data in three different distance ranges, with the number of pulsars used for the fitting procedure indicated in the second column [43]. The uncertainties in the ISMF strength and direction are estimated at ~ 1 μG and $\sim 20°$, respectively (see main text). The last column shows the red arrow used in Figure 5 to represent the best-fit ISMF direction.

Distance range	Number of pulsars	Best-fit ISMF B	ℓ_B	b_B	Arrow in Figure 5
$(0 - 300)$ pc	7	1.9 μG	16°	45°	——→
$(0 - 500)$ pc	18	2.7 μG	6°	28°	——→
$(0.5 - 2)$ kpc	103	2.5 μG	80°	−10°	- - -→

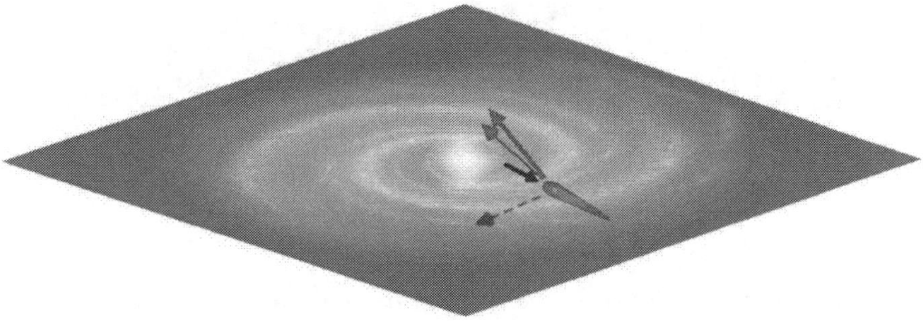

Figure 5. Oblique view of the Galactic disk showing the position of the Sun (yellow dot), the shape of the heliosphere (green comet-like structure), the direction of the interstellar wind against the heliosphere (blue arrow), and the directions of the ISMF vectors that best fit the pulsar data in the three distance ranges listed in Table 1 (red arrows taken from the last column of the table). Figure credit: Philippe Terral.

3.2. Polarization of nearby stars

[44] analyzed the polarization vectors of a group of stars within 40 pc of the Sun and 90° of the nose of the heliosphere, and they derived the ISMF orientation that best matched the polarization data. The result is shown in Table 2 and in Figure 6, where it can be compared to the ISMF orientation inferred from the polarization vectors of stars more distant than 500 pc [11]. Let me emphasize that, in contrast to rotation measures, polarization vectors provide no information on the magnetic field strength (except, very indirectly, with the Chandrasekhar-Fermi method). Moreover, since they are actually headless vectors, they only lead to the field orientation (axis without an arrow), not to the field direction (axis with an arrow).

Journal of Physics: Conference Series **577** (2015) 012008 doi:10.1088/1742-6596/577/1/012008

The conclusions here are similar to those reached in Section 3.1, namely, the ISMF orientation in the close vicinity of the Sun departs significantly from that observed in the large-scale ISM beyond $\sim$ 500 pc of the Sun. The former must be associated with discrete features within the Local Bubble – possibly with the G-cloud that lies just outside the Local Cloud in the general direction of the Galactic center [44]. [44] interpreted their derived ISMF orientation, together with the direction of the interstellar flow past the heliosphere, as evidence that the local ISM is an expanding fragment of the S1 shell of the Loop I superbubble.

Table 2. Galactic longitude, ℓ_B, and latitude, b_B, of the ISMF axis that best fits the polarization vectors of nearby stars [44] and of more distant stars [11]. The last column shows the red bar used in Figure 6 to represent the best-fit ISMF orientation.

Distance range	Best-fit ISMF ℓ_B	b_B	Bar in Figure 6
< 40 pc	$47° \pm 20°$	$25° \pm 20°$	——
> 500 pc	$83° \pm 4°$	$0.4° \pm 0.5°$	- - -

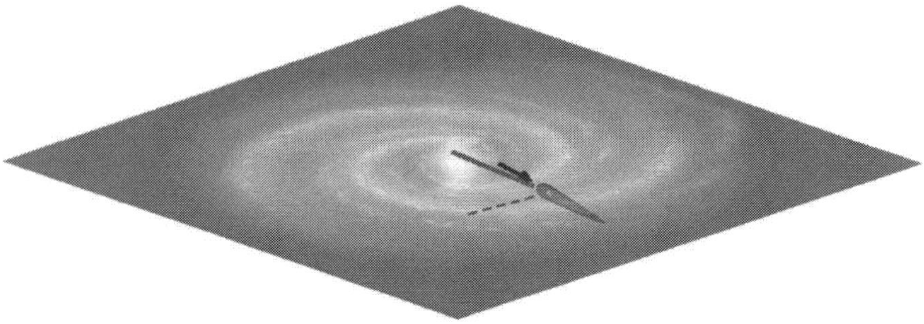

Figure 6. Oblique view of the Galactic disk showing the same basic features as in Figure 5, with the orientations of the ISMF axes that best fit the polarization vectors of the two groups of stars defined in Table 2 (red bars taken from the last column of the table). Figure credit: Philippe Terral.

3.3. Just outside the heliosphere

[45] measured the direction of the interstellar neutral hydrogen flow that penetrates the inner heliosphere and found it to be deflected by $\simeq$ 4° relative to the interstellar helium flow. They interpreted this deflection as resulting from a distortion of the heliosphere caused by the surrounding ISMF and used it to constrain the ISMF orientation. They argued that the ISMF must lie in the plane defined by the H and He velocities, i.e., the plane normal to ecliptic coordinates $(\lambda, \beta) = (167°, -30°)$, on the same side of the He velocity as the secondary H velocity but farther away. They also restricted the range of the ISMF ecliptic coordinates to $\lambda_B = 210° - 240°$ and $\beta_B = 30° - 60°$, assuming an angle of $30° - 60°$ between the ISMF and the interstellar bulk velocity. Finally, they noted that the ISMF projected onto the Galactic plane is roughly tangent to the boundary of the Local Cloud in the direction of the G-cloud, as expected if field lines are compressed between both clouds.

Journal of Physics: Conference Series **577** (2015) 012008 doi:10.1088/1742-6596/577/1/012008

A very different approach relies on the *Interstellar Boundary Explorer* (IBEX), which discovered a circular ribbon of enhanced energetic neutral atom (ENA) emission from the outer heliosphere [46]. The center of the ribbon, toward ecliptic longitude $\lambda = 219.2° \pm 1.3°$ and latitude $\beta = 39.9° \pm 2.3°$ [47], was argued to trace the orientation of the ISMF that shapes the heliosphere [48]. [44] suggested that this field orientation could be representative of the ISMF orientation in the Local Cloud, whose boundary in the considered direction lies only ~ 0.1 pc away from the Sun. It is noteworthy that the ISMF orientation from the IBEX ribbon center falls well within the range estimated from the deflection of the interstellar H flow and lies nearly (within 5°) in the plane defined by the interstellar H and He velocities (see previous paragraph).

The only *in-situ* measurements of ISMFs were made by *Voyager 1*, which is now believed to have crossed the heliopause into the very local ISM [49]. Ever since August 25, 2012, when the spacecraft was at 122 AU of the Sun, it has been measuring ISMFs [50]. For the first month after August 25, 2012, the average magnetic field strength was $B = (4.4 \pm 0.1)\ \mu$G and the average magnetic field direction had azimuthal angle $\lambda_B = 287° \pm 1°$ and elevation angle $\delta_B = 14° \pm 2°$ in RTN coordinates [51]. This field direction is close to the Parker spiral direction, $\lambda_P = 270°$ and $\delta_P = 0°$. Since then, the field strength has been smoothly varying in the range $(3.8 - 5.9)\ \mu$G, and the field direction has been increasingly deviating from the Parker spiral [50]. The *Voyager 1* measurements are consistent with the very local ISMF draping around the heliosphere and being twisted toward the Parker spiral [52].

Table 3. Galactic longitude, ℓ_B, and latitude, b_B, of the ISMF axis inferred from the center of the IBEX ribbon [47], and complete Galactic coordinates, B, ℓ_B, b_B, of the full ISMF vector measured by *Voyager 1* [51]. The last column makes the link with the corresponding red bar or arrow in Figure 7.

Mission	Distance	Inferred/measured ISMF B	ℓ_B	b_B	Bar/arrow in Figure 7
IBEX	$\lesssim 0.1$ pc		$35° \pm 4°$	$57° \pm 1°$	——
Voyager 1	122 AU	$(4.4 \pm 0.1)\ \mu$G	$189° \pm 10°$	$77° \pm 0.2°$	——▸

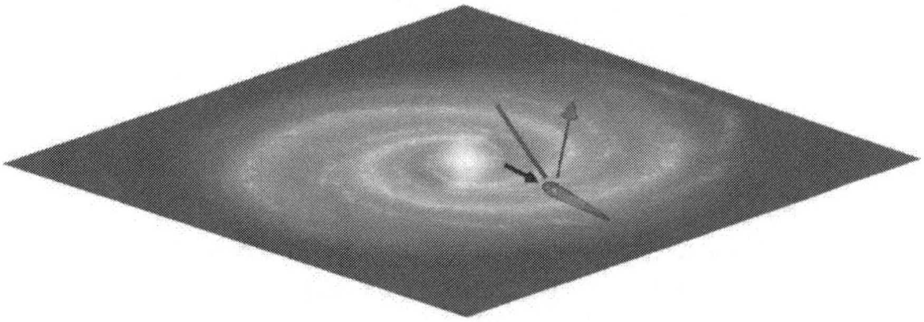

Figure 7. Oblique view of the Galactic disk showing the same basic features as in Figures 5 and 6, with the orientation of the ISMF axis inferred from the center of the IBEX ribbon (thick red bar) and the direction of the full ISMF vector measured by *Voyager 1* (thin red arrow). See Table 3. Figure credit: Philippe Terral.

For comparison purposes, the above ecliptic and RTN coordinates of the very local ISMF were converted to Galactic coordinates in Table 3. The uncertainties in Galactic longitude may appear surprisingly high, but it should be kept in mind that the corresponding angular uncertainties along a great circle are reduced by a factor cos b_B.

4. Summary

Various methods, primarily based on dust polarization, synchrotron emission, Faraday rotation, and Zeeman splitting, converge to show that the ISMF in the Galactic disk has a typical strength of a few μG (except in molecular clouds, where much stronger fields are observed) and a mean direction that is nearly parallel to the Galactic plane and generally close to azimuthal. In the large-scale surroundings of the Sun, the regular, ordered (regular + anisotropic random), and total field strengths are approximately $B_{\rm reg} \simeq 1.5\ \mu$G, $B_{\rm ord} \simeq 3\ \mu$G, and $B \simeq 5\ \mu$G, and the regular field runs clockwise at an angle $\simeq -7°$ or $-8°$ to the azimuthal direction. The substantial difference between $B_{\rm reg}$ and B reflects the presence of important magnetic fluctuations, while the difference between $B_{\rm reg}$ and $B_{\rm ord}$ indicates that part of these fluctuations are anisotropic.

We are now starting to gain a better understanding of magnetic fluctuations in the small-scale vicinity of the heliosphere, both with standard probes (rotation measures of nearby pulsars and polarization vectors of nearby stars) and with more indirect observations (deflection of the interstellar H flow entering the heliosphere and IBEX ribbon of enhanced ENA emission). These independent methods appear to yield compatible results, showing that the local ISMF around the heliosphere points in the first octant of the sky in Galactic coordinates and completely deviates from the nearly azimuthal direction of the large-scale regular ISMF. These severe deviations most likely arise from the pushing, lifting, and twisting of field lines that accompany the expansion of the Local Bubble and the Loop I superbubble and, closer to us, the motions of the Local Cloud and the G-cloud inside the Local Bubble.

Acknowledgments

I would like to express my gratitude to my student, Philippe Terral, who kindly accepted to prepare the plots needed for Figures 5–7, and to the referee, Steven Spangler, who provided me with very valuable and constructive comments.

References

[1] Fermi E 1949 *Physical Review* **75** 1169–1174
[2] Hall J S 1949 *Science* **109** 166–167
[3] Hiltner W A 1949 ApJ **109** 471–+
[4] Hiltner W A 1949 *Science* **109** 165–+
[5] Davis Jr L and Greenstein J L 1951 ApJ **114** 206–+
[6] Kiepenheuer K O 1950 *Physical Review* **79** 738–739
[7] Draine B T and Weingartner J C 1996 ApJ **470** 551–+ (*Preprint* arXiv:astro-ph/9605046)
[8] Lazarian A and Hoang T 2007 MNRAS **378** 910–946 (*Preprint* 0707.0886)
[9] Hoang T and Lazarian A 2008 MNRAS **388** 117–143 (*Preprint* 0707.3645)
[10] Mathewson D S and Ford V L 1970 MmRAS **74** 139–+
[11] Heiles C 1996 ApJ **462** 316–+
[12] Heiles C 2000 AJ **119** 923–927 (*Preprint* arXiv:astro-ph/9910303)
[13] Hildebrand R H 1988 QJRAS **29** 327–351
[14] Benoît A, Ade P, Amblard A, Ansari R, Aubourg É, Bargot S, Bartlett J G, Bernard J P, Bhatia R S, Blanchard A and et al 2004 A&A **424** 571–582 (*Preprint* arXiv:astro-ph/0306222)

[15] Planck Collaboration, Ade P A R, Aghanim N, Alina D, Alves M I R, Armitage-Caplan C, Arnaud M, Arzoumanian D, Ashdown M, Atrio-Barandela F and et al 2014 *ArXiv e-prints* (*Preprint* 1405.0871)

[16] Beuermann K, Kanbach G and Berkhuijsen E M 1985 A&A **153** 17–34

[17] Haslam C G T, Salter C J, Stoffel H and Wilson W E 1982 A&AS **47** 1

[18] Miley G 1980 ARA&A **18** 165–218

[19] Ferrière K 1998 ApJ **497** 759–+

[20] Gaensler B M, Madsen G J, Chatterjee S and Mao S A 2008 PASA **25** 184–200 (*Preprint* 0808.2550)

[21] Beck R 2001 *Space Science Reviews* **99** 243–260 (*Preprint* arXiv:astro-ph/0012402)

[22] Rand R J and Kulkarni S R 1989 ApJ **343** 760–772

[23] Han J L, Manchester R N, Lyne A G, Qiao G J and van Straten W 2006 ApJ **642** 868–881 (*Preprint* arXiv:astro-ph/0601357)

[24] Inoue M and Tabara H 1981 PASJ **33** 603–+

[25] Han J L, Manchester R N and Qiao G J 1999 MNRAS **306** 371–380 (*Preprint* arXiv:astro-ph/9903101)

[26] Rand R J and Lyne A G 1994 MNRAS **268** 497–+

[27] Vallée J P 2005 ApJ **619** 297–305

[28] Brown J C, Haverkorn M, Gaensler B M, Taylor A R, Bizunok N S, McClure-Griffiths N M, Dickey J M and Green A J 2007 ApJ **663** 258–266 (*Preprint* arXiv:0704.0458)

[29] Nota T and Katgert P 2010 A&A **513** A65+

[30] Men H, Ferrière K and Han J L 2008 A&A **486** 819–828 (*Preprint* 0805.3454)

[31] Taylor A R, Stil J M and Sunstrum C 2009 ApJ **702** 1230–1236

[32] Mao S A, Gaensler B M, Haverkorn M, Zweibel E G, Madsen G J, McClure-Griffiths N M, Shukurov A and Kronberg P P 2010 ApJ **714** 1170–1186 (*Preprint* 1003.4519)

[33] Frick P, Stepanov R, Shukurov A and Sokoloff D 2001 MNRAS **325** 649–664 (*Preprint* arXiv:astro-ph/0012459)

[34] Oren A L and Wolfe A M 1995 ApJ **445** 624–641

[35] Han J L, Manchester R N, Berkhuijsen E M and Beck R 1997 A&A **322** 98–102

[36] Ruzmaikin A A, Sokolov D D and Shukurov A M (eds) 1988 *Magnetic fields of galaxies* (*Astrophysics and Space Science Library* vol 133)

[37] Moss D and Sokoloff D 2008 A&A **487** 197–203

[38] Moss D, Sokoloff D, Beck R and Krause M 2010 A&A **512** A61

[39] Heiles C and Troland T H 2005 ApJ **624** 773–793 (*Preprint* arXiv:astro-ph/0501482)

[40] Troland T H and Heiles C 1986 ApJ **301** 339–345

[41] Crutcher R, Heiles C and Troland T 2003 *Turbulence and Magnetic Fields in Astrophysics* (*Lecture Notes in Physics, Berlin Springer Verlag* vol 614) ed Falgarone E and Passot T pp 155–181

[42] Crutcher R M, Wandelt B, Heiles C, Falgarone E and Troland T H 2010 ApJ **725** 466–479

[43] Salvati M 2010 A&A **513** A28 (*Preprint* 1001.4947)

[44] Frisch P C, Andersson B G, Berdyugin A, Piirola V, DeMajistre R, Funsten H O, Magalhaes A M, Seriacopi D B, McComas D J, Schwadron N A, Slavin J D and Wiktorowicz S J 2012 ApJ **760** 106 (*Preprint* 1206.1273)

[45] Lallement R, Quémerais E, Bertaux J L, Ferron S, Koutroumpa D and Pellinen R 2005 *Science* **307** 1447–1449

[46] McComas D J, Allegrini F, Bochsler P, Bzowski M, Christian E R, Crew G B, DeMajistre R, Fahr H, Fichtner H, Frisch P C, Funsten H O, Fuselier S A, Gloeckler G, Gruntman M, Heerikhuisen J, Izmodenov V, Janzen P, Knappenberger P, Krimigis S, Kucharek H, Lee M, Livadiotis G, Livi S, MacDowall R J, Mitchell D, Möbius E, Moore T, Pogorelov N V, Reisenfeld D, Roelof E, Saul L, Schwadron N A, Valek P W, Vanderspek R, Wurz P and Zank G P 2009 *Science* **326** 959–

[47] Funsten H O, DeMajistre R, Frisch P C, Heerikhuisen J, Higdon D M, Janzen P, Larsen B A, Livadiotis G, McComas D J, Möbius E, Reese C S, Reisenfeld D B, Schwadron N A and Zirnstein E J 2013 ApJ **776** 30

[48] Funsten H O, Allegrini F, Crew G B, DeMajistre R, Frisch P C, Fuselier S A, Gruntman M, Janzen P, McComas D J, Möbius E, Randol B, Reisenfeld D B, Roelof E C and Schwadron N A 2009 *Science* **326** 964–

[49] Gurnett D A, Kurth W S, Burlaga L F and Ness N F 2013 *Science* **341** 1489–

[50] Burlaga L F and Ness N F 2014 ApJ **784** 146

[51] Burlaga L F, Ness N F and Stone E C 2013 *Science* **341** 147–

[52] Opher M and Drake J F 2013 ApJ **778** L26 (*Preprint* 1310.0808)

A model for the nose region of the heliosheath

L A Fisk and G Gloeckler
Department of Atmospheric, Oceanic and Space Sciences, University of Michigan, 2455 Hayward St, Ann Arbor, MI 48109-2143,

E-mail: lafisk@umich.edu

Abstract. The *Voyager* 1 spacecraft is currently in the vicinity of the heliopause, which separates the heliosphere from the local interstellar medium. There has been a precipitous decrease in particles accelerated in the heliosphere, and a substantial increase in galactic cosmic rays (GCRs), suggesting easy escape of the former across the heliopause, and entry of the latter. The question is, has *Voyager 1* actually crossed the heliopause and is it now in the interstellar medium? Here the evidence is inconclusive. The direction of the magnetic field observed by *Voyager 1* is essentially unchanged from the direction of the heliospheric magnetic field. However, the plasma density, which is measured from observations of plasma waves, is similar to the expected higher density of interstellar plasma. In this paper, the principal results of an analytic model are presented demonstrating that the higher plasma densities measured by *Voyager 1* are likely to be compressed solar wind. Thus both the magnetic field and the plasma density observations are consistent with *Voyager 1* still remaining well within the heliosheath. The model has a simple test: *Voyager 1* should encounter a current sheet crossing, where the behavior of particles accelerated in the heliosphere and the GCRs will be different from what *Voyager 1* is now observing.

1. Introduction

The *Voyager* 1 spacecraft, hereafter referred to as *V1*, is currently in the vicinity of the heliopause, which separates the heliosphere from the local interstellar medium. There has been a precipitous decrease in particles accelerated in the heliosphere, and a substantial increase in galactic cosmic rays (GCRs) [1,2], suggesting easy escape of the former across the heliopause, and entry of the latter. The question is, has *V1* actually crossed the heliopause and is it now in the interstellar medium? Here the evidence is inconclusive. The direction of the magnetic field observed by *V1* is unchanged from the direction of the heliospheric magnetic field [3], and is in a substantially different direction from the expected direction of the interstellar magnetic field inferred from both *IBEX* observations of energetic neutral atoms (ENAs) [4,5 and references therein] as well as from observations of the polarization of interstellar grains [6]. However, Gurnett et al. [7] observe plasma oscillations at 122.2 AU, 122.5 AU, and 123.9 AU, which are a direct measure of the plasma electron density, and finds densities of 0.055 cm^{-3}, 0.058 cm^{-3}, and 0.083 cm^{-3}, respectively. These densities are comparable to densities expected in the local interstellar medium, and are much larger than the plasma densities observed by the *Voyager 2* (hereafter referred to as *V2*) plasma detector closer to the termination shock [8].

In this conference paper, we present the main arguments and conclusions of a model for the nose region of the heliosheath, which is described in Fisk and Gloeckler [9]. This analytic model is based upon and is consistent with all *Voyager* observations, and demonstrates that the higher plasma densities measured by *V1* are likely due to compressed solar wind. Thus, in the interpretation presented here, both the magnetic field and the plasma density observations are consistent with *V1* still remaining well within the heliosheath, and, as we shall show, *V1* is years away from crossing the heliopause.

We begin with a detailed discussion of the unusual properties and unusual physical processes that occur in the nose region of the heliosheath. We then use these properties and physical processes to develop an analytic model for the flow of the solar wind through the nose region of the heliosheath, demonstrating that the solar wind will be compressed to values equal to those observed by Gurnett et al. [7]. The model allows us to make specific predictions about the distance to the heliopause, and provides a test for the validity of the model: we predict, as discussed in detail in Gloeckler and Fisk [10], that *V1* will soon encounter a current sheet in which the polarity of the magnetic field reverses, clearly indicating that *V1* remains within the heliosheath.

2. The unusual properties of the nose region of the heliosheath

There are two properties of the nose region of the heliosheath on which our model is based: (1) the dominant pressure in the pickup ions and ACRs; and (2) the flow direction of the solar wind.

(1) The particle pressure in the nose region of the heliosheath is primarily the result of the mobile interstellar pickup ions and particles accelerated in the heliosheath (usually referred to as anomalous cosmic rays, ACRs), and is readily modeled and/or observed: (a) The pickup process in the supersonic solar wind is readily modeled and yields the pressure in pickup ions upstream of the termination shock. Observations by the working plasma detector on *V2* find that it is the pickup ions, not the solar wind, that are heated at the termination shock, satisfying the Rankine-Hugoniot relationship, which determines the pickup ion pressure in the heliosheath [8]. This pickup-ion pressure is validated by the *IBEX* observations of energetic neutral atoms resulting from the pickup ions [11]. (b) As for the energetic particles observed in the nose region of the heliosheath, at ~120 AU the differential intensity spectrum of ACRs has a spectral index of -1.5, with an exponential rollover, as is expected from the pump acceleration mechanism of Fisk and Gloeckler [12]. For such a spectrum, the pressure is due primarily to particles with energies that are measured by the Cosmic Ray Subsystem (CRS) detector on *V1* [2], and thus the ACR pressure is readily determined. The combined pressure in the pickup ions (deduced from *IBEX* observations), and in the ACRs (measured by *V1* at distances of ~120 AU) is $\sim 1.2 \times 10^{-12}$ dyne cm^{-2} [11].

The pressure in the pickup ions and ACRs is the dominant pressure in the nose region of the heliosheath. The plasma detector on *V2*, which is closer to the termination shock than *V1*, measures both the ram pressure and the thermal pressure of the solar wind [8]. The pressure in the pickup ions and ACRs is a factor of ~3 larger than the ram pressure, and is a factor of ~150 larger than the thermal pressure of the solar wind. The solar wind is not heated crossing the termination shock, and relative to itself, it remains supersonic in the heliosheath.

(2) Although *V1* does not have a functional plasma detector, Krimigis et al. [13] and Decker et al. [14] have been able to infer the solar wind flow velocity from the convective anisotropy of ~50-keV ions. Their instrument, the Low-Energy Charged Particle (LECP) detector, normally rotates approximately in an R-T plane of the *Voyager* R-T-N coordinate system, where the $+\mathbf{e}_R$ axis points radially outward from the Sun, the $+\mathbf{e}_T$ axis is in the direction of the rotation of the Sun, and $\mathbf{e}_N = \mathbf{e}_R \times \mathbf{e}_T$. From measurements that have been made continuously since *V1* crossed the termination shock, Krimigis et al. [13] and Decker et

Journal of Physics: Conference Series **577** (2015) 012009 doi:10.1088/1742-6596/577/1/012009

al. [14] find that the inferred radial flow speed decreases systematically to essentially zero beyond ~113 AU, with a small azimuthal speed persisting. At distances between ~116 AU and ~120 AU from the Sun, *V1* was rotated at a number of short time intervals to observe the anisotropies of low-energy ions in approximately the $\mathbf{e}_N$ direction [14]. At these distances the anisotropies, and thus the inferred convective flow speed in the polar direction, were found to be small, essentially zero [14]. This result is consistent with that of Stone and Cummings [15], who observe, using the Cosmic Ray Subsystem (CRS) detector that measures somewhat higher energy particles, that the anisotropies in the polar direction become increasingly small with distance into the heliosheath.

3. Implications of the dominant pressure in the pickup ions and ACRs

The hot pickup ions, and particularly the energetic ACRs, have average speeds very much larger than the thermal speeds of the solar wind. The pickup ions and ACRs should flow freely along the magnetic field, and seek to maintain pressure equilibrium in the nose region of the heliosheath, independent of the local conditions in the solar wind.

Thus, in the nose region of the heliosheath we have two distinct, and basically uncoupled gases: the mobile pickup ions and ACRs, which contain the pressure, and the relatively cold thermal solar wind, which contains the mass. The pickup ions and ACRs are responding to the global conditions in the heliosheath as they attempt to maintain pressure equilibrium. The solar wind is flowing into the heliosheath across the termination shock, and its density is responding to local conditions. The solar wind by itself has an equation of state that relates pressure and density, and the same is true for the pickup ions and ACRs. However, it is inappropriate to consider the combined gas of pickup ions, ACRs, and solar wind as having an equation of state since the pressure is determined by global conditions and the density by local conditions.

The need to treat the pickup ions and ACRs separately from the solar wind has important implications for the magnetic field in the nose region of the heliosheath: Consider first the magnetic field associated with the pickup ions and the ACRs. Each pickup ion and ACR particle has a dipole moment, which collectively align to form a magnetization that is proportional to the dominant pressure in the pickup ions and ACRs. Since the pickup ions and ACRs are in pressure equilibrium in the nose region of the heliosheath, we can think of the nose region as a substance with intrinsic dipole moments. The appropriate form for Ampere's Law (in a steady state) is then

$$\mathbf{J} = \frac{c}{4\pi}\nabla \times \mathbf{H}. \tag{1}$$

Here, $\mathbf{H}$ is the magnetic field that includes the magnetization and c is the speed of light.

The current $\mathbf{J}$ in Equation (1) is due to the perpendicular pressure of the pickup ions and ACRs, and including the magnetization and all gradient and curvature drifts, is given by (e.g. [16])

$$\mathbf{J} = \frac{c\mathbf{H} \times \nabla P_{\perp}}{H^2}. \tag{2}$$

Crossing $\mathbf{H}$ into Equation (2) yields the equation for static equilibrium for the pickup ions and ACRs:

Journal of Physics: Conference Series **577** (2015) 012009 doi:10.1088/1742-6596/577/1/012009

$$-\nabla P_{\perp} + \frac{1}{4\pi}(\nabla \times \mathbf{H}) \times \mathbf{H} = 0 . \qquad (3)$$

Note that the **H**-field is entirely determined by the pressure of the pickup ions and ACRs, and does not depend on the solar wind.

It is important to realize in this paper that we use the symbol **H** to denote the magnetic field in the nose region of the heliosheath that is determined by the pickup ions and ACRs, independent of the solar wind. This is standard notation in any basic E&M book (e.g. [17]); the macroscopic magnetic field of an object with intrinsic magnetic dipoles is denoted by **H**. This notation is not standard in conventional plasma physics, where either no distinction is made between an **H**-field and a **B**-field, or the magnetization is considered to be wholly contained in the **B**-field. We reserve the symbol **B** to describe the magnetic field that is convected into the heliosheath and controlled by the solar wind.

The solar wind convects the heliospheric magnetic field across the termination shock and into the heliosheath. The solar wind is coupled to the magnetic field through the Lorentz force, which depends upon **B**. The momentum equation for the solar wind (again in a steady state) is then

$$\rho \mathbf{u} \bullet \nabla \mathbf{u} = -\nabla P_{sw} + \frac{1}{4\pi}(\nabla \times \mathbf{B}) \times \mathbf{B} . \qquad (4)$$

Here, **u** is the solar wind velocity; ρ is the solar wind mass density, and P_{sw} is the solar wind thermal pressure, which remains small compared to the pressure in the pickup ions and ACRs. Similarly, Faraday's Law requires that, in a steady state,

$$\nabla \times (\mathbf{u} \times \mathbf{B}) = 0 , \qquad (5)$$

where the electric field has been set equal to the convective electric field, $\mathbf{E} = -(\mathbf{u} \times \mathbf{B})/c$.

In a conventional plasma, where pressure and density are directly related, **H** and **B** are the same. The pressure in Equation (3) is the same as the pressure in Equation (4) and directly related to the density in Equation (4). In the nose region of the heliosheath **H** and **B** are not the same. The dominant pressure in the pickup ions and ACRs in Equation (3) is determined by the global properties of the nose region and is unrelated to the solar wind pressure or density in Equation (4), which is determined by the local flow properties of the solar wind.

The magnetic field at ~120 AU is observed by *V1* to be primarily in the azimuthal direction [3]. As discussed in Section 2, the radial flow speed of the solar wind is inferred to decrease to near zero in the radial and polar directions. If the heliosheath were a conventional plasma, then from Equation (5) we should expect that the magnetic field strength would increase dramatically to very large values, and yet no such significant change is observed in the field magnitude [3].

The resolution of this contradiction is provided by noting that the *V1* magnetometer observes the **H**-field, which is determined by the pressure in the pickup ions and ACRs, not the **B**-field, which is determined by Equation (5). Indeed, the observed decrease in the radial and polar solar wind flows without a major increase in the observed magnetic field strength is strong evidence for the validity of the approach taken in this paper: the pickup ions and ACRs are a separate gas, responding to global conditions and determining the **H**-field; the solar wind is responding to local conditions and determining the **B**-field.

4. The flow of the solar wind in the nose region of the heliosheath

With the unusual properties and physical conditions in the nose region of the heliosheath described in Section 2, and a few straightforward simplifications, it is possible to calculate the flow of the solar wind analytically, without resorting to a numerical model. The solar wind is treated as a separate gas from the pickup ions and ACRs, coupled to the magnetic field **B**, and governed by standard continuity equations, such as the momentum equation in Equation (4). For our purposes, we use the energy continuity equation,

$$\nabla \bullet \left(\rho \mathbf{u} \left(\frac{u^2}{2} + \frac{5}{2} \frac{P_{sw}}{\rho} \right) - \frac{1}{4\pi} (\mathbf{u} \times \mathbf{B}) \times \mathbf{B} \right) = 0, \qquad (6)$$

where again, **u** is the solar wind velocity; ρ is the solar wind mass density, and P_{sw} is the solar wind thermal pressure. We take the solar wind to be adiabatic, $P_{sw} \propto \rho^{5/3}$.

The solar wind velocity **u** is taken to lie on a cone of constant heliographic latitude, consistent with the observations of energetic particle anisotropies from *V1* discussed in Section 2 that show that the polar flow of the solar wind becomes diminishingly small [14,15]. This simplification is not strictly valid immediately downstream from the termination shock, where polar flows occur, but it is valid beyond about 110 AU, which is the main concern of this paper. The magnetic field is taken to lie in the azimuthal direction. Burlaga et al. [3] observe small deviations of the magnetic field direction from azimuthal, which also do not affect the main conclusions of this paper.

With the flow of the solar wind in the nose region of the heliosheath in the radial and azimuthal directions only, there must be a centerline region, where the flow of the solar wind diverges, one streamline flowing in the positive azimuthal direction and the streamline on the other side of the centerline in the negative azimuthal direction. The actual location of the centerline is unknown. We take the centerline to be offset by ~15° in azimuthal angle from the trajectory of *V1*, which is moving in the direction of motion of the Sun through the local interstellar medium. Such a location of the centerline is consistent with numerical models such as those by Opher et al. [18] in which pressure of the interstellar magnetic field distorts the shape of the heliopause.

As is illustrated in figure 1, the streamlines in the heliosheath on either side of the centerline (the first streamlines) originate from essentially the same location on the termination shock. They create a region that is a vacuum as the two streamlines diverge, each flowing in an opposite direction. However, the magnetic field is azimuthal and crosses the vacuum region. The pickup ions and ACRs will flow easily into the vacuum, along the magnetic field, but so must the solar wind. Thus, the streamlines that are adjacent to the centerline will leak particles as they flow outward, and these particles will fill in the centerline region, making the solar wind density azimuthally independent in the nose region of the heliosheath.

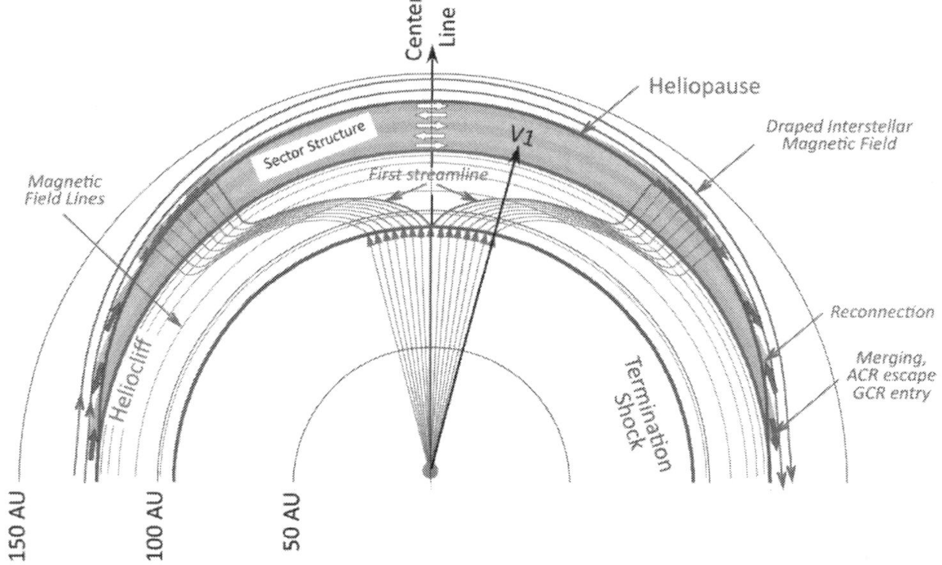

Figure 1. The solar wind streamlines (red curves) and the magnetic field lines (green and blue curves) on a cone of constant heliographic latitude projected onto a rectangular plane (after Fisk and Gloeckler [9]). At the centerline, which is slightly offset from the trajectory of *Voyager 1*, the flow of the solar wind diverges, with one streamline flowing in the positive azimuthal direction and the streamline on the other side of the centerline in the negative azimuthal direction. The two 'first streamlines' and the heliopause define the boundaries of the centerline region. The streamline regions lie to the left and right of the centerline region. Between the heliocliff and the heliopause lies a high-density region where the solar wind flow is highly compressed, reaching the density of the interstellar plasma (~0.1 cm^{-3}), and flows radially outward at a few km s^{-1} toward the heliopause. The azimuthal magnetic field in this high-density region is expected to reverse polarity several times, providing a test of our model. See text.

There can be no mass flow along the magnetic field connecting streamlines on either side of the centerline. By definition, the azimuthal velocity is zero at the centerline. So as streamlines leak particles along the magnetic field, they will create a region with no azimuthal flow, and in a steady state, with density equal to the density on the streamlines. With the magnetic field in the azimuthal direction there is no dependence of the density on azimuthal angle in the vicinity of the centerline region, and the density will increase with radial distance according to

$$\rho u_r r^2 = \rho_{ts} u_{ts} r_{ts}^2 , \tag{7}$$

where u_r is the radial flow speed of the solar wind and *ts* denotes 'evaluated at the termination shock'.

In figure 1 we present the results of our solution to Equation (6), and its consequences for the region beyond the current location of *V1*, and for the location and properties of the heliopause. The calculations for these streamlines are presented in detail in Fisk and Gloeckler [9]. The solar wind streamlines and the magnetic field all lie on a cone of constant heliographic latitude. Figure 1 is the projection of this cone onto a rectangular plane. The solar wind streamlines are in red, originating at the Sun, crossing the termination shock at 94 AU, and then flowing through the heliosheath to the heliopause.

The streamlines shown in figure 1 result from an analytic model. It would clearly be useful to calculate these streamlines in a full numerical simulation of the nose region of the heliosheath. In some ways, the required numerical simulation is simpler than many of the numerical simulations now is use (e.g., [19-22]). In the model presented in this paper, the solar wind is treated as a separate fluid from the pickup ions and ACRs, and simply needs to be modeled so as to have the correct flow directions in the nose region of the heliosheath. Although the pressure in the pickup ions and ACRs does not influence the solar wind flow directly, it can be expected to resist the flow of the solar wind in the polar direction, since such flow would introduce strong pressure gradients in the pickup ions and ACRs, which cannot be relieved along the primarily azimuthal magnetic field. We have then a two-dimensional flow of the solar wind, which is initially supersonic downstream of the termination shock, which is expected to create a centerline region where the solar wind flow diverges; which suffers mass loss from the escape of the solar wind into the centerline region; and which can be cooled by the escape of the thermal energy when the embedded magnetic field crosses the heliopause.

5. Implications of the model of the flow of the solar wind

As is illustrated in figure 1, after the solar wind crosses the termination shock, it first turns in the azimuthal direction, and then turns back into the radial direction, and flows outward to the heliopause, which results in a flow velocity normal to the heliopause. The heliopause can thus not be a tangential discontinuity, but rather must be a rotational discontinuity, since rotational discontinuities have a normal component of the magnetic field and can accommodate flows normal to the discontinuity.

Rotational discontinuities, unlike tangential discontinuities, propagate. The heliopause, however, is stationary relative to the Sun. Thus, to maintain the heliopause stationary in the frame moving with the Sun, the rotation discontinuity must propagate into the interstellar medium at a speed equal to the velocity of the Sun through the local interstellar medium.

The solar wind flow across the heliopause will result primarily in a decrease in temperature, not in mass, since it is the higher energy particles, including the pickup ions, that will most readily escape. The resulting decrease in pressure will not balance the interstellar medium, with the result that the heliosheath will shrink in size with a resulting increase in the solar wind density. The region between where the solar wind flow turns radial and the heliopause is analogous to a balloon that is cooled; the balloon shrinks in size and the density increases.

At the same time the solar wind is escaping, cold dense interstellar gas is entering across the heliopause. The solar wind density should increase only to where the density equals the interstellar density, which we take to be ~0.1 cm^{-3}. In a steady state, then, the density of the solar wind increases to equal the interstellar density, and the pressure in the heliosheath is able to balance the interstellar pressure since no further shrinkage is occurring.

Note that the increased density is not accompanied by an increased pressure, as occurs on the adiabatic streamlines. Rather, the region beyond where the streamlines turn radial is not adiabatic since there is an escape of thermal energy across the heliopause, which cools this region of the heliosheath to achieve the required pressure with an increased density. We have then two separate and distinct regions in the nose region of the heliosheath: a region where the solar wind flow is controlled by the flow parameters at the termination shock and is adiabatic, and a separate region where the solar wind flow properties are determined by the escape of thermal energy across the heliopause and the solar wind flow is not adiabatic.

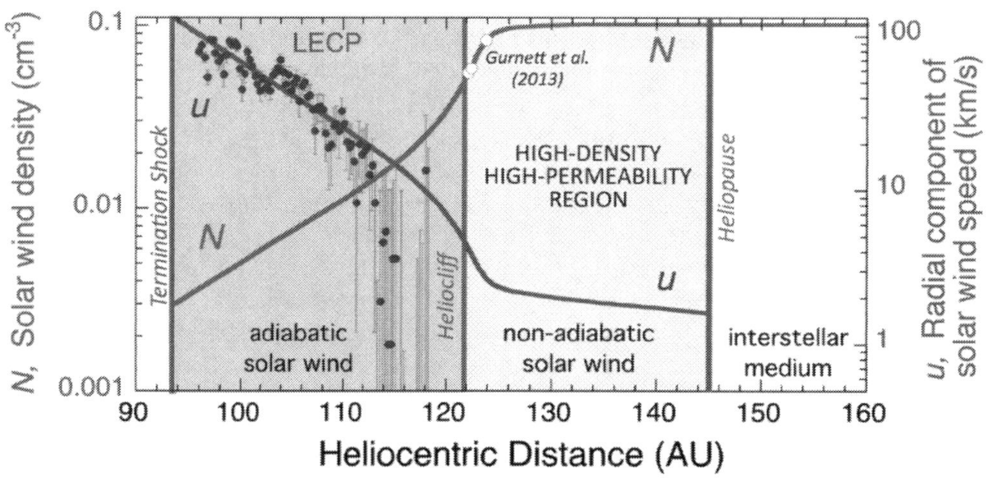

Figure 2. Variations of the radial component of the solar wind speed (blue curve) and solar wind number density (red curve) with heliocentric distance (after Fisk and Gloeckler [9]). *V1* is currently at about 127 AU. The speed profile (blue curve) was chosen to match the average radial speeds deduced from anisotropy measurement of ~ 50-keV ions [13,14] between ~96 AU and ~112 AU (filled blue circles). Plasma densities measured by Gurnett et al. [7] at ~123 AU to 124 AU (red circles), an interstellar plasma density of 0.1 cm^{-3} and a solar wind density at the termination shock of 0.003 cm^{-3} were used in Equation (7), to obtain the speed profile beyond ~122 AU. The two speed profiles were then smoothly connected between 112 AU and 122 AU. This speed profile was used in Equation (11) to calculate the streamlines shown in figure 1. The solar wind is compressed to densities measured by Gurnett et al. [7]. See text for further discussion.

Shown in figure 2 is the radial profile of the solar wind density predicted in the model. The predicted density is consistent with the radial flow speed inferred from the anisotropy observations, which is also shown and yields the densities, and the gradient in the density observed by Gurnett et al. [7]. In this model the density equal to the interstellar density observed by Gurnett et al. [7] results because the solar wind is compressed to where it equals this density. The flow of thermal energy along the field lines that cross the heliopause maintains the solar wind density equal to the interstellar density.

Equation (7) still describes the relationship between the solar wind density and the radial flow speed of the solar wind and can be used to determine u_r in the nonadiabatic region. The radial flow speed becomes small, and is now only a few km s^{-1}, as shown in figure 2.

6. A test of the model

The precipitous decrease in the ACRs, often referred to as the 'heliocliff', should occur where the magnetic field is directly connected to the heliopause. In this model, there is a normal component of the magnetic field across the heliopause, which is a rotational discontinuity. However, we should expect a difference in the nature of the heliocliff, depending on the orientation of the heliosheath magnetic field relative to the interstellar magnetic field. In the case where the orientation is such that the heliosheath and interstellar magnetic fields merge, ACRs and GCRs should flow freely along the magnetic field, the former escaping and the latter entering. In the case where the orientation of the heliosheath field is opposite to the interstellar field, reconnection islands should occur which could limit the escape of ACRs and prohibit the entry of GCRs.

We note that the heliocliff observed by *V1* was marked by the escape of the ACRs, the entry of GCRs, and a sector boundary crossing [3]. There was evidence of increased magnetic field before the heliocliff, but the full escape of the ACRs and entry of GRCs occurred roughly coincident with a sector boundary, which suggests that at this boundary *V1* entered a region where the magnetic field was aligned with the interstellar magnetic field, providing for the abrupt escape of ACRs and entry of GCRs.

The calculations, described in detail in Fisk and Gloeckler [9] and illustrated in figure 1, predict that *V1* is currently about 18 AU from the heliopause. As discussed in Section 5, the solar wind flow speed is very slow beyond the location of V1, such that it will take *V1* more than two magnetic field reversals of the Sun to reach the heliopause. Thus, *V1* is certain to encounter another current sheet, where the magnetic field reverses polarity and we should expect reduced escape of ACRs, and certainly limited entry of GCRs.

Thus, there is a very clear test of the model presented in this paper. If *V1* encounters another current sheet, with different behavior of the ACRs and GCRs, it is clear that *V1* remains in the heliosheath and that *V1* is still years away from crossing the heliopause.

In Gloeckler and Fisk [10] we use *Voyager* observations to estimate when *V1* should encounter the next sector boundary, and find that it could well be within the next 2 years.

When *V1* finally reaches the heliopause there will be no change in the plasma density, which is simply the interstellar density, but the magnetic field should rotate to be more closely parallel or antiparallel to the interstellar magnetic field. There will then be a boundary layer beyond the heliopause, where the magnetic field is still the heliosheath magnetic field, followed by a region of merging or reconnecting magnetic fields. It may well be beyond the lifetimes of the *Voyager* spacecraft before they reach a region that can be said to be truly in the interstellar medium.

7. Concluding remarks

When *V1* passes out of the heliosphere into interstellar space, it will be an event of historical importance, the first human-made object to leave the domain of the Sun. At present the evidence as to whether *V1* is in interstellar space is unclear. There is no doubt that, with the observed precipitous decreases in particles accelerated in the heliosheath, *V1* is in the vicinity of the heliopause. The unchanged magnetic field direction suggests that *V1* remains within the heliosheath [3], whereas the observed plasma densities equal to expected interstellar density suggests that *V1* is in the interstellar medium [7]. Clearly, considering the historical importance of *V1* crossing the heliopause, we need a higher degree of certainty as to whether the crossing has occurred or not.

In this paper, we present an analytic model that is based upon and is consistent with all *Voyager* observations, and in which the higher plasma densities inferred by *V1* are due simply to compressed solar wind. Thus, in the interpretation presented here, both the magnetic field and the plasma density observations are consistent with *V1* still remaining well within the heliosheath, and, as we showed, *V1* is years away from crossing the heliopause. Of most importance, the model presented here has a simple test: We predict that *Voyager 1* will encounter a current sheet crossing, where the behavior of ACRs and GCRS will be different from what *V1* is now observing. Should such a current sheet crossing occur there should be no doubt that *V1* remains within the heliosheath.

Acknowledgements
This work was supported in part by NASA Grant NNX10AF23G and by NSF Grant AGS-1043012. We acknowledge and are very appreciative of the *Voyager 1* and *2* teams, who provided us with an opportunity to examine their data.

References

[1] Krimigis S M, Decker R B, Roelof E C, Hill M E, Armstrong T P, Gloeckler G, Hamilton D C and Lanzerotti L J 2013 *Sci.* **341** 144
[2] Stone E C, Cummings A C, McDonald F B, Heikkila B C, Lal N and Webber W R 2013 *Sci.* **341** 150
[3] Burlaga L F, Ness N F and Stone E C 2013 *Sci.* **341** 147
[4] McComas D J, et al. 2009 *Sci.* **326** 959
[5] McComas D J, et al. 2012 *ApJS* **203** 1
[6] Frisch P C 2011 *AIP Conf. Proc. 1436, Physics of the Heliosphere: A 10 Year Retrospective* ed J Heerikhuisen, G Li, N Pogorelov and G Zank (NY: AIP) p 295
[7] Gurnett D A, Kurth W S, Burlaga L F, and Ness N F 2013 *Sci.* **341** 1489
[8] Richardson J D 2008 *GRL* **35** L23104 doi:10.1029/2008GL036168
[9] Fisk L A and Gloeckler G 2014 *ApJ* **789** 41
[10] Gloeckler G, and Fisk L A 2014 *GRL* in press
[11] Gloeckler G, and Fisk L A 2010 *AIP Conf. Proc. 1302, Pickup Ions Throughout the Heliosphere and Beyond* ed J A le Roux, V Florinksi, G P Zank, and A J Coates (NY: AIP) p 110
[12] Fisk L A and Gloeckler G 2013 *ApJ* **776** 79
[13] Krimigis S M, Roelof E C, Decker R B, and Hill M E 2011 *Nature* **474** 359
[14] Decker R B, Krimigis S M, Roelof E C, and Hill M E 2012 *Nature* **489** 124
[15] Stone E C, and Cummings A C 2011 *Proc. of the 32nd Intl. Cosmic Ray Conference* (vol 12) (Beijing: IUPAP) p 29
[16] Gurnett D A, and Bhattacharjee A 2005 *Introduction to Plasma Physics* (Cambridge: Cambridge Univ Press)
[17] Jackson J D 1999 *Classical Electromagnetics* (New York: John Wiley and Sons Inc)
[18] Opher M, Stone E C, and Gombosi T 2007 *Sci.* **316** 875
[19] Pogorelov N V, Borovikov S N, Zank G P, and Ogino T 2009 *ApJ* **696** 1478
[20] Izmodenov V V, Malama Y G, Ruderman M S, Chalov S V, Alexashov D B, Katushkina O A, and Provornikova E A 2009 *Space Sci. Rev.* **146** 329
[21] Washimi H, Zank G P, Hu O, Tanaka T, Munakata K and Shinagawa H 2011 *MNRAS* **416** 1475
[22] Opher M, Drake J, Swisdak M, Schoeffler K M, Richardson J D, Decker R B and Toth G 2011, *ApJ* **734** 71

Journal of Physics: Conference Series **577** (2015) 012010 doi:10.1088/1742-6596/577/1/012010

Connecting the interstellar magnetic field at the heliosphere to the Loop I superbubble

P C Frisch[1], A Berdyugin[2], H O Funsten[3], A M Magalhaes[4], D J McComas[5], V Piirola[2], N A Schwadron[6], D B Seriacopi[4], and S J Wiktorowicz[7]

[1]Dept of Astronomy and Astrophysics, University of Chicago, Chicago, IL 60637, USA
[2]Finnish Centre for Astronomy with ESO, University of Turku, Finland
[3]Los Alamos National Laboratory, Los Alamos, NM
[4]Inst de Astronomia, Geofisica e Ciencias Atmosfericas, Universidade de Sao Paulo, Brazil
[5]Southwest Research Inst, San Antonio, TX and University of Texas, San Antonio, TX
[6]Space Science Center, University of New Hampshire, Durham, NH
[7]Dept of Astronomy, University of California at Santa Cruz, Santa Cruz, CA

E-mail: [1]`frisch@oddjob.uchicago.edu`

Abstract. The local interstellar magnetic field affects both the heliosphere and the surrounding cluster of interstellar clouds (CLIC). Measurements of linearly polarized starlight provide the only test of the magnetic field threading the CLIC. Polarization measurements of the CLIC magnetic field show multiple local magnetic structures, one of which is aligned with the magnetic field traced by the center of the "ribbon" of energetic neutral atoms discovered by the Interstellar Boundary Explorer (IBEX). Comparisons between the bulk motion of the CLIC through the local standard of rest, the magnetic field direction, the geometric center of Loop I, and the polarized dust bridge extending from the heliosphere toward the North Polar Spur direction all suggest that the CLIC is part of the rim region of the Loop I superbubble.

1. Origin of local interstellar magnetic field

The Sun resides in a region of space with very low interstellar densities [1] that is offset by ~ 400 pc from the density maximum of the Orion spiral arm (also known as the Orion Spur, [2]). The interstellar magnetic field (ISMF) in this interarm region is characterized by an ordered magnetic field with a strength 1.4 ± 0.3 μG that is directed toward the local radiant at $\ell \sim 88^\circ$ in the galactic plane [3–5]. The older clusters of the Scorpius-Centaurus Association (ScoOB2), located ~ 120 pc from the Sun, are also offset from the density maximum of the Orion arm [2].

A superbubble was created by the ScoOB2 stellar winds and supernova during three epochs of star formation over the past 15 Myrs [6–9]. A classic expanding superbubble will sweep up the ISMF during expansion, expanding asymmetrically in directions parallel to the ISMF for initially uniform gas, and with the compressed magnetic field lines parallel to the rim of the shell in directions perpendicular to the magnetic field [10–12]. The epochs of star formation in ScoOB2 swept the expanding superbubble into the interarm region through which the heliosphere moves [Fig. 1, 8; 9; 13]. The interstellar signatures of the Loop I superbubble have been analyzed based on tracers of neutral gas, polarized starlight, and radio continuum emission [6; 7; 14–20].

Journal of Physics: Conference Series **577** (2015) 012010 doi:10.1088/1742-6596/577/1/012010

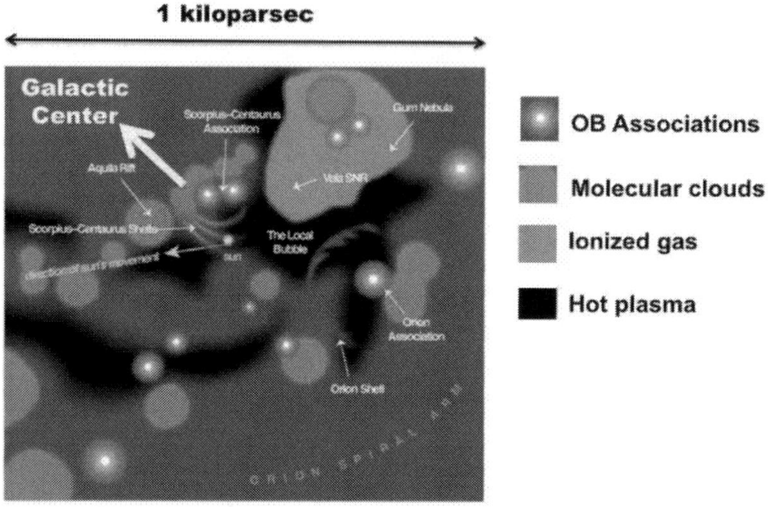

Figure 1. The distribution of molecular clouds (orange) traced by the CO 12 cm line is shown for regions within ~ 1 kpc of the Sun. Stellar associations (bright starry dots) form from these molecular clouds, generating superbubble shells that compress interstellar dust and gas during expansion (gray and red shell-like arcs). The solar motion through the LSR (yellow arrow) has carried the Sun through the low density interior of the Local Bubble for the past several million years. The figure is from [25].

The original superbubble expanded into a giant dust cloud, the remnants of which form the Aquila Rift molecular cloud, and initiated the sequential star formation that produced the upper Scorpius subgroup [6; 7]. Recent stellar evolution in the upper Scorpius subgroup generated the supernova that produced the bright X-ray source associated with the North Polar Spur radio emission. Observations and models of the X-ray plasma and radio shell show that it is a supernova remnant with an age of ~ 2 Myrs that was reheated possibly within the past 100,000 years [21]. Hydrogen column densities and dust opacity jump at $\sim 80 - 100$ pc in the North Polar Spur segment of Loop I ($330^\circ < \ell < 30^\circ$, $0^\circ < b < 40^\circ$, [21; 22]). Cosmic rays generated in the remnant light up the magnetic field compressed in the superbubble shell with synchrotron emission, producing the giant non-thermal radio continuum feature, Loop I, that dominates the northern hemisphere sky [14]. The magnetic field associated with Loop I dominates starlight polarization in the northern hemisphere [ISMF, 14; 18; 23], and the polarized emission of interstellar dust grains [24].

A coherent picture of the Loop I ISMF can found using three sources of polarized light, starlight polarization, polarized dust emission,. and polarized synchrotron emission. Starlight is linearly polarized by asymmetric dust grains that have larger opacities along the axis that is perpendicular to the magnetic field, so that optical polarization vectors are parallel to the magnetic field direction (e.g. [26]). The polarized emission of dust grains is parallel to the most optically opaque axis, and therefore perpendicular to the ISMF direction for aligned grains [24; 27]. The polarization of synchrotron emission is aligned with the acceleration vector of electrons and ions propagating along the ISMF, so that it is perpendicular to the ISMF direction for galactic cosmic rays gyrating around the ISMF. The property that the optical polarization vectors are perpendicular to the Loop I synchrotron polarizations (e.g. compare polarization maps in references [23] and [14]), and also perpendicular to polarized infrared light from dust

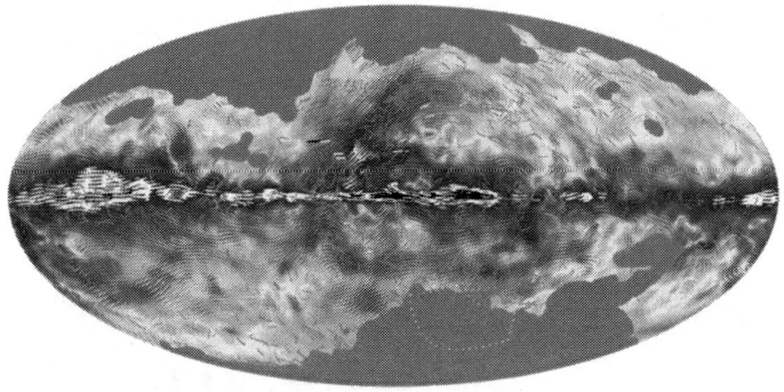

Figure 2. Stellar polarization data are overplotted on the magnetic field direction based on polarized dust emission measured by Planck [24]. The green triangle shows the direction of the ISMF traced by the center of the IBEX Ribbon arc [33]. A high-latitude dust arc is found in this direction. The blue bars show the polarizations of distant (> 40 pc) stars that trace the maximum of the Loop I intensity. Red bars show polarizations for stars within 40 pc. Green bars show the filament stars that are located toward the inner ridge of the nearby portion of Loop I (Fig. 3). In the northern parts of Loop I and in the first galactic quadrant, where the ISMF extends to the solar location (Fig. 4), the ISMF direction traced by the optical and infrared polarizations are consistent. Yellow regions appear in sightlines of highest dust emissivity.

associated with Loop I [24; 27], shows that the position angles of polarized starlight trace the magnetic field directions in the global diffuse interstellar medium, and that the linearly polarized starlight has a plane of polarization that is parallel to the magnetic field direction. This property enables the mapping of the local ISMF direction using high-sensitivity polarization measurements, $\leq 0.01\%$ [28–32]. Fig. 2 shows starlight polarizations plotted over the ISMF direction obtained from measurements of polarized dust emission by Planck [24; 27].

Optically polarized starlight reveals the ISMF associated with Loop I [23]. Polarization surveys by Mathewson and Ford [23], Santos et al. [18] and Berdyugin et al. [20] have mapped starlight polarizations that trace the ISMF of Loop I. Polarization strengths jump at ~ 80 pc where $\ell \sim 35°$, and at ~ 150 pc where $\ell \sim 270°$, indicating that the densest parts of Loop I extend closest to the Sun for longitudes $\sim 35°$.

2. Location of the heliosphere in the rim of Loop I

The Loop I regions of highest column density are located beyond ~ 100 pc, but spatially smoothed E(B-V) data show that the Loop I dust shell is apparent within 50–100 pc at high latitudes and in the first galactic quadrant (Fig. 3). The void in the dust that is centered near $\ell \sim 320°$, $b \sim -20°$ shows that the Loop I superbubble interior merges into the Local Bubble interior in the fourth galactic quadrant ($\ell = 270° - 360°$). The only interstellar cloud found near the center of the void is the G-cloud [34], where the color excess corresponding to the low observed column densities (E(B-V)< 0.001 mag) is too small for detection.

Several types of data suggest that the Sun is embedded in the rim of the evolved Loop I superbubble. The first is that the geometry of the arcs associated with Loop I places the Sun at the bubble edge. Assuming that synchrotron arcs and H° arcs associated with the North Polar Spur ($\ell \sim 10° - 30°$) form a single superbubble places the center of Loop I at $\ell = 320°$, $b = 12°$, distance 120 pc, with a radius of ~ 115 pc [17; 35]. An alternative geometric model is consistent with the series of high-latitude arcs visible in the H° data [15; 16]. The associated

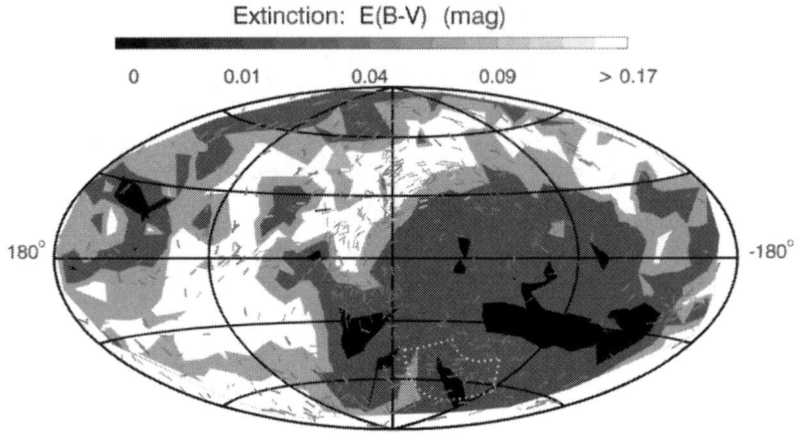

Figure 3. Smoothed contours of color excess E(B-V) for stars 50–100 pc show that dust associated with the inner rim of Loop I extends to inside of 100 pc in the North Polar Spur region. Optical polarization vectors for stars within 40 pc (red) show similar position angles in some regions when compared to more distant stellar polarizations ($<$ 150 pc, blue) that trace the Loop I ISMF. The filament stars are plotted in green. The green triangle identifies the IBEX Ribbon ISMF direction. The void in ISM in the fourth galactic quadrant forms where the interiors of the Loop I superbubble and Local Bubble merge. The plot is centered on $\ell = 0$, and galactic longitude increases to the left. Figure details are given in [31].

radio continuum arcs have been modeled as two shells by Wolleben [36], using the intensity of polarized radio continuum emission (also see [20] and Fig. 6). The shell geometry suggests that the Sun is most likely in the rim of the shell identified as "S1" [36], which is centered at $\ell = 346^\circ \pm 5^\circ$, $b = 3^\circ \pm 5^\circ$ and 78 ± 10 pc. [29; 31; 36; 37].

The kinematical properties of nearby interstellar gas provide the second type of information suggesting that the Sun is located in the rim of Loop I. Interstellar gas within 30 pc flows past the Sun with an upwind direction in the Local Standard of Rest (LSR) that is directed toward the center of the Loop I superbubble [1; 38]. For the assumption that the different cloudlets in the flow move with a rigid body motion, the heliocentric vector flow of nearby ISM past the Sun is away from $\ell = 12.4^\circ$, $b = 11.6^\circ$, at a velocity of -28.1 ± 4.6 km s^{-1} [1; 38]. Using the LSR motion derived from Hipparcos data by Schonrich et al. (2010), and including the uncertainties of both the solar apex motion and the heliocentric interstellar motion, this heliocentric vector flow corresponds to the upwind LSR velocity of ℓ,b= $338.4^\circ \pm 15.2^\circ$, $-5.3^\circ \pm 8.3^\circ$ and velocity 17.2 ± 4.6 km s^{-1} for the bulk CLIC motion. This is $11^\circ \pm 19^\circ$ from the S1 shell center. Alternate velocity vectors for local gas yield similar results. For example, a local cloud velocity derived from a restricted data set that omits outlying and redundant velocity components produced a heliocentric upwind velocity of $\ell = 5.8^\circ \pm 0.8^\circ$, $b = 12.8^\circ \pm 0.7^\circ$, velocity -25.5 ± 0.3 km s^{-1} [39], which then corresponds to the LSR velocity $\ell = 323.6^\circ \pm 1.7^\circ$, $b = -16.2^\circ \pm 1.4^\circ$ and a velocity of -17.0 ± 0.6 km s^{-1}. Thus, regardless of the specific assumptions about the individual components used for calculating the bulk flow, the upwind direction is always toward the low density void around the Loop I center.

The third set of data relating the CLIC with the perimeter of Loop I is plotted in Fig. 4, where the polarization strengths are shown to increase steadily with distance inside of a cone of radius 40° that is centered on the IBEX ribbon ISMF direction (also see related figures in

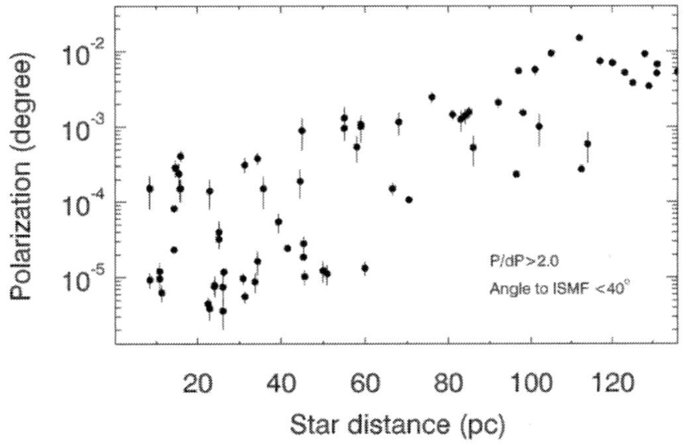

Figure 4. Polarizations are plotted against the distances of stars that are within 40° of the magnetic field direction indicated by the center of the IBEX Ribbon (see text). The steady increase in polarization with distance shows that the aligned dust grains extend to the solar location. The closest interstellar polarizations are toward stars within 5 pc, indicating that the stream of polarized dust toward Loop I must extend close to the heliosphere in the first galactic quadrant.

[28] and [31]). The IBEX ribbon ISMF is directed toward the North Polar Spur (Figs. 2, 3), so this polarization bridge shows that both an ISMF and a dust stream extend from the solar location out to the interstellar density jump at 100 pc in the North Polar Spur direction. The connection that we find between the ISMF direction at the heliosphere and the dominant local field direction from the polarization data [31; 32] then links the very local ISMF to the ISMF of Loop I.

This cluster of local interstellar cloudlets (CLIC) is decelerating, as shown by blue-shifted velocity components in both the upwind and downwind directions [1; 38]. Representative of this deceleration is the supersonic collision between the LIC and the BC in the direction of Sirius [37; 41].

3. Interstellar magnetic field shaping the heliosphere

The discovery of a "ribbon" of energetic neutral atoms (ENAs) by the Interstellar Boundary Explorer (IBEX) spacecraft [42–44] has provided a remarkable opportunity to understand the relation between the dominant magnetic structure in the sky, Loop I, and the ISMF that shapes the heliosphere. Optical polarization data that trace the ISMF in the low-density local interstellar medium (ISM) gives us a new tool for probing the physical properties and history of nearby interstellar clouds. We have undertaken a project to map the local ISMF direction, and are finding that the ISMF that shapes the heliosphere is associated with the rim of the Loop I superbubble [30–32].

Although there is no generally accepted physical mechanism for the formation of the IBEX ribbon [45], most existing models indicate that it appears in sightlines that are perpendicular to the ISMF as it drapes over the heliosphere upwind of the heliopause [44]. The ribbon is extraordinarily circular, with an ellipticity of < 0.3, and is centered at galactic coordinates of $\ell = 36.7° \pm 2.1°$, $b = 56.0° \pm 0.6°$ (after converting the J2000 ecliptic position in [33] to galactic coordinates). The class of models that have successfully reproduced the ribbon configuration predict that the ribbon forms ~ 50 AU upwind of the heliopause, and that the ISMF direction is within $\sim 15°$ of the center of the ribbon arc [46]. Because of the importance of interstellar pressure terms to the shape and dimensions of the outer heliosphere, the geometric configuration of the ribbon is highly sensitive to the physical properties of the surrounding interstellar cloud [46–49].

The polarity of the local ISMF is not defined close to the Sun by astronomical data. It has been measured by Voyager 1 in the heliosphere depletion region, and the field was found to be directed upwards out of the ecliptic plane [50]. Observations of the Faraday rotation measures of four pulsars in the third galactic quadrant region of low interstellar densities also find that the polarity of the ISMF over ~ 100 pc scales is directed upwards through the galactic and ecliptic planes [51].

4. Structure in the local interstellar magnetic field

We have determined the magnetic field direction in the local interstellar medium where gas column densities are low, $N(H^o) < 10^{18.7}$ cm^{-2}, dust is unreddened, E(B-V)< 0.001 mag, and interstellar optical polarizations are weak, $< 0.01\%$ [30–32]. These studies use new polarization data collected in the southern and northern hemispheres as well as data in the literature. The present analysis focuses on data for stars within 40 pc and 90° of the heliosphere nose in order to confine the study to local ISM in the upwind hemisphere of the sky.

We have developed a method [30; 31] for obtaining the ISMF direction that takes advantage of the fact that a linear polarization position angle of zero degrees indicates that the starlight polarization vector is aligned with the ISMF. Testing the data sample for all possible field directions then yields the "true" field direction corresponding to the lowest value of the weighted mean of the sine of those position angles. The weighting factors allow the use of lower sensitivity polarization data. The polarization data used for this study are plotted in Fig. 3 as the red and green polarization vectors. The polarizations plotted in green consist of a distinct group of stars, here termed the "filament stars" (see below).

Frisch et al. [31] applied this analysis method to a polarization dataset and found a best-fitting magnetic field direction toward $\ell = 47° \pm 20°$, $b = 25° \pm 20°$. The angle between this ISMF direction and the IBEX ribbon field direction is $32° \pm 28°$.

In our new study [32], using an expanded dataset, we find that at least two distinct magnetic field directions are traced by the polarization data [32]. The first structure is traced by the group of stars termed here the 'filament stars'; it is well-defined both by the elongated spatial grouping (green polarizations in Fig. 3), and by the linear decrease of the polarization position angles with distance. This magnetic structure is sampled by stars as close as ~ 5 pc. More details and possible origins of this filament are given in [32].

When the filament stars are omitted from the calculations of the best-fitting ISMF, the ISMF that dominates the fit has the direction $\ell = 36°$, $b = 49°$ ($\pm 16°$). The values of the merit function for this fit are plotted for magnetic field directions in all positions on the sky (Fig. 5). This ISMF direction represents the dominant magnetic structure in the hemisphere around the heliosphere nose, according to our fitting algorithm. The direction is in excellent agreement with the ISMF direction from the IBEX ribbon center to within the uncertainties. The angle between this dominant ISMF direction and the LSR velocity vector of the bulk flow of the local ISM past the Sun is $74° \pm 24°$. A nearly perpendicular angle between the CLIC flow velocity and the CLIC magnetic field is consistent with the expected configuration of a magnetic field that has been compressed in an expanding superbubble shell.

The strength of the ISMF can not be obtained from optical polarization data, but other data and models indicate that it is ~ 3 μG. If an ISMF strength of 3 μG is assumed in the Heerikhuisen et al. models of the ribbon [46], the ribbon center is offset from the direction of the ISMF at infinity by $\sim 4°$. The offset becomes larger for weaker field strengths. Photoionization

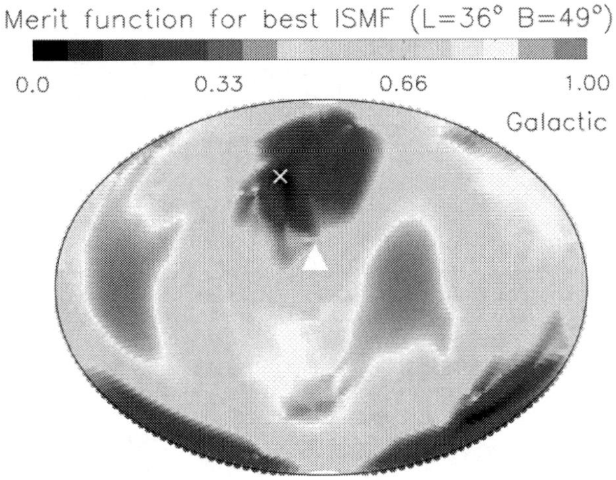

Figure 5. The spatial variations of the merit function for the best-fitting ISMF direction obtained from our polarization data set of qualifying stars (e.g.within 40 pc and 90° of the heliosphere nose) are plotted in galactic coordinates, using an Aitoff projection centered on the galactic center. The white "X" and triangle indicate the ISMF and heliosphere nose directions, respectively. The polarization data set underlying this figure omits the filament stars that trace a different magnetic field direction (see [32] for details). The best-fitting ISMF direction is obtained by minimizing a merit function, which in essence minimizes the ensemble of polarization position angle sines. The best-fitting ISMF direction to the polarization data is within $7.6° \pm 16.2°$ of the magnetic field direction indicated by the IBEX ENA Ribbon [33].

equilibrium models of the Local Interstellar Cloud (LIC) around the heliosphere find a field strength of ~ 3 μG [52]. The pressure of the plasma in the inner heliosheath traced by IBEX ENA measurements, compared to interstellar pressures, indicates a strength for the ISMF of ~ 3.3 μG [53].

The ISMF is a conduit for TeV galactic cosmic rays that flow into the heliosphere [41]. Cosmic rays have gyroradii less than 750 AU in the local ISM for energies ≤ 10 TeV. Stellar polarization data show that the ISMF traced by the IBEX ribbon extends into the space beyond the immediate heliosphere vicinity. This supports results showing that the observed TeV cosmic ray anisotropies are consistent with cosmic ray diffusion along the ISMF traced by the IBEX ribbon [41].

The geometrical relation between the local magnetic field, the positions and kinematics of local interstellar clouds, and the Loop I S1 superbubble, suggest that the Sun is located in the boundary of this evolved superbubble. The quasi-perpendicular angle between the bulk kinematics and magnetic field of the local ISM indicates that a complete picture of low density interstellar clouds needs to include information on the interstellar magnetic field. Fig. 6 shows the projection of the Wolleben S1 shell [36] onto the galactic plane, with bulk motion of the ISM flowing past the Sun overplotted as the thick light gray arrow, and the fifteen clouds derived in the Redfield and Linsky [34] analysis plotted as individual color-coded arrows. The S1 shell geometry is consistent with the inclusion of the CLIC within the shell.

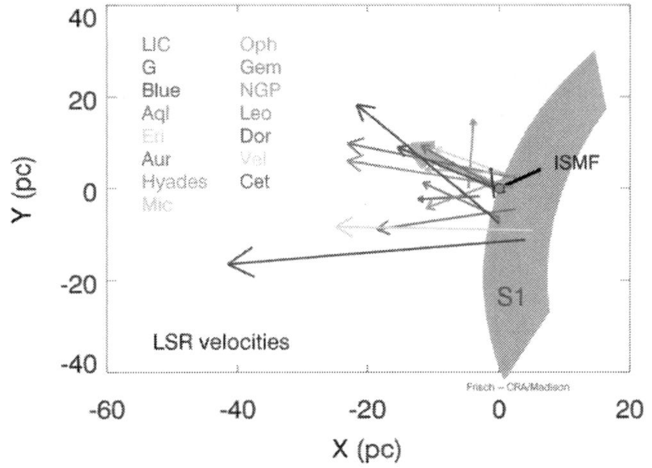

Figure 6. The Loop I S1 shell [36] and nearby cloud [34] LSR velocities [37] are shown projected onto the galactic plane. The black line shows the ISMF direction from the IBEX ENA Ribbon center [33]. Omitting filament stars from the calculation gives an ISMF direction from the polarization data and IBEX Ribbon that agree. The large gray arrow shows the bulk flow of local ISM through the LSR [1]. This ISM flow is nearly perpendicular to the the ISMF and S1 shell. The Sun is located at the orange dot.

5. What is an interstellar cloud?

Eugene Parker once asked "what is an interstellar cloud" [54]. The low density partially ionized interstellar gas in which the heliosphere is now embedded ($n(H^{\circ}) \sim 0.2$ cm^{-3}, $n(H^{+}) \sim 0.06$ cm^{-3}, $T \sim 6300$ K [52]) was once identified as "intercloud medium" [55] to distinguish it from the dusty or opaque clouds that attenuate starlight. Beyond about 7 AU, the mass density of interplanetary space is dominated by neutral interstellar hydrogen. In some sense this interstellar material forms a heliospheric interstellar cloud.

Interstellar material can be characterized by volume density, column density, ionization, kinematics, turbulence, temperature, composition, and/or magnetic field. The tradition of identifying interstellar clouds by velocity began with an early all-sky survey of interstellar absorption components toward bright stars where multiple Doppler-shifted components were resolved [56]. Since then, the Doppler-shifted velocity of a cloud has proven the easiest form of cloud identification. The shortcoming of flagging clouds by absorption line velocity is that this diagnostic depends on the instrument. Adjacent absorption components typically are blended in velocity space, and the number of identified clouds has been found to increase nearly exponentially as instrumental spectral resolution improves [54; 57]. The question of "what is a cloud" then becomes even more problematical for the decelerating flow of ISM past the Sun.

This tradition of identifying clouds by velocity has been followed for the local ISM, but the concept that clouds are individual entities is transcended by the inclusion of an interstellar magnetic field. We have found that the bulk flow of local interstellar gas past the Sun has a direction that is nearly perpendicular to the direction of the local ISMF determined from optical polarization data [31; 32; 37; 41]). This connection between cloud kinematics and magnetic field extends the concept of interstellar clouds beyond velocity components, so as to include large scale structural features in the ISMF such as the Loop I superbubble. Furthermore, the ISMF shaping the heliosphere corresponds to the ISMF traced by the polarizations of nearby starlight.

A discussion of the multiple models for the "cloud-structure" of the local ISM is outside of the scope of this paper. However the comparison between the polarization data (e.g. Fig. 3) and the 15-cloud structure modeled by Redfield and Linsky ([34], also see [1]) produces the interesting result that no significant nearby polarizations are yet found in the upwind direction toward the G-cloud. This appears to be related to the destruction of grains by interstellar shocks [32]. The relative velocities between adjacent cloud components support the presence of local shocked ISM, both between the LIC and G-cloud [58; 59], the Blue Cloud and the LIC [41], and other nearby sightlines.

6. Conclusions

Measurements of nearby interstellar gas and dust indicate that the local ISM is associated with the rim of the evolved Loop I superbubble. The bulk motion of nearby interstellar gas through the LSR indicates an upwind direction toward the center of Loop I. Both polarized starlight and the center of the IBEX ribbon indicate that the magnetic field is perpendicular to the bulk velocity of local ISM. The ISMF compressed in the rim of an expanding superbubble shell that is sweeping up interstellar material is expected to be perpendicular to the expansion velocity of the shell. The increase of polarizations with distance indicates that the source of some of the CLIC dust is part of an interstellar dust stream that connects to the nearest side of Loop I toward the North Polar Spur region.

Acknowledgements: This work has been partially supported by the NASA Explorer Program through funding of the IBEX project.

References

[1] Frisch P C, Redfield S and Slavin J 2011 *ARA&A* **49**
[2] Elias F, Alfaro E J and Cabrera-Caño J 2009 *MNRAS* **397** 2–13
[3] Rand R J and Kulkarni S R 1989 *ApJ* **343** 760–772
[4] Han J 2009 *IAU Symposium* **259** 455–466
[5] Heiles C 1996 *ASP Conf. Ser. 97: Polarimetry of the Interstellar Medium* p 457
[6] de Geus E J 1992 *A&A* **262** 258–270
[7] Crawford I A 1991 *A&A* **247** 183–201
[8] Frisch P C 1995 *Space Sci. Rev.* **72** 499–592
[9] Maíz-Apellániz J 2001 *ApJ* **560** L83–L86
[10] Vallee J P 1984 *A&A* **136** 373–377
[11] MacLow M and McCray R 1988 *ApJ* **324** 776–785
[12] Ferriere K M, Mac Low M and Zweibel E G 1991 *ApJ* **375** 239–253
[13] Frisch P C 1981 *Nature* **293** 377–379
[14] Berkhuijsen E M 1971 *A&A* **14** 359–386
[15] Heiles C, Chu Y , Troland T H, Reynolds R J and Yegingil I 1980 *ApJ* **242** 533–540
[16] Heiles C 1998 *ApJ* **498** 689–703
[17] Heiles C 1998 *IAU Colloq. 166: The Local Bubble and Beyond* pp 229–238
[18] Santos F P, Corradi W and Reis W 2011 *ApJ* **728** 104
[19] Hartmann D and Burton W 2012 *Atlas of Galactic Neutral Hydrogen* (Cambridge U Press)
[20] Berdyugin A, Piirola V and Teerikorpi P 2014 *A&A* **561** A24
[21] Iwan D C 1980 *ApJ* **239** 316–327
[22] Reis W and Corradi W J B 2008 *A&A* **486** 471–484
[23] Mathewson D S and Ford V L 1970 *MmRAS* **74** 139–182
[24] Planck Collaboration, 2014 *ArXiv e-prints* (*Preprint* 1405.0871)
[25] Frisch P C 2000 *American Scientist* **88** 52–59
[26] Lazarian A 2007 *J. Quant. Spec. Radiat. Transf.* **106** 225–256
[27] Planck Collaboration 2014 *ArXiv e-prints* (*Preprint* 1405.0873)

[28] Bailey J, Lucas P W and Hough J H 2010 *MNRAS* **405** 2570–2578

[29] Frisch P C 2010 *ApJ* **714** 1679–1688

[30] Frisch P C, Andersson B, Berdyugin A, Funsten H O, Magalhaes M, McComas D J, Piirola V, Schwadron N A, Slavin J D and Wiktorowicz S J 2010 *ApJ* **724** 1473–1479

[31] Frisch P C, Andersson B G, Berdyugin A, Piirola V, DeMajistre R, Funsten H O, Magalhaes A M, Seriacopi D B, McComas D J, Schwadron N A, Slavin J D and Wiktorowicz S J 2012 *ApJ* **760** 106

[32] Frisch P C, Andersson B G, Berdyugin A, Piirola V, Funsten H O, Magalhaes A M, Seriacopi D B, McComas D J, Schwadron N A, Slavin J D and Wiktorowicz S J 2014 *ApJ, in preparation*

[33] Funsten H O, DeMajistre R, Frisch P C, Heerikhuijsen J, Higdon D M, Janzen P, Larsen B, Livadiotis G, McComas D J, Möbius E, Reese C, Reisenfeld D B, Schwadron N A and Zirnstein E J 2013 *ApJ* **776** 30

[34] Redfield S and Linsky J L 2008 *ApJ* **673** 283–314

[35] Berkhuijsen E M, Haslam C G T and Salter C J 1971 *A&A* **14** 252–262

[36] Wolleben M 2007 *ApJ* **664** 349–356

[37] Frisch P C and Schwadron N A 2014 *Astro Soc Pacific Conf Series*, **484** 42

[38] Frisch P C, Grodnicki L and Welty D E 2002 *ApJ* **574** 834–846

[39] Gry C and Jenkins E B 2014 *A&A* **567** A58

[40] Schonrich R, Binney J and Dehnen W 2010 *MNRAS* **403** 1829–1833

[41] Schwadron N A, Adams F C, Christian E R, Desiati P, Frisch P, Funsten H O, Jokipii J R, McComas D J, Moebius E and Zank G P 2014 *Science* **343** 988–990

[42] McComas D J, Allegrini F, et al. 2009 *Science* **326** 959–462

[43] Funsten H O, Allegrini F, et al. 2009 *Science* **326** 964–967

[44] Schwadron N A, Bzowski M, et al. Crew G B, Gruntman M, Fahr H, Fichtner H, 2009 *Science* **326** 966

[45] McComas D J, Lewis W S and Schwadron N A 2014 *Reviews of Geophysics* doi:10.1002/2013RG000438

[46] Heerikhuisen J, Zirnstein E, Funsten H, Pogorelov N and Zank G 2014 *ApJ* **784** 73

[47] Heerikhuisen J and Pogorelov N V 2011 *ApJ* **738** 29

[48] Frisch P C, Heerikhuisen J, Pogorelov N V, DeMajistre B, Crew G B, Funsten H O, Janzen P, McComas D J, Moebius E, Mueller H, Reisenfeld D B, Schwadron N A, Slavin J D and Zank G P 2010 *ApJ* **719** 1984–1992

[49] Ratkiewicz R, Strumik M and Grygorczuk J 2012 *ApJ* **756** 3

[50] Burlaga L F, Ness N F and Stone E C 2013 *Science Express* 10.1126/science.1235451

[51] Salvati M 2010 *A&A* **513** A28

[52] Slavin J D and Frisch P C 2008 *A&A* **491** 53–68

[53] Schwadron N A, Allegrini F, et al. 2011 *ApJ* **731** 56–77

[54] Frisch P C 2001 *The Interstellar Medium of Our Galaxy* (Kluwer) pp 647–676

[55] Rogerson J, York D, Drake J, Jenkins E, Morton D and Spitzer L 1973 *ApJ* **181** L110–L114

[56] Adams W S 1949 *ApJ* **109** 354–379

[57] Welty D E and Hobbs L M 2001 *ApJS* **133** 345–393

[58] Frisch P C and York D G 1991 *Extreme Ultraviolet Astronomy* p 322 (Pergamon)

[59] Grzedzielski S and Lallement R 1996 *Space Sci. Rev.* **78** 247–258

Has Voyager 1 really crossed the heliopause?

G Gloeckler and L A Fisk

Department of Atmospheric, Oceanic and Space Sciences, University of Michigan, 2455 Hayward St, Ann Arbor, MI 48109-2143 USA

gglo@umich.edu

Abstract. The *Voyager* 1 spacecraft is currently in the vicinity of the heliopause, which separates the heliosphere from the local interstellar medium. There has been a precipitous decrease in particles accelerated in the heliosphere, and a substantial increase in galactic cosmic rays (GCRs). The evidence is unclear, however, as to whether *Voyager* 1 has crossed the heliopause into the local interstellar medium, or remains within the heliosheath. In this paper we propose a test that will determine whether *Voyager 1* has crossed the heliopause: If *Voyager 1* remains in the heliosheath, the high plasma densities must be due to compressed solar wind, with the consequence that *Voyager 1* will encounter another current sheet, where the polarity of the magnetic field reverses. *Voyager 1* observations can be used to predict that the next current sheet crossing is likely to occur during 2015. A prediction is also provided as to what the *Voyager 2* plasma detector will measure in the next few years.

1. Introduction

The *Voyager 1* spacecraft, hereafter referred to as *V1*, is currently in the vicinity of the heliopause, which separates the heliosphere from the local interstellar medium. Particles accelerated in the heliosphere have disappeared, while galactic cosmic rays (GCRs) have reached maximum intensity [1,2], suggesting easy escape of the former across the heliopause and entry of the latter. But has *V1* actually crossed the heliopause and is it now in the interstellar medium? Here the evidence is inconclusive. The direction of the magnetic field observed by *V1* is unchanged from the direction of the heliospheric magnetic field [3], and is in a substantially different direction from the expected direction of the interstellar magnetic field inferred from both *IBEX* observations of energetic neutral atoms (ENAs) [4,5, and references therein], as well as from observations of the polarization of interstellar grains [6]. However, *Gurnett et al.* [7] observe plasma oscillations at 122.2 AU, 122.5 AU, and 123.9 AU, which are a direct measure of the plasma electron density, and find densities of 0.055 cm^{-3}, 0.058 cm^{-3}, and 0.083 cm^{-3}, respectively. These densities are comparable to densities expected in the local interstellar medium, and are much larger than the plasma densities observed by the working plasma detector on *Voyager 2* (hereafter referred to as *V2*) closer to the termination shock [8].

A simple test will determine whether *V1* remains in the heliosheath or has already crossed the heliopause into the local interstellar medium. The test is based on two principles:

(*i*) One of the most observable features of the heliosphere, which would clearly identify that *V1* remains in the heliosheath, is a current sheet crossing. The heliosphere contains a single large current sheet, which separates two hemispheres of opposite magnetic polarity. As the solar cycle evolves, the current sheet becomes highly tilted relative to the solar equatorial plane, and appears to rotate through the poles of the Sun, accomplishing the polarity reversal of the dipole component

Journal of Physics: Conference Series **577** (2015) 012011 doi:10.1088/1742-6596/577/1/012011

of the solar magnetic field. Warps in the current sheet will be compressed as the current sheet is convected into the distant heliosheath, where the radial solar wind speed becomes very small, as is inferred from observations of particle anisotropies [9,10]. Magnetic reconnection will then occur, resulting in two regions of opposite polarity, with reduced magnetic strength due to the reconnection. The single current sheet that separates these two regions marks the time where the field reversal of the Sun occurred at the heliographic latitude and radius of *V1*.

(ii) If *V1* remains in the heliosheath, the large densities observed by *Gurnett et al.* [7] must be the result of compressed solar wind. If the solar wind is compressed, it must be the result of a reduced flow velocity of the solar wind, particularly a reduced radial flow velocity. *V1* is thus moving faster than the outward flow of the solar wind, and will encounter additional current sheets, where the polarity of the magnetic field reverses.

The likely distance at which *V1* will encounter the next current sheet can be estimated from *V1* observations, and thus provides a clear test for whether *V1* remains in the heliosheath.

Fisk and Gloeckler [11] have developed a detailed model for the nose region of the heliosheath, the region of the heliosheath in the direction of motion of the Sun through the local interstellar medium, the regions that the *Voyagers* are exploring. The model is based on the assumption that the solar wind can be compressed. The model yields flow patterns for the solar wind that are consistent with the convective solar wind velocities inferred from >40 keV anisotropy observations, the technique used to determine the solar wind velocity in the absence of a working plasma detector on *V1* [9,10]. The model also yields the likely distance to the heliopause, as well as the likely properties of the heliopause. For our purposes here, the details of the model of *Fisk and Gloeckler* [11] are not needed, other than to note that it is possible to construct a complete model for the nose region of the heliosheath in which the solar wind is compressed.

We begin by presenting the arguments for why it is both possible and indeed to be expected that the solar wind is compressed at the current location of *V1*. We then use *Voyager* observations to predict when *V1* is likely to encounter another current sheet. Guided by *V1* observations and drawing on the work of *Fisk and Gloeckler* [11] we also predict what the *Voyager 2* (*V2*) plasma detector should measure during the next few years. In Concluding Remarks we discuss the different behavior of energetic particles that should be expected when *V1* enters a region of opposite magnetic polarity.

2. The argument for compressed solar wind

The nose region of the heliosheath is a subsonic region, and thus the sum of the particle and magnetic pressure is expected to be constant. It is sometimes argued that since particle density and pressure are related to each other through an equation of state, the solar wind density should also be constant in the nose region of the heliosheath, i.e. the solar wind cannot be compressed.

However, the nose region of the heliosheath is not a conventional plasma. The particle pressure in the nose region of the heliosheath is the result of the mobile suprathermal particles, such as interstellar pickup ions, and particles accelerated in the heliosheath (usually referred to as anomalous cosmic rays (ACRs)). The velocity distributions of these mobile particles are readily observed and/or modeled: (1) The pressure in pickup ions in the supersonic solar wind upstream of the termination is easily determined using standard models of the pickup ion process. Observations by the working plasma detector on *V2* find that it is the pickup ions and halo solar wind [12], not the bulk solar wind, that are heated at the termination shock, satisfying the Rankine-Hugoniot relationship, which determines the pickup ion and halo pressure in the heliosheath [13]. This component of the ion pressure is validated by the *IBEX* observations of energetic neutral atoms resulting from the pickup ions [14]. (2) Energetic particles (e.g. ACR oxygen) observed in the nose region of the heliosheath at ~117 AU have a power law differential intensity spectrum with a spectral index of -1.5, with an exponential rollover, as is expected from the pump acceleration mechanism of *Fisk and Gloeckler* [15]. For such a spectrum, the pressure is due primarily to particles with energies that are measured by the Cosmic Ray Subsystem (CRS) detector on *V1* [2], and thus the ACR pressure is readily determined. The combined perpendicular pressure in the halo solar wind, pickup ions (deduced from *IBEX* observations) and in the ACRs (measured by *V1* at distances of ~117 AU) plus magnetic field and bulk solar wind pressure is estimated to be 1.47×10^{-12} dyne cm^{-2}, close to our earlier value of $\sim 1.2 \times 10^{-12}$ dyne cm^{-2} [14].

The pressure in the halo solar wind, pickup ions and ACRs is the dominant pressure in the nose region of the heliosheath. The plasma detector on *V2*, which is closer to the termination shock than *V1*, measures both the ram pressure and the thermal pressure of the solar wind [8]. The pressure in the pickup ions and ACRs is a factor of ~3 larger than the ram pressure, and is a factor of ~350 larger than the thermal pressure of the solar wind. The bulk solar wind is not heated crossing the termination shock, and relative to itself, it remains supersonic in the heliosheath.

The hot pickup ions, and particularly the energetic ACRs, have average speeds very much larger than the thermal speeds of the solar wind. The pickup ions and ACRs should flow freely along the magnetic field, and seek to maintain pressure equilibrium in the nose region of the heliosheath, independent of the local conditions in the solar wind.

Thus, in the nose region of the heliosheath we have two distinct and basically uncoupled gases: the mobile halo particles, pickup ions and ACRs, which contain the dominant pressure, and the relatively cold bulk solar wind, which contains most of the mass. The pickup ions and ACRs are responding to the global conditions in the heliosheath as they attempt to maintain pressure equilibrium. The bulk solar wind is flowing into the heliosheath across the termination shock, and its density is responding to local conditions. The bulk solar wind by itself has an equation of state that relates pressure and density, and the same is true for the gas of halo particles, pickup ions and ACRs. However, it is inappropriate to consider a single gas of halo particles, pickup ions, ACRs, and the bulk solar wind as having an equation of state since the pressure is determined primarily by global conditions and the density primarily by local conditions. There is thus no restriction on compressing the bulk solar wind.

We should in fact expect that the bulk solar wind must be compressed at distances at least out to ~122 AU, simply from the solar wind velocity measurements. Although *V1* does not have a functional plasma detector, *Krimigis et al.* [9] and *Decker et al.* [10] have been able to infer the solar wind flow velocity from the convective anisotropy of >40 keV ions, and find with their Low-Energy Charged Particle (LECP) detector that the inferred radial flow speed decreases systematically to a very small value beyond ~113 AU, with a small azimuthal speed persisting. At distances between ~116 AU and ~120 AU from the Sun, *V1* was rotated at a number of short time intervals to observe the anisotropies of low-energy ions in a direction normal to the usual azimuthal rotation of LECP [10]. At these distances the anisotropies, and thus the inferred convective flow speed in the polar direction, were found to be small, with an average of essentially zero [10]. This result is consistent with that of *Stone and Cummings* [16], who observe, using the Cosmic Ray Subsystem (CRS) detector that measures somewhat higher energy particles, that the anisotropies in the polar direction become increasingly small with distance into the heliosheath.

The *V1* solar wind observations can only be made when there are low-energy energetic particles present, from which the convective anisotropies can be determined. The measurements cease at ~122 AU, when the particles accelerated in the heliosheath escape. However, inside this distance, where the solar wind flow velocity becomes small in all directions, the continuity of mass equation requires that the solar wind is compressed, and increases substantially. Beyond 122 AU there are no solar wind measurements on *V1*, but it is reasonable to assume that however small, the solar wind continues to flow radially outward, since otherwise the solar wind density would increase without bound.

3. Prediction of when *V1* will encounter another current sheet

In this section we present our prediction as to when *V1* will encounter another current sheet crossing. Our prediction is based primarily on the radial dependence of the average radial component of the solar wind speed, $\langle u_r(r)\rangle$, for which no direct measurements exist on *V1*. We determine $\langle u_r(r)\rangle$ from several indirect measurements, which taken together highly constrain when *V1* should encounter the next current sheet.

3.1. The average radial speed of the solar wind from 1 AU to ~113 AU

Starting at 1 AU, $\langle u_r(r)\rangle$ gradually decreases [17,18,19], due to the process of picking up recently ionized interstellar neutral gas, to a somewhat smaller value immediately upstream of the termination shock at r_{ts}, the distance at which the solar wind ram pressure (determined by the upstream speed) roughly balances the component of the pressure of the Local Interstellar Cloud (LIC) normal to the heliopause (e.g. [18]).

Upon crossing the termination shock at r_{ts}, $\langle u_r(r)\rangle$ decreases by the termination shock compression ratio, which is measured to be ~3 by both *V1* and *V2*. The compression ratio will likely be somewhat lower at solar maximum (a lower upstream speed) and somewhat higher at solar minimum (a higher upstream speed).

Further downstream of r_{ts}, out to ~118 AU, $\langle u_r(r)\rangle$ can be reliably estimated from the LECP observations [9,10]. The radial component of the solar wind speed is inferred from anisotropy measurements of >40 keV ions and, as shown in figure 1, shows quasi-periodic variations, $\delta u_r(t,r)$, with time of about ±20 km/s over distance scales of several AU around a decreasing value of $\langle u_r(r)\rangle$. Thus, *V1* measures $\langle u_r(r)\rangle \pm \delta u_r(t,r)$.

Note that $\langle u_r(r)\rangle$ decreases linearly in figure 1 to where the variations, $\delta u_r(t,r)$, are larger than $\langle u_r(r)\rangle$, and can even result in negative radial solar wind flows. In this region, beyond ~113 AU, where the variations exceed the mean, we cannot use LECP observations directly and require another technique to infer $\langle u_r(r)\rangle$. For $\langle u_r(r)\rangle$ between 105 and 112.5 AU we can reliably use a linear fit to the LECP data.

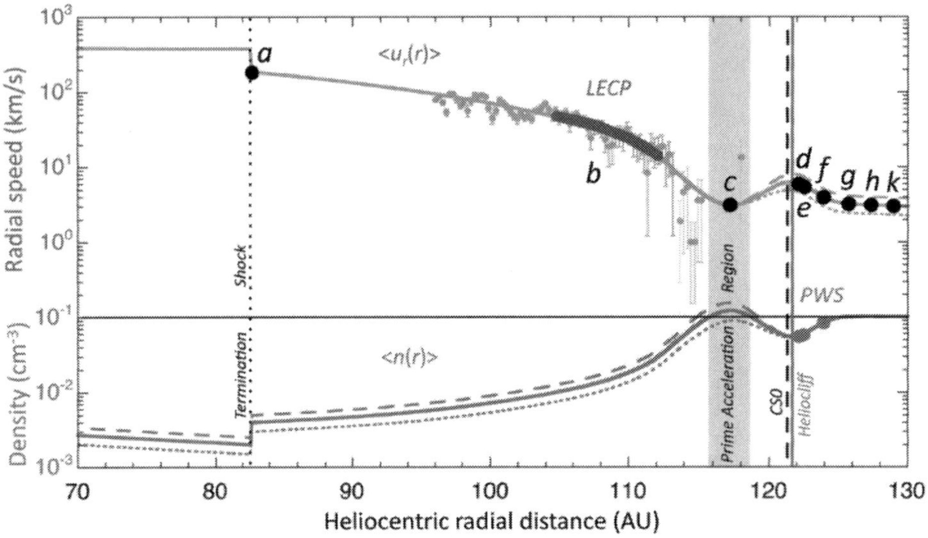

Figure 1. Variation of the average radial speed component, $\langle u_r(r)\rangle$, and the associated average number density, $\langle n(r)\rangle$, of the solar wind with heliocentric distance r. The termination shock is at heliocentric distance r_{ts}, which depends on and will vary with the solar wind speed upstream of the shock. Equation (2), with n_{ts}, the density just downstream to the termination shock, relates $\langle n(r)\rangle$ to $\langle u_r(r)\rangle$. A smooth variation of $\langle u_r(r)\rangle$ with r (no abrupt changes) is achieved using a cubic spline fit through point *a* just downstream of the TS, line segment *b* (the linear fit to the LECP data from 105 AU to 112.5 AU), point *c* at ~117 AU (in the prime acceleration region), points *d*, *f*, and *g* (speeds derived from plasma densities measured by *Gurnett et al.* [7]), and points *h* and *k* (speeds derived from Local Interstellar Medium densities). The speed at *c* is adjusted until the CS0 trajectory goes through point $X^*(r^*,t^*)$, shown in Figure 2 (bottom panel). The dashed, solid, and dotted curves correspond to values of n_{ts} of 0.005, 0.004, and 0.003 cm^{-3}, respectively. The distance, r^* and time t^* of the last current sheet crossed by *V1* were reported and labeled CS0 by *Burlaga et al.* [3]. It is worth remarking that the solar wind is already highly compressed in the heliosheath, reaching a very high density at ~117 AU. At the heliocliff the density has a local minimum, about a factor of two below the Local Interstellar Medium density.

3.2. The average radial speed of the solar wind from ~113 to 121.31 AU

As the current sheet responsible for the reversal in the magnetic field polarity is convected outward with the solar wind, it will encounter *V1* twice. First, in the supersonic solar wind, where the current sheet is moving faster than *V1*, the current sheet overtakes *V1* at time t and distance r. Then, since $<u_r(r)>$ decreases with r (see figure 1) to values below the speed of *V1* (~ 17 km/s), *V1* will overtake the same current sheet at a later time t^* and further distance r^*. Fitting the related current sheet crossings observed by *V1* places strong constraints on the radial dependence of $<u_r(r)>$ and can be used to determine $<u_r(r)>$ in the region beyond 113 AU.

We use the last current sheet crossing observed by *V1*, labeled CS0 by *Burlaga et al.* [3], which occurred at t^*_{CS0} = 2012.57 and r^*_{CS0} =121.31 AU. This is the current sheet crossing that occurred just before the 'heliocliff', where there was a precipitous decrease in the ACRs. We have identified the first crossing of CS0, when it overtook *V1*, to have occurred on DOY 135 in 2000 (14 May 2000, t_{CS0} = 2000.367) when *V1* was at 77.48 AU. As can be seen in figure 2 (top panel), the first crossing occurred when the current sheet rotated through the latitude of *V1* and the dominant polarity changed; a close examination of the daily averages of the magnetic field direction showed an abrupt change in direction at hour 19 of 14 May 2000 (http://vgrmag.gsfc.nasa.gov/data.html).

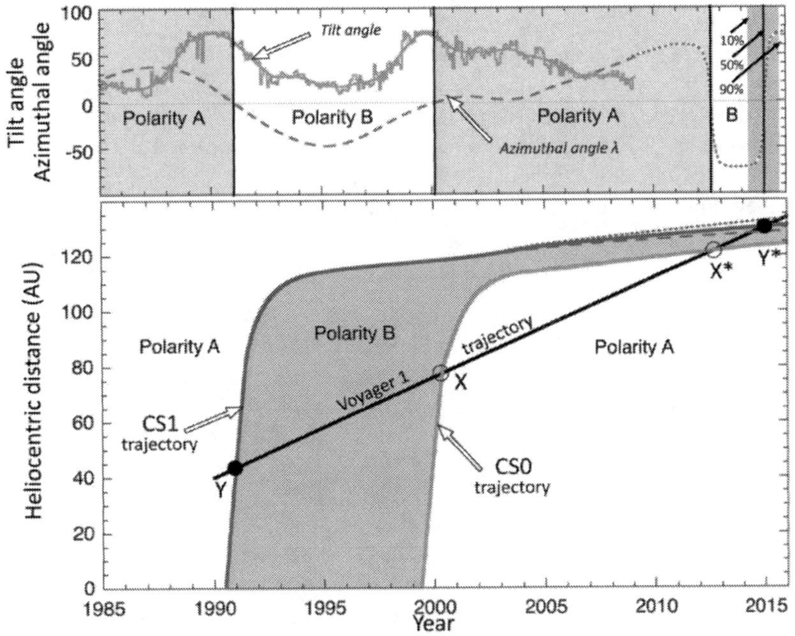

Figure 2. (top) Variations with time of the maximum extent of the tilt angle in the northern hemisphere [20] during a Carrington rotation (step curve) and the 28% weighted average of the azimuthal angle, λ, (in the RTN coordinate system) of the magnetic field (dashed curve) observed by MAG on *Voyager 1*. The solid curve is the 5% weighted average of the step curve. The dotted curve is an extrapolation of the average azimuthal magnetic angle beyond 2009. The dominant polarity reverses when λ (dashed or dotted curve) crosses zero. This reversal should coincide with the rotation of the current sheet past 90° observed at 1 AU. It should be noted that estimates of large tilt angles have large uncertainties, making it difficult to find the exact time when rotation past 90° occurred. The times of the next current sheet crossing by *V1* are indicated by the vertical lines labeled 10% (probability), 50% and 90%. (bottom) Trajectories of current sheet crossings CS0 (observed) and CS1 (predicted). Even though numerous current sheet crossings were observed by *V1* from 1991 to 2012.57 (between points *Y* and *X**), none were seen after the CS0 crossing at *X**. The most plausible explanation is that magnetic reconnection occurs beyond ~122 AU, where the solar wind speed is very small, as discussed in the text.

The trajectory (distance versus time) of CS0 is calculated using the following simple equation:

$$t(r) = t_0 + \int_{r_0}^{r} \frac{dr}{\langle u_r(r) \rangle} . \quad (1)$$

We assume that the solar wind speed at 1 AU is 455 km/s, an appropriate solar maximum value, consistent with the 453 km/s speed measured by *Ulysses* at the *V1* latitude at approximately the same time (2 December 1999) (http://omniweb.gsfc.nasa.gov/coho/). We assume that out to 112.5 AU $<u_r(r)>$ is determined as described in section 3.1. Then, by requiring that the trajectory must pass through both (t_{CS0}, r_{CS0}) and (t^*_{CS0}, r^*_{CS0}), we find a mean radial speed of $<u_r(r)>$ = 4.3 km/s in the region from ~113 to 121.31 AU. Note that this calculation is particularly sensitive to values of $<u_r(r)>$ between 113 and 121 AU since with this slow speed, the dwell time of the current sheet in this region is long.

It is unlikely that $<u_r(r)>$ is constant at 4.3 km/s from ~113 to 121.31 AU, since this would result in an abrupt change in $<u_r(r)>$ at ~113 AU. In subsequent calculations we will make the transition of $<u_r(r)>$ smooth across ~113 AU (see caption to figure 1). The trajectory of CS0 is shown in figure 2 (bottom panel). As required, it goes through points X (t_{CS0}, r_{CS0}) and X^* (t^*_{CS0}, r^*_{CS0}).

3.3. The average radial speed of the solar wind beyond 121.31 AU

Using observations of plasma oscillations, *Gurnett et al.* [7] report very accurate measurements of the plasma electron density during three short time intervals when *V1* was at 122.155, 122.465 and 123.910 AU, respectively. The measured densities are shown in figure 1. The corresponding radial solar wind speed, assuming that the density, n, measured by *Gurnett et al.* [7] is due to compressed solar wind, can be found from the continuity of mass equation,

$$\langle u_r(r) \rangle = \frac{n_{ts} r_{ts}^2 u_{ts}}{n(r) r^2} . \quad (2)$$

In the model of *Fisk and Gloeckler* [11], the density of the solar wind at radial distances beyond where *Gurnett et al.* [7] observed is assumed to have a value equal to the interstellar density. The solar wind density is maintained at this value due to the nature of the heliopause and the pressure in the LIC. We take the solar wind density beyond ~124 AU to be constant at ~0.1 cm^{-3} [e.g. 18], in which case, in this region, $<u_r(r)>$ decreases as r^{-2}.

3.4. The profile of the average radial speed of the solar wind

Taking the results of sections 3.1-3.3 together, and providing smooth transitions across the boundaries between the regions considered, yields the profile of the average radial speed of the solar wind shown in figure 1, out to 130 AU. The parameters used in determining this profile can be varied, but only within fairly narrow limits, as we consider below. Also shown in figure 1 is the density profile calculated from equation (2). Note that there is a region just inside the heliocliff where the density reaches a maximum. It is interesting to note that the density maximum occurs at the location of prime acceleration of ACRs, where they reach their maximum energies [2].

3.5. The prediction of when V1 will encounter the next current sheet

As discussed in section 3.2, current sheet CS0, which *V1* observed at t^*_{CS0} = 2012.57 and r^*_{CS0} =121.31 AU, was the same current sheet that overtook *V1* about 12 years earlier at t_{CS0} = 2000.367 and r_{CS0} = 77.48 AU. It follows then that the next current sheet that *V1* should encounter is the current sheet that overtook *V1* during the previous change in the polarity of the solar magnetic field, which, as can be seen in figure 2 (top panel), occurred at t_{CS1} = 1990.996 when *V1* was at r_{CS1} = 43.54 AU.

Using the date and location of this initial encounter of *V1* with CS1, and the average radial speed of the solar wind in figure 1 (to which we can make small adjustments in the parameters, particularly in the density at the termination shock, n_{ts}), we can estimate when *V1* will next encounter CS1. A lower limit on when the encounter will occur is obtained by noting that *V1* has not yet overtaken CS1, and

Journal of Physics: Conference Series **577** (2015) 012011 doi:10.1088/1742-6596/577/1/012011

thus $t^*_{CS1} > \sim 2014.4$, which implies that $<u_r(r)>$ at 122.155 AU is greater than ~4.65 km/s. This speed condition is satisfied with a solar wind density just downstream of the termination shock of $n_{ts} = 0.003$ cm^{-3}, which is reasonable for solar maximum conditions. To estimate the upper limit on $<u_r(r)>$ we take the solar wind density just downstream of the termination shock to be 0.005 cm^{-3} which yields $<u_r(r)>$ at 122.155 AU of 7.85 km/s. Finally, a density just downstream of the termination shock of 0.004 cm^{-3} (a reasonable value for solar maximum conditions) yields $<u_r(r)>$ at 122.155 AU of 6.13 km/s.

Shown in figure 2 (bottom panel) are three trajectories of CS1 with values for n_{ts} of 0.003, 0.004, and 0.005 cm^{-3}, the dashed, solid, and dotted curves, respectively. The times when the trajectories of *V1* and CS1 cross are 2014.43, 2015.07, and 2015.87, respectively.

Based on our analysis, the probability of *V1* overtaking CS1 on 26 January 2015, and 14 November 2015 is 50% and 90% respectively.

3.6. Future solar wind measurements by the Voyager 2 plasma instrument

In the Fisk & Gloeckler [11] model of the nose region of the heliosheath the flow of the solar wind is primarily in the radial and azimuthal directions. There is a centerline where the solar wind flow must split, streamlines going to the left and to the right of the centerline. *Voyager 1* is relatively close to the centerline, but *V2* is exploring the heliosheath further away from the centerline. With its working plasma instrument, solar wind parameters will be measured directly in the nose region of the heliosheath at distances comparable to those *V1* has now traversed. It is interesting to anticipate what type of plasma *V2* will see as it approaches and then passes the heliocliff boundary. In particular, we will likely have definitive information on whether or not the heliocliff is the heliopause.

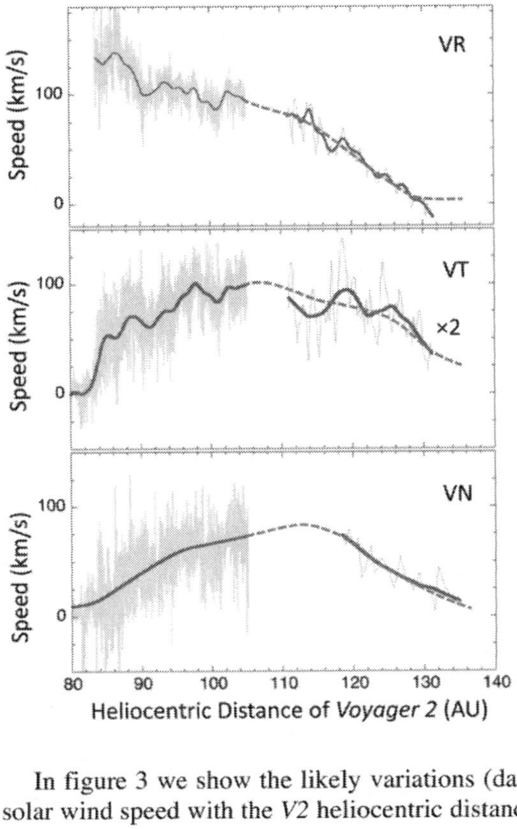

Figure 3. Variations with *Voyager 2* heliocentric distance of the RTN components of the bulk solar wind speed (top to bottom: V_R, V_T and V_N). The pink curves are *V1* estimates of the speed components [9,10,16] plotted at the *V1* distance plus 15 AU. The plasma instrument should be able to measure the N (polar) and T (azimuthal) components of the solar wind, especially during the periodic roll maneuvers of the spacecraft. The radial speed component should be well measured up to ~125 AU at which point the radial speed will be comparable to the speed of *V2*. Beyond that only a portion of the distribution will be observable, making evaluation of the radial speed more problematic. See text for additional discussion. (*V2* plasma data (blue curves) courtesy of John Richardson).

In figure 3 we show the likely variations (dashed red curves) of the three RTN components of the solar wind speed with the *V2* heliocentric distance. Our estimates are based on three assumptions: (1)

the heliocliff boundary is blunt, which pushes the heliocliff out to ~137 AU in the *V2* direction, (2) at a given heliocentric distance the azimuthal speed increases with the angle from the centerline, and (3) the interstellar medium takes control of the general trend of the speed as it decreases at an upstream distance of ~20 AU from the heliocliff, allowing us to use the *V1* speed measurements in determining how the three components will decrease beyond ~ 117 AU in the *V2* direction.

Measurements that are not available from *V1* are the density and temperature or thermal speed of the solar wind in the various regions of the heliosheath. If we had such direct measurements on *V1* we would not be debating whether or not *V1* is now in interstellar space. We use estimates of pressure and density of the bulk solar wind, halo particles, pickup ions and ACR as a function of the *V1* heliocentric distance to deduce variations of the solar wind density and its thermal speed with *V2* distance. The results are summarized in figure 4. We anticipate that the bulk solar wind will be compressed (its density will increase) as it slows down, and its thermal speed (temperature) will start also to gradually increase beyond ~110 AU. Maximum density and temperature will be reached near the heliocliff. The sudden drop in temperature is the result of the disappearance of mobile particles, leaving only the dense solar wind.

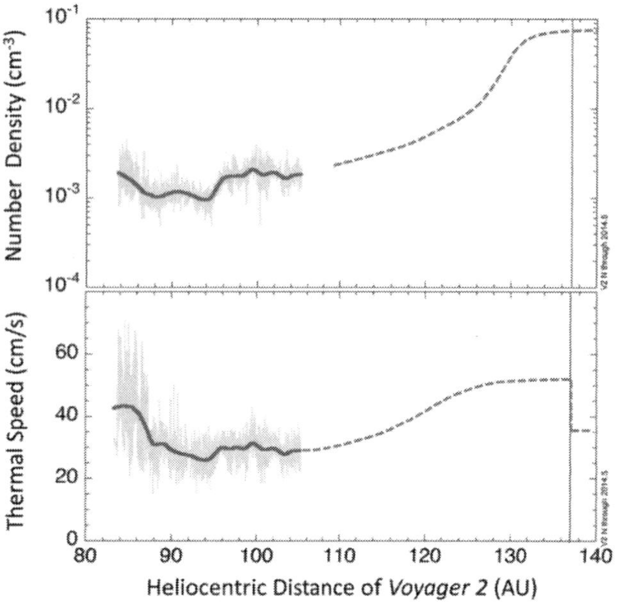

Figure 4. Variations with *Voyager 2* heliocentric distance of the number density (top panel) and thermal speed (bottom panel) of the solar wind. The red dashed curves are predicted future measurements by the plasma instrument on *V2*. See text for additional discussion. (*V2* plasma data (blue curves) courtesy of John Richardson).

4. Concluding remarks

In this paper we have provided a test of whether *V1* has crossed the heliopause or still remains in the heliosheath. If *V1* remains in the heliosheath, the high density measurements of *Gurnett et al.* [7] must be due to compressed solar wind, which should result from a very slow outward radial flow speed, from which it is possible to predict that *V1* will most likely encounter another current sheet crossing during 2015.

The approach taken in this paper, that it is possible to compress the solar wind, is based upon the detailed model of *Fisk and Gloeckler* [11]. In this model there is also a prediction that ACRs and GCRs will behave differently when the magnetic field polarity reverses. The model predicts that when the heliosheath magnetic field is oriented in the same general direction as the interstellar magnetic field, the two fields can merge beyond the heliopause, yielding easy escape of ACRs and easy entry of GCRs. This is apparently the case with the magnetic polarity that *V1* is now observing. When the heliosheath magnetic field is oriented generally opposite to the interstellar magnetic field, reconnection should occur beyond the heliopause, reconnection islands will form, and escape of the ACRs, and certainly entry of GCRs will be restricted.

Thus, if *V1* encounters another current sheet, it will provide strong observation evidence that the heliopause has not yet been crossed. If the region beyond the current sheet, with opposite magnetic polarity, is accompanied by less escape of ACRs and restricted entry of GCRs, it will be strong observational evidence for the model of *Fisk and Gloeckler* [11].

Finally, we made predictions of what the *V2* plasma instrument should be seeing as *V2* approaches and passes the heliocliff. If these direct measurements of plasma parameters by *Voyager 2* turn out to be as predicted in figures 3 and 4, then these observations will lend further credence to the heliosheath model of *Fisk and Gloeckler* [11].

Acknowledgments

This work was supported in part by NASA Grant NNX10AF23G and by NSF Grant AGS-1043012. We thank John Richardson for proving up-to-date *Voyager 2* plasma data through their web address: http://web.mit.edu/space/www/voyager.html

References

[1] Krimigis S M, Decker R M, Roelof E C, Hill M E, Armstrong T P, Gloeckler G, Hamilton D C and Lanzerotti L J 2013 Search for the exit: Voyager 1 at Heliosphere's border with the galaxy *Sci.* **341** 144–147

[2] Stone, E C, Cummings A C, McDonald F B, Heikkila B C, Lal N and Webber W R 2013 Voyager 1 observes low-energy galactic cosmic rays in a region depleted of interstellar ions *Sci* **341** 150–153

[3] Burlaga L F, Ness N F and Stone E C 2013 Magnetic field observations of Voyager 1 entered the heliosheath depletion region *Sci.* **341** 147–150

[4] McComas D J, et al. 2009 Global observations of the interstellar interaction from the Interstellar Boundary Explorer (IBEX) *Sci.* **326** 959–962

[5] McComas D J, et al. 2012 The first three years of IBEX observations and our evolving heliosphere *Astrophys. J. Supp.* **203** doi:10.1088/0067-0049/203/1/1

[6] Frisch,P C 2011 How local is the local interstellar magnetic field *Physics of the Heliosphere: A 10 Year Retrospective* (*AIP Conf. Proc.* vol. 1436) ed J Heerikhuisen et al. (New York: AIP) pp 295–301

[7] Gurnett D A, Kurth W S, Burlaga L F and Ness N F 2013 In situ observations of interstellar plasma with Voyager 1 *Sci.* **341** 1489–1492

[8] Richardson J D 2008 Plasma temperature distributions in the heliosheath *GRL* **35** L23104 doi:10.1029/2008GL036168

[9] Krimigis S M, Roelof E C, Decker R B and Hill M E 2011 Zero outward flow velocity for plasma in a heliosheath transition layer *Nature* **474** 359–361

[10] Decker R B, Krimigis S M, Roelof E C and Hill M E 2012 No meridional plasma flow in the heliosheath transition region *Nature* **489** 124–127

[11] Fisk L A and Gloeckler G 2014 On whether or not Voyager 1 has crossed the heliopause *Astrophys. J.* **789** doi:10.1088/0004-637X/789/1/41

[12] Gloeckler G, Fisk L A, Mason G M, Roelof E C and Stone E C 2012 Analysis of suprathermal tails using hourly averaged proton velocity distributions at 1 AU *Physics of the Heliosphere: A 10 Year Retrospective* (*AIP Conf. Proc.* vol. 1436) ed J Heerikhuisen et al. (New York: AIP) pp 136-143

[13] Richardson J D 2008 Plasma temperature distributions in the heliosheath *GRL* **35** L23104 doi:10.1029/2008GL036168

[14] Gloeckler G and Fisk L A 2010 Proton velocity distributions in the inner heliosheath derived

from energetic hydrogen atoms measured with Cassini and IBEX *Pickup Ions Throughout the Heliosphere and Beyond* (*AIP Conf. Proc.* vol. 1302) ed J A le Roux et al. (New York: AIP) pp 110–116

[15] Fisk L A and Gloeckler G 2013 The global configuration of the heliosheath inferred from recent Voyager 1 observations *Astrophys. J.* **776** doi:10.1088/0004-637X/776/2/79

[16] Stone E C and Cummings A C 2011 Voyager observations in the heliosheath: A quasi-stagnation region beyond 113 AU *Proc. of the 32nd Intl. Cosmic Ray Conf.*. vol 12 (Beijing: IUPAP) p 29

[17] Richardson J D, Liu Y, Wang C and McComas D J 2008 Determining the LIC H density from the solar wind slowdown *Astron. Astrophys.* **491**(1) 1–5

[18] Gloeckler G, Fisk L A and Geiss J 1997 Anomalously small magnetic field in the local interstellar cloud *Nature* **386** 374–377

[19] Lee M A 1997 Effects of cosmic rays and interstellar gas on the dynamics of a wind *Cosmic Winds and the Heliosphere* ed J R Jokippi et al. (Tucson: University of Arizona Press) p 857

[20] Hoeksema J T 1995 The large-scale structure of the heliosheath current sheet during the Ulysses epoch *Space Sci. Revs.* **72** 137–148 http://wso.stanford.edu/Tilts.html

A new perspective on the interstellar cloud surrounding the Sun from UV absorption line results

Cecile Gry[1] and Edward B. Jenkins[2]

[1] Aix Marseille Université, CNRS, LAM (Laboratoire d'Astrophysique de Marseille) UMR 7326, 13388, Marseille, France

[2] Department of Astrophysical Sciences, Princeton University Observatory, Princeton, NJ 08544, USA

E-mail: cecile.gry@lam.fr, ebj@astro.princeton.edu

Abstract. We offer a new, more inclusive, picture of the local interstellar medium, where it is composed of a single, monolithic cloud that surrounds the Sun in all directions. Our study of velocities based on Mg II and Fe II ultraviolet absorption lines indicates that the cloud has an average motion consistent with the velocity vector of gas impacting the heliosphere and does not behave like a rigid body: gas within the cloud is being differentially decelerated in the direction of motion, and the cloud is expanding in directions perpendicular to this flow, much like the squashing of a balloon. The outer boundary of the cloud is in average 10 pc away from us but is highly irregular, being only a few parsecs away in some directions, with possibly a few extensions up to 20 pc. Average H I volume densities vary between 0.03 and 0.1 cm-3 over different sight lines. Metals appear to be significantly depleted onto grains, and there is a steady increase in this effect from the rear of the cloud to the apex of motion. There is no evidence that changes in the ionizing radiation influence the apparent abundances. Additional, secondary velocity components are detected in 60% of the sight lines. Almost all of them appear to be interior to the volume holding the gas that we identify with the main cloud. Half of the sight lines exhibit a secondary component moving at about - 7.2 km/s with respect to the main component, which may be the signature of an implosive shock propagating toward the cloud's interior.

1. Introduction

Our aim is to propose a simplified picture of the local interstellar medium (LISM), where it is composed mainly of a single, continuous cloud, which envelops the Sun in all directions. This outlook differs from the current, commonly accepted view that the LISM is a collection of identifiable cloud structures, all spatially and kinematically distinct from one another, participating in a common bulk motion (cf Frisch, Redfield & Slavin, 2011 for a review). In the most complete version of this model, Redfield & Linsky (2008) interpreted the different velocity components detected in absorption in the spectra of nearby stars in terms of 15 coherently moving clouds, all moving at different but similar velocities. In their picture the heliosphere is not immersed in the local interstellar cloud (LIC), but is situated near its edge, in the transition region between the LIC and another cloud referred as the G cloud. With such a configuration, whereby the Sun is surrounded by several distinct and separate clouds, we would expect to find some lines of sight going through the gaps between the clouds, which should create at least a few cases showing no absorption. However such a line of sight has never been found: every stellar spectrum including a sensitive enough absorption line in the UV shows at least one interstellar component. We note also that three of the clouds (LIC, G and a third one denoted NGP for North Galactic Pole) almost complement each other to cover most of the sky with no overlap, and are described with similar velocity vectors. A question arises: Do the data support a picture that these clouds could be unified into a single cloud? For instance, the G cloud had been introduced by Lallement & Bertin (1992) to account for a slight velocity discrepancy in the direction of α Cen. Now that we have a larger sample of nearby sight lines and a better knowledge of the ISM flow in the heliosphere, we can revisit the question, "Is the G cloud still needed as a distinct cloud?" We therefore decided to re-examine the LISM UV-absorption database with a new focus, i.e. making the assumption that the ISM close to the solar system can be viewed as a unique, continuous medium that envelops the Sun in all directions and, in the simplest approximation, produces one absorption component in every line of sight. With this assumption in mind, we examine how this conforms with a picture of a homogeneous cloud, moving with a coherent velocity vector as a rigid body.

2. The data

We re-examine the Mg II and Fe II results from UV lines observed with HST (GHRS and STIS) at high resolution (R$\simeq$ 100 000, or $\Delta V \simeq 3\,\mathrm{km\,s^{-1}}$). We use the data compiled in Redfield & Linsky (2002) Table 3 (for Mg II) and Table 4 (for Fe II), providing the velocity, broadening parameter, and column density for each component. To guarantee the homogeneity in sensitivity and preserve the integrity of the velocity calibration, we do not include data from ground-based spectroscopy; therefore the CaII or NaI results are not included in this analysis. Note that we do not reanalyze the data; we adhere to the published results listed by Redfield & Linsky (2002). However we have checked the velocity scales for data that had new wavelength calibrations, resulting in 6 slight velocity changes.
Our resulting database includes 111 components, in a total of 59 lines of sight, of which 54 are observed in Mg II, 32 are observed in Fe II, and 27 are observed in both elements. New observations in additional directions were reported at this conference (Redfield et al, this volume) and appeared later in a paper by Malamut et al. (2014). The future addition of these new results should help to refine the conclusions presented here.

3. Identification of a single circum-heliospheric cloud

3.1. Step 1: identification of the component most likely to originate in the circum-heliospheric cloud

In every line of sight (l_i,b_i) we select the absorption component closest in velocity to $V_{i,helio}$, i.e. the projection onto the sight line of the velocity vector of the ISM flow interacting with the heliosphere as measured by Ulysses and previous experiments (Möbius et al. 2004) or by IBEX (McComas et al. 2012). We call this component "Component 1". There is by definition one Component 1 in every line of sight.

3.2. Step 2: Mean velocity vector that best fit all Components 1

We perform a least squares fitting of the Components 1 sample to a single vector in the sky in the heliocentric reference frame. The best-fit gives the following mean vector:

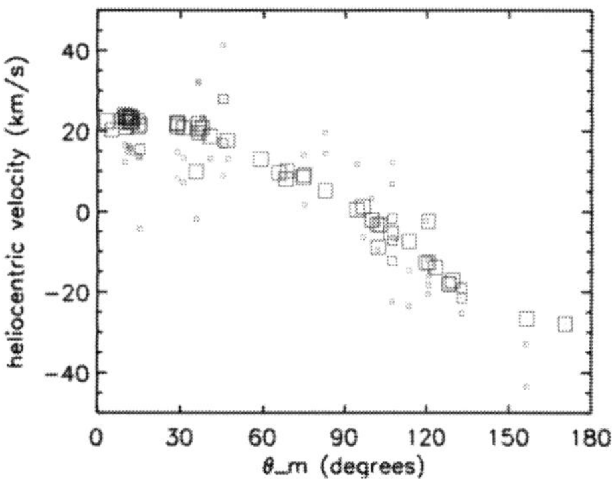

Figure 1. Measured velocities of all components plotted against θ_m, the angular distance to the apex of the Components 1 mean vector. Large symbols indicate components that are dominant in their line of sight in terms of column density. Medium-size symbols depict cases where two components in a line of sight have identical column densities within their error bars. The smallest symbols indicate components that are clearly secondary. The adopted reference column densities are those of Fe II, when available, and Mg II otherwise.

$$l_m = 185.84 \pm 0.83°;\ b_m = -12.79 \pm 0.67°;\ V_m = 25.53 \pm 0.26\text{km/s}$$

Our measured velocity vector is indeed very close to both solar system measurements: it has the same direction as the IBEX vector, but the velocity amplitude is closer to that of the Ulysses measurement. We define for each line of sight the angle θ_m, representing its angular distance to the apex of the above mean velocity vector. On Figure 1 our Components 1 stand out as forming a conspicuous $V_m \cos\theta_{i,m}$ curve, where $V_m = 25.53$ km s^{-1}. They also appear to be the dominant component (shown as large squares) in most sight lines.

3.3. Step 3: Components 1 velocity deviations from the best-fit velocity vector

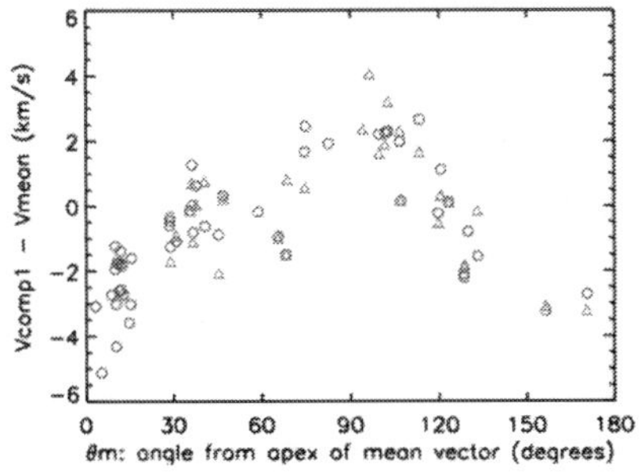

Figure 2. Velocity residuals of Components 1, the components closest to V_{helio} , relative to the $V_m\cos\theta_{i,m}$ relationship for the best-fit vector derived from the heliocentric Mg II velocities. Circles: Mg II, triangles: Fe II

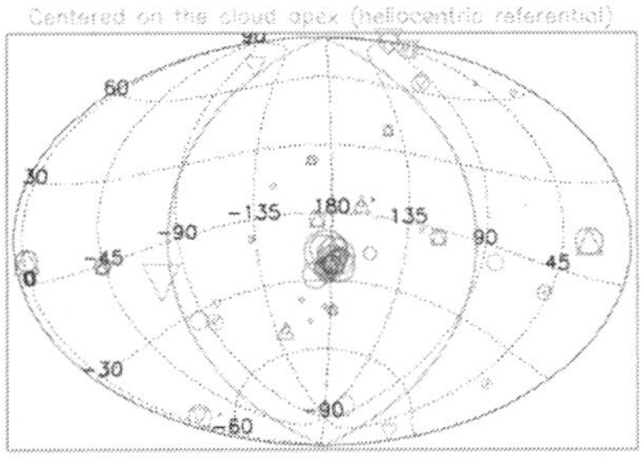

Figure 3. Components 1 velocity deviations from the best-fit vector, in Galactic coordinates but centered on the apex of motion. Red/blue: positive/negative deviations. Symbol size: magnitude of the deviations. Red curve: circle 90° away from the direction of motion.

Figure 2 exhibits a clear trend of the velocity deviations with θ_m: they are negative toward the apex and anti-apex directions and positive in directions perpendicular to the motion. When the trend is taken into account as a systematic velocity gradient, the remaining dispersion is comparable to the velocity determination uncertainties ($\simeq$ 1.1 km s^{-1}). Figure 3 shows that the large positive deviations (red symbols) are indeed roughly distributed along a great circle at 90°away from the direction of motion, showing that the expansion perpendicular to the flow vector is seen in all azimuthal angles.

3.4. Conclusion: a single cloud

We think that the simple, symmetrical, single hump pattern exhibited in Fig. 2, and the coherent spatial distribution of the residuals, indicate this is a fundamental, second-order dynamical effect associated with an otherwise coherently moving cloud. We conclude that, provided we relax the assumption that the cloud is rigid, our Components 1 sample can indeed be interpreted as the projected velocities of a unique interstellar cloud, detected in all directions, with a mean velocity close to the ISM flow into the heliosphere: the Local Cloud. This cloud accounts for 53 % of the components in number. It accounts for 70 % of the column density in the first 50 pc. It includes most of the LIC, G, NGP, Blue, Leo, Aur, Cet, clouds and half of the Eri and Gem clouds components in the Redfield & Linsky (2008) model.

4. Characteristics of the cloud

4.1. Dynamics within the cloud

We developed a simple, idealized model for the velocity deviations. We start with a spherical volume centered on the observer and estimate the velocity perturbations when the sphere is differentially decelerated. To maintain its original volume (and internal pressure) the cloud is deformed into an oblate ellipsoid that has its minor axis aligned with the deceleration vector. We derive the velocity perturbations $\Delta V(\theta_d)$ away from a solid-body motion, expressed as a function of θ_d, which is the angle away from the minor axis of the ellipsoid:

$$\Delta v(\theta_d) = C\left[(1-e^2)^{-1/6}(1-e^2\cos^2\theta_d)^{1/2} - 1\right],$$

where e is the eccentricity of the ellipsoid and C is a constant. Provided the variable e is not large ($e < 0.8$), the shape of the curve tracing Δv as a function of θ_d is invariant and its amplitude is governed only by the product of C and e.

Minimizing the quadratic differences of the observed velocity deviations with respect to $\Delta V(\theta_d)$, we derive the direction of the minor axis (i.e. the origin of the angle θ_d):

$$l_d = 174 \pm 5°;\ \ b_d = -12 \pm 4°,$$

which is relatively close to the apex of the mean velocity vector found in the previous section. If the differential deceleration and distortions are due to the cloud interaction with an external medium or a magnetic field, the minor axis found for the spatial deformation is very likely to indicate a fundamental direction for the cloud's internal motions in response to these external forces, independent of the chosen reference frame. For this reason, we adopt (l_d, b_d) as the reference direction for the cloud.

4.2. Extent and density of the cloud

Mg II and Fe II column densities exhibit no correlation with distance: this means that the cloud is small compared to target distances. On the other hand, the column densities show a high dispersion and there is no obvious trend with direction: we conclude that the cloud is very irregular in shape.

To study the extent and density of the cloud we examine the H I compilation and results from Wood et al (2005), with particular attention to their detections of astrospheres. Since an astrosphere arises from the interaction of the stellar wind with neutral atoms, its presence for a particular line of sight means that the target star is embedded in neutral gas. In the H I sample an astrosphere is detected in 8 single-component lines of sight, where we conclude that the Local Cloud fills up the entire sight line, and where the ratio N(HI)/d reliably measures the average density. The latter varies in these sight lines from 0.03 to 0.1 cm^{-3}, yielding a mean neutral density in the Local Cloud of 0.049 cm^{-3}, in agreement with the solar system measurements of interstellar pick-up ions and anomalous cosmic rays of n(HI)= 0.055 ± 0.029 cm^{-3} (Gloeckler et al. 2009). Inspection of Figure 4 shows that (1) the Local Cloud extends up to 20 pc in a few sight-lines and to less than 15pc in most of them, (2) if we adopt n(HI) = 0.05 cm^{-3}, the mean extent of the cloud is <N(HI)/n(HI)> = 9.3 ± 7.6 pc, (3) in no directions does the Sun lie near the cloud edge: the minimum distance to the boundary is somewhere between 1.3 and 4.7 pc, depending on the density value chosen, and (4) highest column densities are found in directions perpendicular to the cloud deformation minor axis.

4.3. MgII and FeII abundances

As illustrated in Figure 5 a correlation exists between the abundances of MgII and FeII and the angle θ_d, which suggests the presence of an abundance gradient within the cloud.

We first check if the observed gradient can be explained by ionization effects. The LISM is known to be partially ionized and its ionization is maintained by a strong UV flux dominated by the two stars ϵ and β CMa, located respectively at θ_d = 65°and 51°. Shielding by the cloud material makes the ionizing flux decrease with cloud depth and this means that the ionization should be maximum in the direction of the two stars and decrease away from it. However, there is no visible correlation of the Mg II and Fe II abundances with the angular distance to the star ϵ CMa. On the other hand, in regions were photons more energetic than 13.6 eV can penetrate the cloud, an element can be more or less ionized than H, depending on its relative photoionization to recombination rate ratio $\Gamma(X)/\alpha(X)$ compared to $\Gamma(H)/\alpha(H)$ (Jenkins 2004). As a consequence, we would expect to find Fe II more ionized and Mg II less ionized than H, making N(Fe II)/N(H I) lower and N(Mg II)/N(H I) higher than in fully neutral gas. However,

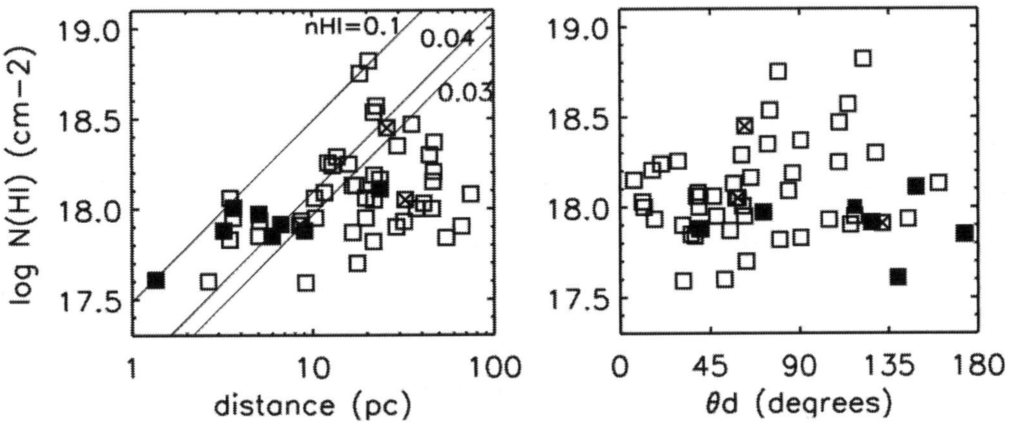

Figure 4. H I column densities. □: Components 1. ■: targets presenting evidence for an astrosphere and only one H I component, i.e. sight lines completely filled with the Local Cloud. Crossed □: targets where the presence of the astrosphere is uncertain. Diagonal lines show lines of constant densities n(H I) = 0.03, 0.04, 0.1 cm^{-3} .

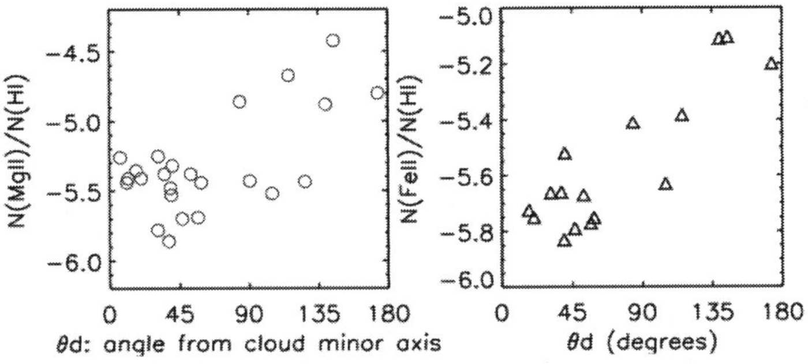

Figure 5. N(Mg II)/N(H I) and N(Fe II)/N(H I) for sight lines in common to both H I and Mg II, Fe II samples, plotted against θ_d, the angle from the cloud's deformation minor axis pointing near the apex of motion.

we will see that the opposite is true since Mg II appears more depleted than expected. Therefore there is no evidence that ionization influences the apparent abundances.

The trends displayed in Figure 5 are thus most probably related to the gas-phase abundances of the two elements. The magnitude of the depletion is gradually increasing from the upwind (θ_d = 180°) to the downwind direction (θ_d = 0°). This had been noticed already by Redfield & Linsky (2008) and interpreted as a physical difference between the LIC and the G cloud, whereas in our picture it is the sign of an abundance gradient within the cloud. The metal abundances are roughly 4 times (0.6 dex) higher in the rear of the cloud than in the front. We note that there is a constant depletion ratio of $\simeq$ 1.5 (0.17 dex) between Mg and Fe, independent of θ_d, which is significantly lower that the mean interstellar value. The fact that we observe more depletion near the head of the cloud, compared to that at the rear (i.e., upwind direction), indicates that a progressive erosion of the grain population may either be taking place, perhaps as a consequence of whatever process is causing the differential deceleration, or may alternatively have occurred in relation to the origin of the cloud as proposed by Kimura et al (2003).

5. Nature of the other components

Out of 59 sight lines, 34 (58%) show at least one other component in addition to Component 1, 13 (22%) show at least two other components, one shows three and another shows five additional components. We examine the characteristics of these other components to try and pinpoint their nature.

5.1. Secondary nature of the other components.

Figure 6 shows that in any line of sight Component 1, i.e., the component whose velocity is most consistent with the velocity vector of gas impacting the heliosphere, is generally also the dominant component. Out of the 4 lines of sight where another component is larger than Component 1, three of them (ι Cap, υ Peg and G191B2B) are longer than 50 pc, meaning that the larger component may be situated far away. For the one short sightline toward 61 Cyg A (3.5 pc), the profile fitting is very uncertain and could in fact result in a reversed column density ratio for the two components if different b-values were chosen (Wood & Linsky 1998). Therefore the components we have kinematically identified as the Local Cloud represent the bulk of the matter in the first 50 pc.

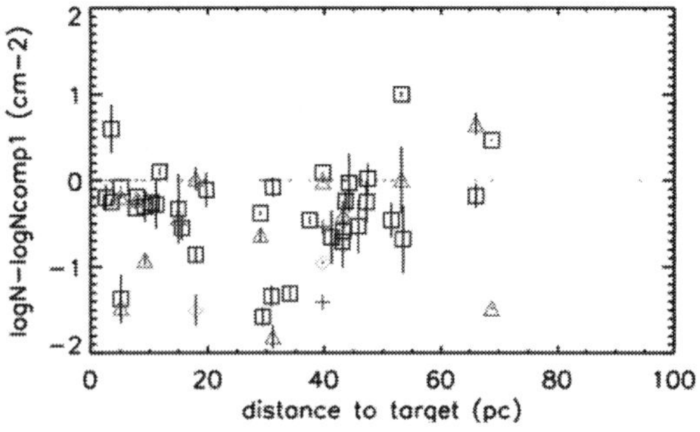

Figure 6. Column density ratios (in log) between the other components and their corresponding Component 1: log N/Ncomp1 against distance. The horizontally arranged dots are components 1. They show the distances of all lines of sight. The other components, numbered in order of increasing $|\Delta V|$ relative to Component 1 are as follows: Black squares are components 2, red triangles components 3, green diamonds components 4 and blue crosses components 5 and 6.

5.2. Proximity of the other components.

We note in Figure 7 that 1) longer sight lines do not show more components than very short ones and 2) even the shortest ones exhibit multiple components: 5 sight lines shorter than 5 pc show 2 or even 3 components. This implies that most of the secondary components exist at distances short of those to the edges of the cloud, and that some of them could be situated within the same volume as that occupied by the local cloud.

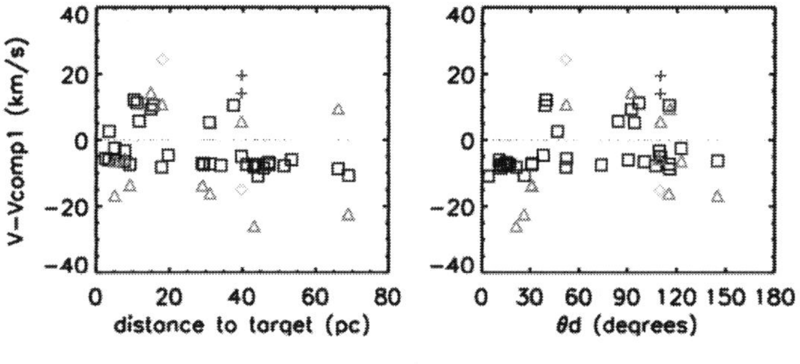

Figure 7. Velocity shifts of the other components relative to Components 1, against distance and θ_d. The meanings of different symbols are the same as those in figure 6. Notice the preponderance of horizontally aligned symbols just below zero: these are what we refer to as the "Cetus Ripple" components.

5.3. Consistency in the velocity shifts of the other components relative to the Local Cloud

Figure 7 also illustrates the fact that half of the secondary components have a consistent $\Delta V = -7.2 \pm 1.5$(rms) km s^{-1} relative to the Local Cloud component and about half of the sight lines include such a component. Figure 8 shows that the components exhibiting the -7.2 km s^{-1} shift also cover approximately half of the sky. When fitting these components as a cloud having a coherent motion relative to the Local Cloud, we get a velocity vector originating from Cetus ($l_{CR} = 124° \pm 4°$, $b_{CR} = -67° \pm 3°$), which is what motivated us to name them the Cetus Ripple components. However, their velocity shifts plotted against the angular distance to the best-vector origin do not follow the expected cosine relationship, but appear instead very constant over θ. This shows that the Cetus Ripple components do not

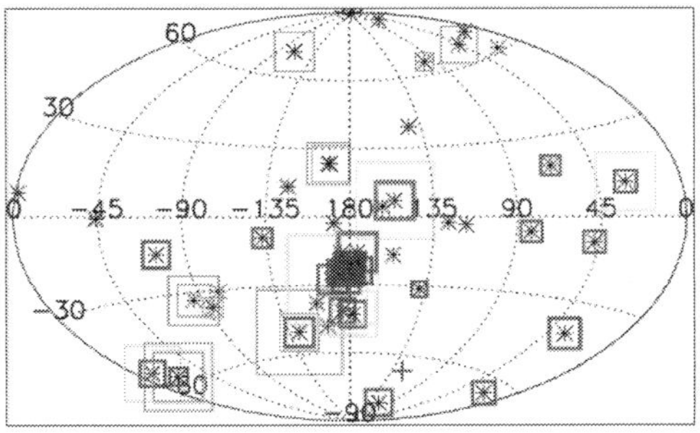

Figure 8. Velocity shifts of the other components relative to Component 1. Black asterisks: all targets. Squares: components other than Components 1. Red/blue: positive/negative shifts. Dark blue: Cetus Ripple components. Symbol size: amplitude of the shift. The blue cross indicates the rough center of the hemisphere occupied by the Cetus Ripple components.

present a coherent linear motion, but instead we conclude that their motion is characteristic of an implosion converging near our location.

5.4. A plausible model for the Cetus Ripple as a shock progressing toward the cloud's interior

Considering all known characteristics of the pre-shock gas (i.e. the Local Cloud), we show that it is possible to define a shock velocity (between 20 and 26 km s^{-1} depending on the chosen value for the magnetic field) that yields post-shock conditions matching remarkably well our observations for the Cetus Ripple components in terms of velocity, column density and velocity dispersion (see Gry & Jenkins 2014 for details). The origin of this shock seems compatible with a sudden increase in thermal pressure by the surrounding gas present in the Local Bubble.

5.5 Possible origins for the remaining components

Closer than 50 pc, the remaining components which are due neither to the local cloud nor to the Cetus Ripple constitute a minor fraction of the matter. They may arise from secondary shocks or small velocity waves induced by turbulent motions, or from gas escaping the local cloud. Larger components likely to arise from separate clouds seem to exist beyond 50 pc.

6. Summary and conclusions

We have shown that the UV absorption line database is compatible with the picture of the LISM as one single, monolithic cloud that surrounds the Sun in all directions.

Its extent is within 9.3±7.6 pc in most directions, and the Sun is not close to the edge of the cloud.

Motions within the cloud deviate from that of a homogeneous rigid body. The second-order kinematics inside the cloud suggests that it may be deforming from an apparent progressive deceleration in the direction of motion.

H I densities within the cloud vary between 0.03 and 0.1 cm^{-3}, and the metal abundances show a gradient in the direction of motion.

This Local Cloud dominates the absorption in the first 50 pc. Most secondary absorption components detected in 60% of the sight lines appear to be located inside the boundary of the Local Cloud.

Half of the secondary components have a velocity shift of -7.2 km s^{-1} relative to the Local Cloud and may result from an imploding shock created by an imbalance of pressure with the surrounding medium.

References

Frisch, P.C., Refield S. & Slavin J.D., 2011, Annu. Rev. Astron. Astrophys. 2011. 49, 237
Gloeckler, G., Fisk, L. A., Geiss, J., et al. 2009, Space Sci. Rev., 143, 163
Gry, C. & Jenkins, E.B., 2014, A&A 567, A58
Jenkins, E. B. 2004, in Carnegie Obs Ap Serie, Vol. 4, Origin and Evolution of the Elements, ed. A. McWilliam & M. Rauch
Kimura, H., Mann, I. & Jessberger, E. K. 2003, ApJ 582, 846
Lallement, R. & Bertin, P., 1992, A&A 266, 479
Malamut, C., Redfield, S., Linsky, J.L., Wood, B. E., and Ayres, T. R. 2014, ApJ 787, 75

McComas, D. J., Alexashov, D., Bzowski, M., et al, 2012, Science 336, 1291
Möbius, E., Bzowski, M., Chalov, S., 2004, A&A 436, 897
Redfield, S. & Linsky, J.L., 2002, ApJS 139, 439
Redfield, S. & Linsky, J.L., 2008, ApJ 673, 283
Wood, B.E., Redfield, S., Linsky, J.L., Mueller, H.R. & Zank, G.P., 2005, ApJS 159, 118
Wood, B.E. & Linsky, J. L., 1998, ApJ, 492, 788

A more comprehensive description of this work can be found in Gry & Jenkins (2014).

Modelling the March 2012 solar events and their impacts at Voyager 1 in the vicinity of the heliopause

D S Intriligator[1], W Sun[1, 2], W D Miller[1], M Dryer[1], C Deehr[2], W R Webber[3], J M Intriligator[1, 4], and T M Detman[1]

[1] Carmel Research Center, Santa Monica, CA 90406, USA
[2] University of Alaska, Fairbanks, AK 99775, USA
[3] New Mexico State University, Las Cruces, NM 88003, USA
[4] Bangor University, Bangor, Wales, LL57 2TT, UK

E-mail: devriei@aol.com

Abstract. Using our three dimensional (3D) time-dependent kinematic HAFSS model we propagate the March 2012 solar events from the Sun to Voyager 1 (V1) to investigate if these solar events could be responsible for the V1 April-May 2013 Plasma Wave Spectrometer (PWS) measurements of enhanced 2-3 kHz signals. These PWS measurements provided the basis for the confirmation that V1 entered the interstellar medium (ISM). These enhanced PWS signals also were associated with significantly increased plasma densities consistent with expectations for the ISM. The origin of these enhanced plasma densities was attributed to the April-May 2013 arrival at the heliopause of the disturbances associated with the March 2012 solar events. Using HAFSS we propagate from the Sun the March 2012 solar wind ambient background and the solar wind impulsive events for comparisons with in-situ spacecraft measurements at Earth, V1 (at 34 degrees North), and Voyager 2 (at 30 degrees South). The in-situ measurements of the March 2012 events were obtained over a wide range in radial distance, longitude, and latitude. This emphasizes the importance of using a 3D time-dependent model that originates at the Sun. From our analyses we conclude that the March 2012 solar events could have been responsible for the April-May 2013 V1 enhanced plasma wave signals and their associated increased plasma densities measured by the PWS indicating V1 was in the ISM.

1. Introduction

The Voyager 1 (V1) energetic particle measurements from April-August 2012 provided information [1-3] that the V1 spacecraft (s/c) may have been near the heliopause (HP). The LECP [1] and CRS [4,2] energetic particle measurements apparently even indicated that on August 25, 2012 V1 was crossed by the HP [4]. However, the V1 magnetometer data [3] appeared to be not as definitive, since the magnetic field magnitude increased as expected for a crossing of the HP, but the expected accompanying change in the magnetic field direction was not seen. The V1 Plasma Spectrometer (PLS) had not been able to make plasma measurements since 1980, so the expected increase in plasma density across the HP and into the ISM could not directly be observed. Subsequently [5] found evidence in the Plasma Wave Spectrometer (PWS) measurements of enhanced signals in the 2-3 kHz range consistent with the substantial increase in plasma density expected if V1 entered the ISM. [5] suggested that these enhanced PWS plasma wave signals were associated with the V1 PWS detecting the impacts of the March 2012 solar events on the HP.

The purpose of this present paper is to investigate whether solar events in March 2012 gave rise to interplanetary plasma disturbances (e.g., shocks, merged interaction regions (MIRs), etc.) that could

Journal of Physics: Conference Series **577** (2015) 012013 doi:10.1088/1742-6596/577/1/012013

have been responsible for the observed enhanced signals in the 2-3 kHz PWS measurements. We present our initial results from 3D time-dependent kinematic simulations of the March 2012 solar events by the HAFSS (Hakamada-Akasofu-Fry Source Surface) method, which simulates the slowly varying solar wind background starting at the Sun and the impulsive time-dependent solar wind events associated with solar activity (e.g., coronal mass ejections (CMEs).

In the HAFSS model perturbations are imposed on the boundary conditions at the Sun to simulate solar activity. The tuning of both the background solar wind state and the impulsive solar wind from solar events is iterative, but the background model tuning only needs to be done once for the period of interest. In previous analyses we developed methods that make shock simulation inputs converge very quickly. Our tuning uses strictly quantitative verification of model results against spacecraft data. Only quantitative testing of such models in a 3D time-dependent context against spacecraft data can both further validate and improve the models. See [6] for a more detailed description of the HAFSS model.

In March 2012 NOAA solar Active Region (AR) 1429 emitted a series of disturbances and CMEs. Table 1 summarizes the characteristics of this March 2012 solar activity as obtained from the NOAA Space Weather Prediction Center daily Report of Solar Geophysical Activity. In Table 1 [7] the first column is the number of the event; the second, the four-digit year; the third, the four-digit month and day; the next column is the two-digit hour and the two-digit minutes of the event; the next two columns are the solar latitude and longitude of the event; VS is the shock speed (in km/s) of the event's CME, shock, etc. We have two columns labelled VS. The first VS column lists the "tuned" shock speed values that we estimated from comparisons between the initial HAFSS simulations and the ACE data. The column labelled (VS) - with parentheses - lists the original SWPC/USAF estimates. The difference between these VS and (VS) values is indicative that at times there is a large uncertainty in the reported SWPC/USAF (VS) estimates. The TAU value (in hours and minutes) is the assumed duration of the piston driving time of the coronal shocks above the flare site. Thereafter the shape of the shock is determined by the HAFSS simulation and not by an assumed cone angle. VSW is the ambient solar wind speed (in km/s); OPT is the optical magnitude of the event; and XRAY is the x-ray magnitude of the event.

Table 1. March 2012 Solar Events and HAFSS Model Inputs

NNNN	YYYY	MMDD	HHMM	LAT	LON	VS	(VS)	TAU	VSW	OPT	XRAY
0001	2012	0305	0002	N17	E029	800	(2273)	0100	750	XXX	X5.4
0002	2012	0305	0409	N17	E041	840	(1340)	0100	750	2B	XXX
0003	2012	0307	0024	N17	E027	1450	(2293)	0015	750	XXX	XXX
0004	2012	0309	0343	N17	W003	350	(1285)	0300	750	XXX	XXX
0005	2012	0310	1200	N17	W024	1110	(1037)	0100	750	XXX	XXX

2. HAFSS Comparisons with OMNI data at Earth at 1 AU

The first comparisons of the HAFSS simulations with spacecraft data for the March 2012 events are shown in Figure 1. It shows comparisons at 1 AU between the in-situ OMNI spacecraft data (the black line) in the vicinity of Earth and two types of HAFSS model results. The blue line shows the modelling results at Earth using only the background (ambient) solar wind inputs. The red line indicates the results of HAFSS simulations that combine the background solar wind inputs and the impulsive solar wind inputs from the solar events in Table 1. Solar wind speed (upper panel) and density (lower panel) are shown in Figure 1.

The solar wind speed comparisons indicate relatively good agreements between the full HAFSS modelling results (the red line) and the in-situ OMNI measured speeds (the black line). The HAFSS simulations show good agreement with the observed OMNI arrival times of all three events and the speed magnitudes of the first and third events, but HAFSS overestimates the speed increase of the second event. Overall, the speed profile from the HAFSS results is quite similar to the OMNI speed

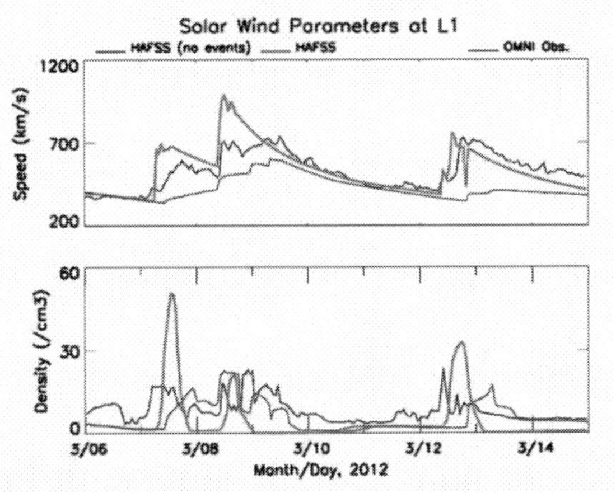

Figure 1. Comparison of HAFSS 3D time-dependent simulation time series results with L1 OMNI observations near Earth (see text). The red line indicates the HAFSS simulation using as inputs the background solar wind and the solar wind from the impulsive solar events (Table 1). The blue line shows the HAFSS results at Earth using only the background (ambient) solar wind inputs. The OMNI data are shown by the black line.

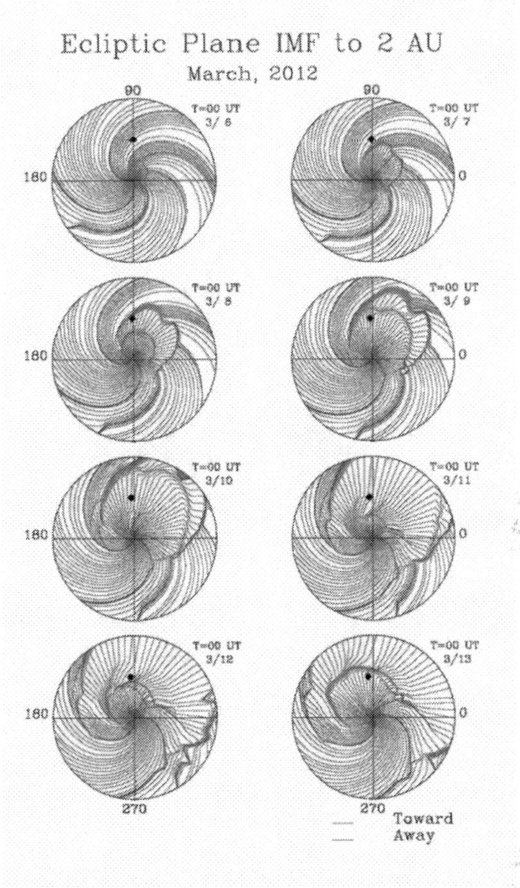

Figure 2. HAFSS ecliptic plane simulation plots of the IMF from the Sun to 2 AU.

data (black line). The plasma density comparisons show fairly good agreements for the arrival times of the events, but HAFSS overestimates the magnitude of the density increase for the first event.

In Figure 2 we show the HAFSS simulation of a series of plots indicating the results of HAFSS calculations of the changing interplanetary magnetic field (IMF) configuration in the ecliptic plane out to 2 AU as a function of time between March 6 and March 13, 2012. The black dot indicates the location of Earth. The blue lines show the configuration of the IMF having an orientation "toward" the Sun. The red lines show the "away" orientation of the IMF. The IMF changes in the time sequence of the figures clearly show the expansion of the March 2012 disturbances as they propagate from the Sun to 2 AU.

3. HAFSS Comparisons at V2 at 30 degrees South

Figure 3 is a plot of the HAFSS simulated IMF in the ecliptic plane out to 140 AU as of April 1, 2013. Figure 3 thus estimates the results for more than one year, from March 2012 to April 1, 2013, of the propagation of the disturbances. The format in Figure 3 is similar to the individual IMF plots in Figure 2. From Figure 3 it is evident that the magnetic field effects of these disturbances primarily propagate

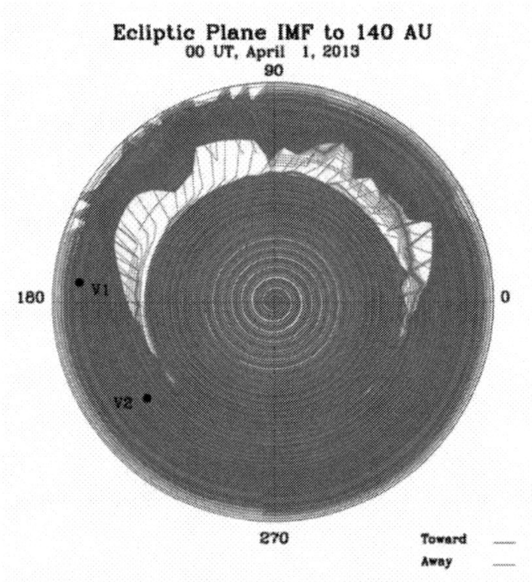

Figure 3. HAFSS ecliptic plane plot of the magnetic field from the Sun to 140 AU. The black dots indicate the locations of V1 at 174 degrees and 123 AU and of V2 at 217 degrees and 101 AU.

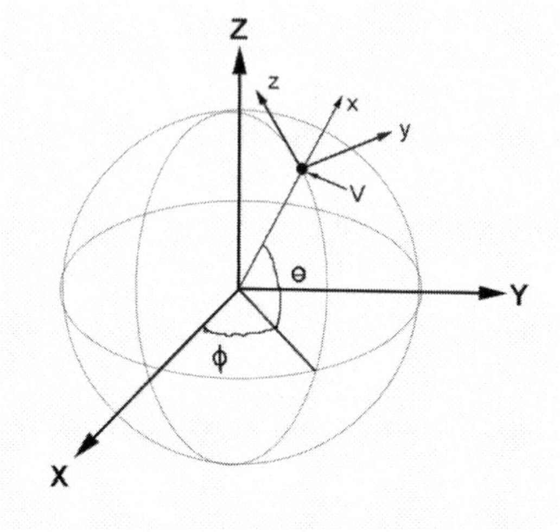

Figure 4. Coordinate transformation, where
$x = X \cos\Theta \cos\Phi + Y \cos\Theta \sin\Phi + Z \sin\Theta$
$y = -X \sin\Phi + Y \cos\Phi$
$z = -X \sin\Theta \cos\Phi - Y \sin\Theta \sin\Phi + Z \cos\Theta$

in the upper hemisphere (0-90-180 degrees). Thus, there is a large asymmetry in longitude of the IMF disturbance as it propagates to greater radial distances from the Sun.

Although Figure 3 shows the IMF in the ecliptic plane, Voyagers 2 and 1 are not located in that plane. Voyager 2 is located at 30 degrees South and Voyager 1 is located at 34 degrees North. In order to more accurately determine the effects of the March 2012 events at each of these spacecraft we employed a coordinate transformation and applied it to each of these spacecraft. Figure 4 shows the coordinate transformation. In the HAFSS code all calculations of speed, density, and magnetic field are in Heliographic Equatorial (HEQ) coordinates so that longitude in the EC plots or other tilted plane plots should be in HEQ coordinates. In Figure 4 XYZ is in the HEQ coordinate system.

A new coordinate is designed as the xyz system at point V (location of Voyager). The tilt angle of the xy plane to the XY plane is the Θ angle. The difference of longitude between XYZ and xyz system is the angle Φ.

Figure 5 is similar in format to Figure 3 and shows the HAFSS calculated plasma density in the ecliptic plane out to 140 AU as of April 1, 2013, due to the effects of the March 2012 solar events. It is evident that there is a strong similarity between the HAFSS simulated propagation of the magnetic field in Figure 3 and the HAFSS simulated propagation of the plasma density in Figure 5.

Figure 6 is similar in format to Figure 3 and shows the HAFSS calculated plasma density on April 1, 2013 out to 140 AU in the V2 plane, which is tilted to 30 degrees South, and is in an *xyz* coordinate system in which the 0 axis is the radius from the Sun to V2; likewise, Figure 7 shows on April 1, 2013, the HAFSS estimated plasma density to 140 AU in the meridian plane at the longitude of V2 at 217 degrees. These figures thus improve on Figures 3 and 5 by estimating how the March 2012 disturbances propagated from the Sun in planes containing the actual location of V2.

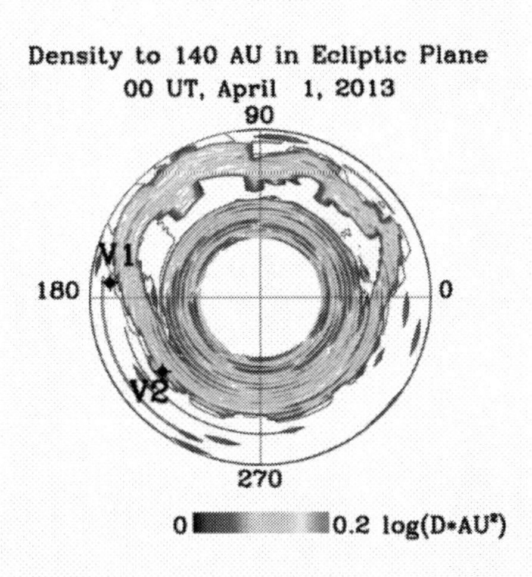

Figure 5. HAFSS ecliptic plane simulation plot of the plasma density from the Sun to 140 AU on April 1, 2013.

Figure 8 is similar to Figure 1 and compares the time series of the HAFSS plasma results (blue and red lines) with the Voyager 2 PLS plasma observations (black dots). The comparisons are shown for the plasma speed (upper panel) and plasma density (lower panel). In these plots the longer-term trends are genuine predictions. The shorter, nearly periodic fluctuations are artefacts of the current state of the calculation method. The measured plasma speeds are substantially less than the HAFSS modelled

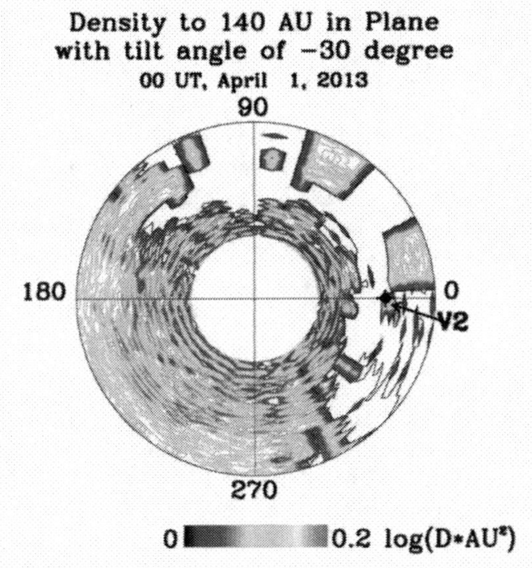

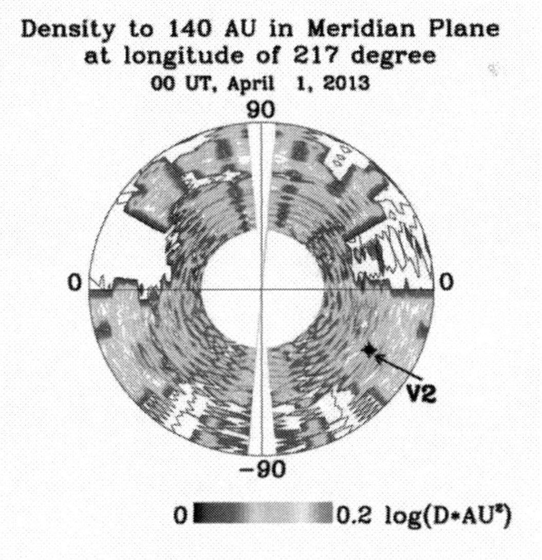

Figure 6. HAFSS plasma density simulations from the Sun to 140 AU in the V2 plane tilted at -30 degrees.

Figure 7. HAFSS plasma density simulations from the Sun to 140 AU in the V2 meridian plane at 217 degrees longitude.

results, which is consistent with the fact that, unlike our 3D MHD time-dependent HHMS-PI model [8,9] that takes account of the decelerating effect of pickup protons on both bulk speed and shock speed, HAFSS does not. Neither HAFSS nor HHMS-PI takes account of the effects of the termination shock or of the heliosheath or of the heliospheric current sheet on the plasma speed and density.

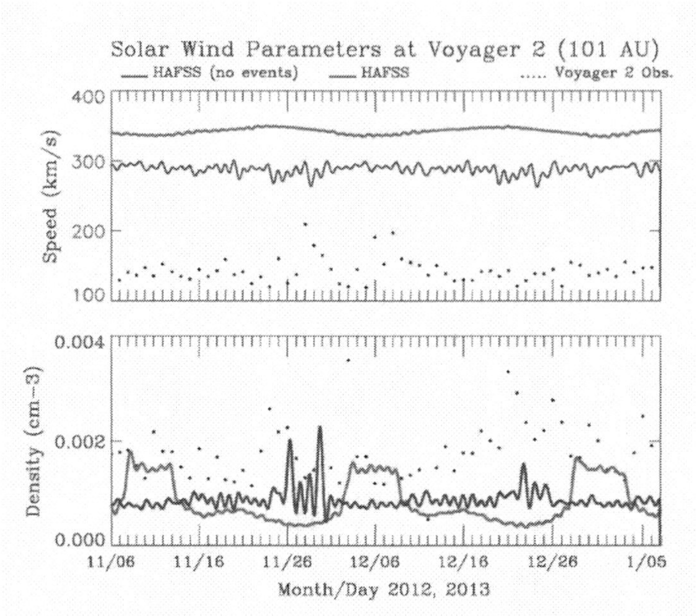

Figure 8. Same as Figure 1 for HAFSS simulations at V2 (at 101 AU) from November 6, 2012 to January 6, 2013.

The density plots in Figure 8 show that there is a similarity between the two modelled plasma density profiles (lower panel) and the data. As noted above, the longer-term trends (e.g., the square waves in the red line and the more irregular rises in the blue line) are genuine predictions and can be seen to correspond roughly to the timing of the peaks in the measured densities. The higher frequency smaller amplitude fluctuations in both lines of HAFSS density predictions are artefacts of the current calculation method. We also note the differences between the HAFSS results based on solely the background solar wind inputs (blue line) and those based on the background solar wind plus the impulsive solar wind inputs (red line). The latter appear to be more similar to the data. Another primary conclusion from the density panel is the overall agreement between the predicted and the actual density magnitudes, which is much closer than the agreement between the magnitudes of the predicted and the actual speeds.

4. HAFSS Comparisons at V1 at 34 degrees North

Figures 9 and 10 are similar in format to Figures 6 and 7, respectively. Figure 9 shows the HAFSS simulation of the plasma density propagation on April 1, 2013 from the Sun out to V1 and beyond to 140 AU in the V1 plane, which is tilted to 34 degrees North and is in an *xyz* coordinate system in which the 0 axis is the radius from the Sun to V1. Likewise, Figure 10 shows on April 1, 2013 the HAFSS simulation of the plasma density propagation, in the meridian plane at the longitude of V1 at 174 degrees.

Figure 11 is similar in format to Figures 1 and 8, and shows the HAFSS simulation of the plasma speed and density in late March and early April 2013 for the March 2012 solar events propagating from the Sun to V1. As in Figure 8, in these plots the longer-term trends are genuine predictions while the shorter, nearly periodic fluctuations are artefacts. However, there are no PLS plasma observations at V1 since the V1 PLS instrument failed in 1980. Judging from the comparisons with V2 in Figure 8, it seems likely that if PLS data had been available, then comparing the HAFSS modelled profiles of the plasma speed (upper panel) and the actual variations in the measured plasma speed (including increases in the magnitude of the plasma speed starting in March/April 2013) we would have primarily found a qualitative correspondence in the plasma speed profiles, with the HAFSS speed magnitudes substantially higher, as was the case at V2. Similarly as in Figure 8, it is tempting to consider that the profiles of the modelled plasma densities (lower panel) also might have been indicative of the variations in the measured plasma density if the V1 PLS instrument were operational.

5. Discussion

Previously it already was tempting to conclude, as suggested by [5], that the effects of the March 2012 solar initiated activity could have propagated to and beyond V1 at ~ 123 AU. Now our HAFSS simulations in Figure 11 can be taken as additional evidence that the March 2012 events are plausible candidates for the origin of the 2-3 kHz enhanced signals detected by the V1 PWS during April/May 2013. The speed and density estimates in the HAFSS simulations in Figure 11 show a substantial enhancement in speed at the end of March, followed by higher densities in the first few days of April 2013, only a few days before the actual beginning of the strong PWS activity on April 9. Considering that, as noted above, HAFSS does not take into account either the decelerating effect of pickup protons or any interaction with the termination shock or the heliosheath or the heliospheric current sheet this is remarkably good agreement with the observations. Thus, we are more confident that the March 2012

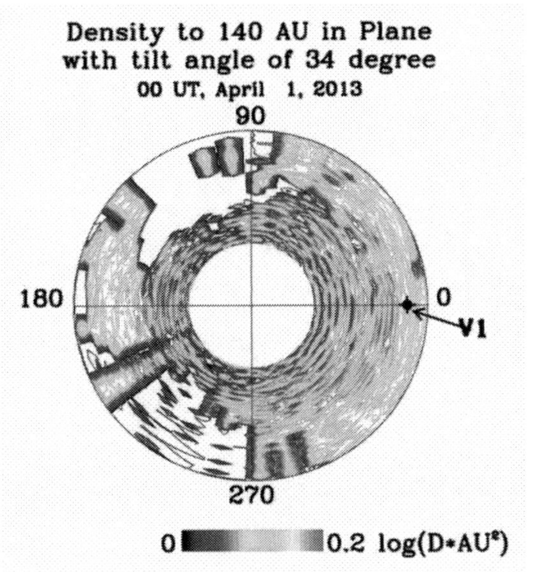

Figure 9. HAFSS plasma density simulations from the Sun to 140 AU in the V1 plane tilted at +34 degrees.

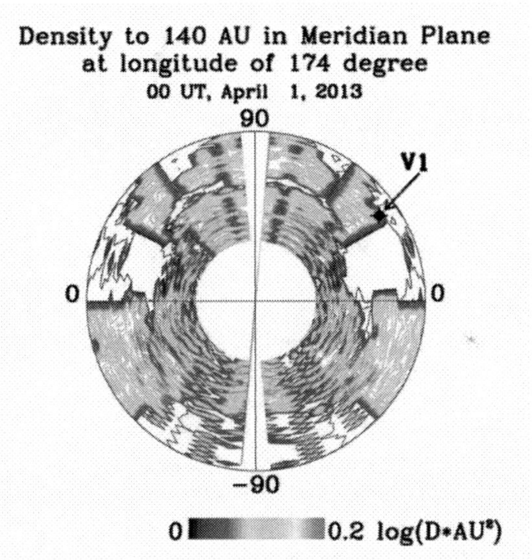

Figure 10. HAFSS plasma density simulations from the Sun to 140 AU in the V1 meridian plane at 174 degrees longitude

solar events could have been responsible for the April-May 2013 2-3 kHz enhanced signals observed by the V1 PWS and their associated increased plasma density measured by the PWS indicating that V1 was in the ISM.

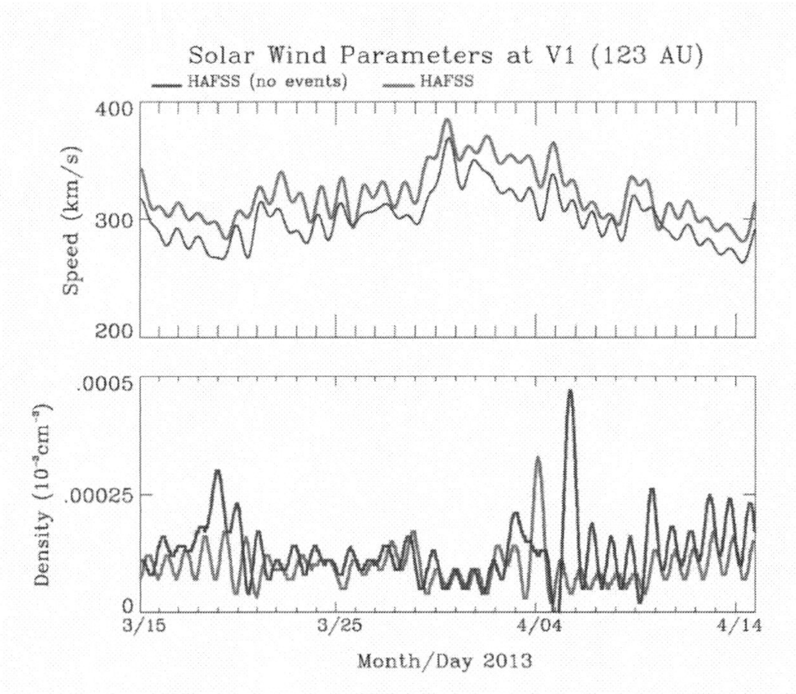

Figure 11. Same as Figure 1 for HAFSS simulations at V1 (at 123 AU) from March 15 to April 15, 2013.

HAFSS is ideally suited for these analyses since it simulates the slowly evolving background solar wind inputs and the impulsive solar wind inputs due to solar event activity. Three-dimensional simulations, including the obviously needed coordinate transformations and the event initiations at or very close to the Sun, are essential for studies of this kind. The large asymmetries in the solar wind flow emphasize the importance of these capabilities. Two-dimensional and certainly 1D simulations that start at Earth or elsewhere are definitely not appropriate in this context.

6. Acknowledgments

We thank the Voyager and OMNI teams, the NSSDC, and NOAA's SWPC for their data. Our work was supported in part by a NASA grant and by Carmel Research Center.

7. References

[1] Krimigis S M et al., 2013, *Sci*, **341**, 144-147
[2] Stone E C et al., 2013, *Sci*, **341**, 150-153
[3] Burlaga, L F, et al., 2013, *Sci*, **341**, 147-150
[4] Webber W R, and McDonald F B, 2013, *Geophys Res. Lett.*, **40**,1665-1668
[5] Gurnett D A et al., 2013, *Sci*, **341**, 1489-1492
[6] Intriligator D S, Sun W, et al., 2008, Particle Acceleration and Transport in the Heliosphere and Beyond (Eds. Li G, Hu Q,Verkhoglyadova O, Zank G, et al.), *AIP* **1039**, 375-383
[7] Intriligator D S, Sun W, et al., 2005, *J. Geophys. Res.*, **110**, A09S10, doi:10.1029/2004JA010939
[8] Detman T M, Intriligator D S, et al., 2011, *J. Geophys. Res.*, **116**, A03105, doi:10.1029/2010JA015803
[9] Intriligator, D S, Detman T M, Gloeckler G, et al., 2012, *J. Geophys. Res.*, **117**, A06104, doi:10.1029/2011JA017424.

Spatial Retention of the IBEX Ribbon Source Particles – Further Work on the Dominant Turbulence Model

Philip A. Isenberg

Space Science Center and Department of Physics, University of New Hampshire, Durham, NH 03824, USA

E-mail: phil.isenberg@unh.edu

Abstract. We present an overview of the "dominant turbulence" picture of pickup proton isotropization and its recent application to the spatial retention of these particles in the nearby interstellar medium as an explanation for the IBEX ribbon structure. We report the progress of further modeling in this area, expanding the published model to 3D and adding a contribution of energetic neutral atoms from the inner heliosheath to the ribbon volume. We find that the ribbon flux expected at IBEX from this model is still below the observed value, but the discrepancy is much smaller than that of our previous result.

1. Introduction

The "IBEX ribbon" is a narrow region of enhanced energetic neutral atom (ENA) flux detected by the Interstellar Boundary EXplorer (IBEX) spacecraft. This enhancement is ~ 20° wide, and emits a factor of 2 - 3 larger ENA flux than the surrounding regions. The energy spectrum is broad, but peaks at ~ 1 keV. Its position on the sky is well-defined as a circle around the heliosphere satisfying $\hat{\mathbf{r}} \cdot \hat{\mathbf{b}} = 0$, where $\hat{\mathbf{r}}$ is the heliospheric radial direction and $\hat{\mathbf{b}}$ is the direction of the local interstellar magnetic field according to current best estimates.

Since its discovery in 2009, a number of hypotheses have been advanced to explain the ribbon (see McComas et al. [1, 2] or Schwadron et al. [3] for extensive lists and discussion). However, at this point none of these hypotheses have convincingly modeled all the observed features. The most advanced scenario is the "secondary ENA" model first proposed by McComas et al. [1] and further elaborated by Heerikhuisen et al. [4] and Möbius et al. [5] (see also Chalov et al. [6] for a somewhat different treatment). In this picture, the ribbon is the product of a multi-step process that starts with the well-known pickup of inflowing interstellar hydrogen in the solar wind. The charge-exchange interaction creating the pickup protons in the solar wind also produces a radially expanding wind of hydrogen atoms streaming at the solar wind speed, ~ 1 keV. These ENAs pass through the heliosheath and become ionized and picked up in the cool, dense interstellar medium (ISM). These secondary pickup protons gyrate around the ISM magnetic field until some of them are neutralized by yet another charge-exchange. At this time, those new ENAs that were moving toward the Earth when they were neutralized will continue that straight-line motion and can be detected at IBEX.

The main difficulty with this scenario comes from its explanation of the ribbon spatial structure. The original idea required the pickup protons in the ISM to remain near their initial pitch angle in phase space until they were re-neutralized. The initial pitch-angle cosine of radially streaming solar wind ENAs is $\mu_o = \hat{\mathbf{r}} \cdot \hat{\mathbf{b}}$, and only those pickup protons with instantaneous motion back along the

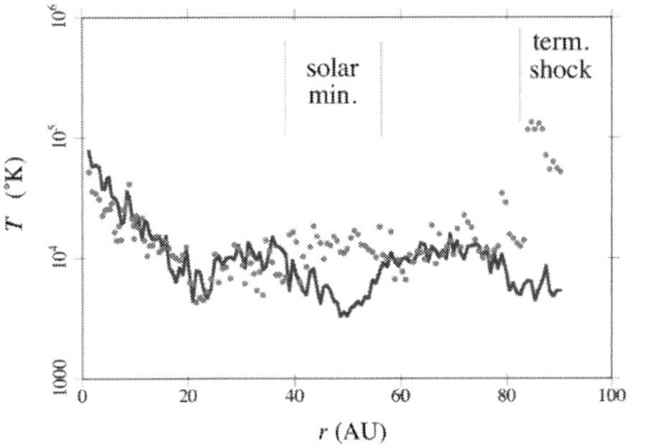

Figure 1. Application of the DT formalism to the outer solar wind. Red dots show the core proton temperature, averaged over 3 solar rotations, as measured at Voyager 2. Blue line shows the predictions of a DT model [17], using solar wind measured at 1 AU as input. During solar minimum, the ecliptic solar wind measured at 1 AU is not representative of the higher latitude solar wind encountered at Voyager.

radial direction can be detected at IBEX. For these particles to coincide with $\hat{\mathbf{r}} \cdot \hat{\mathbf{b}} \approx 0$, the secondary ENAs cannot change their pitch angles very much before they are neutralized. If this requirement is somehow accomplished, the estimated flux from such a configuration is consistent with that observed by IBEX [5].

However, the ring distributions corresponding to $\hat{\mathbf{r}} \cdot \hat{\mathbf{b}} \approx 0$ are known to be highly unstable to pitch-angle scattering by ambient and self-generated waves [7, 8]. The time scale for significant pitch angle scattering of these rings is several days [9, 10], but the charge-exchange time is much longer, about 2 years [9, 11]. If these pickup protons become substantially isotropized before they can be neutralized by charge-exchange, the resulting ENAs would come from all directions and there would be no distinct ribbon structure.

To deal with this issue, Schwadron & McComas [12] suggested a modification to this scenario. Rather than somehow suppressing the pitch-angle scattering of the ISM pickup protons, they proposed that the pickup protons did scatter to isotropy and generated strong turbulent fluctuations in the process. One expects that the turbulence generated by a ring of protons at $\hat{\mathbf{r}} \cdot \hat{\mathbf{b}} = 0$ would be stronger than that generated by pickup protons starting at other pitch angles. In this way, the spatial region defined by $\hat{\mathbf{r}} \cdot \hat{\mathbf{b}} \approx 0$ would become more turbulent, transport of 1 keV protons along the magnetic field would become more difficult there, and these protons could be preferentially retained in that region as source particles for a ribbon.

Such a retention process requires a physical mechanism to distinguish between conditions of strong pitch-angle scattering, leading to isotropization and diffusively inhibited transport of the pickup protons, and weak pitch-angle scattering, which allows the pickup protons to effectively stream along the field away from the ribbon. In a recent paper, Isenberg [13] presented a quantitative mechanism based on the "dominant turbulence" (DT) approximation. This concept was first developed to describe the pickup-proton-driven turbulence in the outer solar wind, whose dissipation can heat the core solar wind protons [14, 15]. In subsequent works, incorporating increasing levels of detail [16-19], this quantitative description has been shown to successfully match the solar wind temperature measurements at Voyager 2 out to the solar wind termination shock (Figure 1). Isenberg [13] applied this mechanism to a simplified model of the heliosphere and the surrounding ISM, and showed that a qualitatively reasonable ribbon structure could be produced. However, the ENA flux emitted by this simple ribbon model was considerably less than that implied by the IBEX measurements, so more detailed models are required.

In this paper, we will extend the 2D model of [13] to 3D and present preliminary results incorporating the addition of ENAs from the inner heliosheath. In §2 we briefly review the DT formalism as applied to the ISM, and in §3 we outline the basics of the earlier [13] model. In §4, we describe the extensions made here and the results obtained. In §5, we discuss these results and present our conclusions.

2. The dominant turbulence assumption

Pickup protons interact with their surrounding plasma primarily through cyclotron-resonant scattering by parallel-propagating MHD waves. Four modes of these waves are available: right-circularly polarized fast mode waves and left-circularly polarized ion-cyclotron waves, each propagating in either direction along the magnetic field. The waves will scatter the protons toward isotropy in the plasma reference frame, and the final shape of the fully-scattered distribution depends on the relative intensities of the four resonant wave spectra. In general, a given pickup proton is able to resonate with two modes at a time. Scattering by one mode ("unstable") will cause the proton to lose energy in the plasma frame, and scattering by the other mode ("stable") will add energy to the proton. These energy changes are matched by changes in the wave intensities, growing or damping, such that each interaction conserves the total energy of waves and particles.

One standard treatment is to assume that the damped wave modes, those that give energy to the protons, disappear quickly, so the scattering is controlled by the interaction with the growing modes. This is called the "bispherical" assumption, since it results in proton shell distributions composed of two spherical sections, and it typically leads to the prediction of intense resonant waves. This assumption was developed to model the ion pickup around comets, where the production rate of new ions is very large, so the damping of the stable waves is very strong [20, 21]. The prediction of intense growth of the unstable waves has been borne out by observations in these cases.

However, the bispherical assumption cannot be used to treat pickup proton isotropization in circumstances where the production of new protons is slow, such as in the outer solar wind and the ISM. In this case, the damping of stable waves is weak and nonlinear turbulent processes can be fast enough to maintain the presence of all four resonant modes. When all the resonant modes take part in the scattering process, the fully-scattered pickup protons will be more isotropic than in the bispherical prediction and the net energy given to the fluctuations by isotropization of new pickup protons is significantly smaller than the bispherical value.

This distinction was central to the successful modeling of the core proton temperature measured by Voyager 2 in the outer solar wind. The fact that the bispherical assumption completely overestimates the core proton heating [22] strongly indicates that some level of stable waves are continually present to contribute to the scattering. Our DT hypothesis, which assumes that turbulence maintains equal intensities of these four modes on the production timescale for new pickup protons, has done a very good job of matching the observed temperatures, as shown in Figure 1.

The quantitative application of this approximation is based on the total energy of the fully-scattered pickup proton shell, representing the gain or loss of fluctuation energy by the isotropization of newly ionized hydrogen. The specific DT assumptions take the intensities of the four resonant wave modes to be equilibrated by nonlinear wave-wave interactions on a time scale faster than others in the system, and assumes that the spectra of each of these modes is kept as a power law in wavenumber

$$I_{\pm}(k) \sim |k|^{-5/3}. \qquad (1)$$

Newly ionized protons are scattered to a closed-shell distribution, $v(\mu)$, on a kinetic time scale faster than the macroscopic fluid time scales of ISM evolution. The shape of this distribution is given by (see [15])

$$\frac{dv}{d\mu} = \left[\sum_j V_j \frac{I_j(k_r)}{|\mu v - W_j|}\left(1 - \frac{\mu V_j}{v}\right)\right]\left[\sum_j \frac{I_j(k_r)}{|\mu v - W_j|}\left(1 - \frac{\mu V_j}{v}\right)^2\right]^{-1}, \qquad (2)$$

where the sums are taken over all the cyclotron resonances at that value of $v_{\parallel} = \mu v$. The functions V_j and W_j are the resonant wave phase and group speeds, respectively, obtained from the cold plasma dispersion relation, and the individual resonant wavenumbers are given by the cyclotron resonance condition as

$$k_r = \frac{\Omega}{\mu v - V_j}. \qquad (3)$$

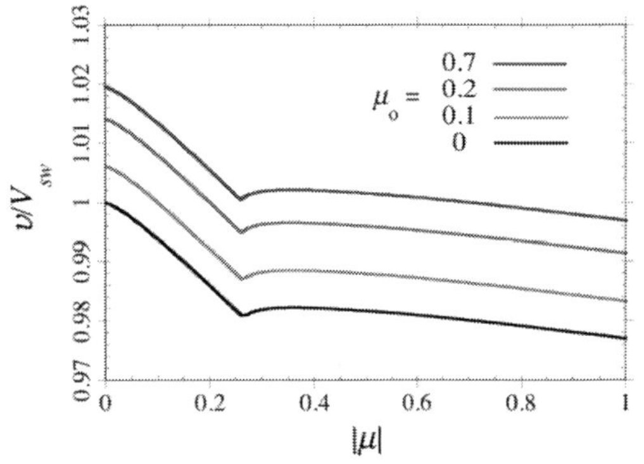

Figure 2. Shapes of the fully-scattered pickup proton shell, $v(\mu)$, for different initial pitch angles, μ_o, taking V_A/V_{sw} = 1/18. At the initial pitch angle, the speed of the pickup proton is $v(\mu_o) = V_{sw}$.

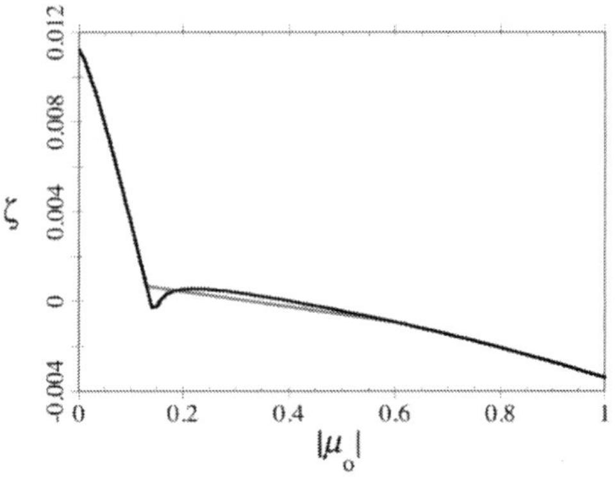

Figure 3. Fractional energy released to turbulent fluctuations by isotropization of a pickup proton with initial pitch angle μ_o. The current model uses the straight red segment to replace a portion of this function.

With the relative intensities given by (1), equation (2) can be integrated to give the fully scattered distribution, which only depends on the value of V_A/V_{sw} and the initial pitch angle of the ring-beam. (We consider here that the ENAs, and therefore the newly ionized protons have speeds equal to V_{sw}.)

In Figure 2, we show the shape of the fully-scattered shell, $v(\mu)$, for a typical Alfvén speed in the ISM, V_A = 25 km s^{-1} and for protons picked up at V_{sw} = 450 km s^{-1}. The shape does not vary much for different initial pitch angles, but the condition that $v(\mu_o) = V_{sw}$ means that the total energy of the shell is larger when the newly ionized ENA has substantial motion along the magnetic field. The fractional change in proton energy which results from the scattering to a closed shell is given by the integral over these shapes. We define the function $\zeta(\mu_o)$ to be the fractional energy released to the resonant fluctuations by scattering process, such that the energy added to turbulence by each new ionization is $E_w = 0.5\ \zeta\ m\ V_{sw}^2$. This function is plotted in Figure 3 for the same parameter values as in Figure 2.

We note that the function ζ passes through zero, so scattering to near-isotropy from some initial pitch angles will add energy to the turbulence, while scattering from other angles will drain energy (if any is present) from the turbulence. This property provides the physical mechanism that helps to distinguish the pickup protons which are diffusively retained in the plasma which ionized them from those protons which continue to stream away from the region of the ribbon. In our modeling thus far, we have used a simplified form of the ζ function, where a portion of the curve is replaced by the straight red segment shown in Figure 3.

3. Ribbon structure from spatial retention

In [13], we applied this DT formalism to a simple model of the ISM as the first step toward constructing a spatial retention model for the IBEX ribbon. We took the ISM to be a uniform magnetized plasma approaching the Sun with a velocity $\mathbf{V_{ISM}}$, ignoring any effects of flow deflection or field draping at the spherical heliopause. We considered the ionization, pickup and isotropization of secondary ENAs from the solar wind in the plane containing $\mathbf{V_{ISM}}$, the uniform field $\mathbf{B_{ISM}}$, and the Sun (see Figure 4). We defined a 2D Cartesian system, where x is the coordinate along the magnetic field and y points away from the Sun perpendicular to the field. We considered only particles of energy ~ 1 keV, taking the speed of the ENAs and their pickup proton products to be V_{sw} = 450 km s^{-1}.

As detailed in [13], we derived model equations for the pickup proton density $n\ (x, y)$

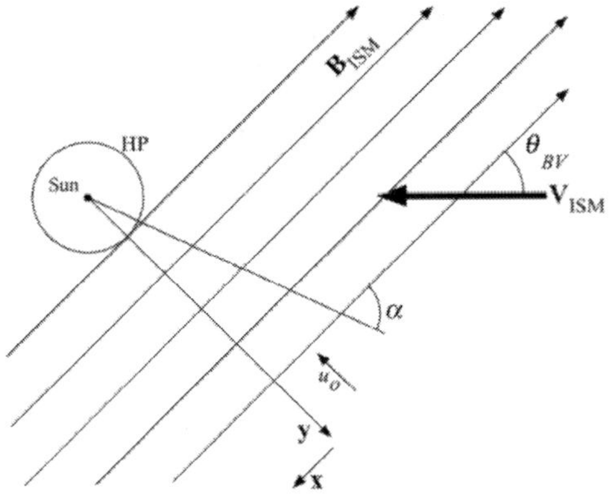

Figure 4. Model diagram of the nearby ISM in this paper. The ISM plasma streams to the left at $\mathbf{V_{ISM}}$ = 23.2 km s^{-1}, carrying a constant magnetic field oriented at θ_{BV} = 45° to the flow. The heliopause (HP) is taken to be a sphere, which reflects incident pickup protons. The ISM is uniform, with proton density n_p = 0.07 cm^{-3} and hydrogen density N_H = 0.16 cm^{-3}, as in [13].

$$-u_o \frac{\partial n}{\partial y} = \frac{\partial}{\partial x}\left(D\frac{\partial n}{\partial x}\right) + P - L\,, \qquad (4)$$

the turbulent intensity in Elsässer units $Z^2 = <\delta v^2> + <\delta b^2/4\pi\rho>$,

$$-u_o \frac{\partial Z^2}{\partial y} = -\frac{2Z^4}{\lambda_{\|}V_A} + Q\,, \qquad (5)$$

and the turbulent correlation length, $\lambda_{\|}$, which quantifies the dissipation scale

$$-u_o \frac{\partial \lambda_{\|}}{\partial y} = \frac{2Z^2}{V_A} - (\lambda_{\|} - \lambda_{res})\frac{Q}{Z^2}S(Q)\,, \qquad (6)$$

as the ISM plasma flows toward the Sun with normal component u_o. In these equations, D is the spatial diffusion coefficient along the field, determined from quasilinear theory to be ~ $\lambda_{\|}^{2/3}/Z^2$. The charge-exchange production and loss rates of pickup protons are given by P and L, respectively, and are functions of position. The characteristic wavelength of the pickup-proton waves is $\lambda_{res} = 2\pi V_{sw}/\Omega_p$. The turbulence is modeled phenomenologically by designating the total intensity and the correlation length, as discussed by Breech et al. [23] and Oughton et al. [19]. The turbulent driving from the pickup proton scattering is denoted by $Q = \zeta\, V_{sw}^2\, P/n_p$, where n_p is the ISM proton density, and it appears as a signed source term in the turbulence equations. Please see [13] for a more extensive explanation.

To specify the criterion between pickup protons that are scattered fast enough to complete their nearly isotropic shells, and those that are not, we require the scattering rate to be faster than the production rate for new pickup protons. This reduces to a limit on the turbulent intensity, such that if

$$Z^2 > (1 + r_A)\frac{\sigma n_p}{\Omega} V_A{}^2 V_{sw} \qquad (7)$$

(where σ is the charge-exchange cross-section and r_A is the Alfvén ratio of the turbulence) we take the pickup protons to be nearly isotropic and coupled to the plasma. Otherwise, we assume that the new pickup protons are able to stream away and we set $n = 0$ there.

In [13], we solved the system of equations (4) - (6) from an outer boundary at y = 2100 AU in to the position of the heliopause, taken as a circle of radius 150 AU. At the outer boundary, we set the density and the turbulent intensity to zero, so the model ribbon is purely a local phenomenon, generated out of the ENAs produced in the solar wind. The pickup protons tend to be spatially retained around the region where $\hat{\mathbf{r}} \cdot \hat{\mathbf{b}} = 0$, despite their diffusion along the field away from this region. Integrating the sunward ENAs emitted by these nearly isotropic pickup protons along the line of sight to IBEX, we found that the ENA flux was enhanced in a qualitatively ribbon-like structure. However, the enhanced region spanned ~ 30° at the half-maximum point, wider than the observed ribbon, and the peak flux was only 22.4 (cm^2 sr s keV)$^{-1}$, substantially lower than the ~ 120 (cm^2 sr s keV)$^{-1}$ inferred by Schwadron et al. [3].

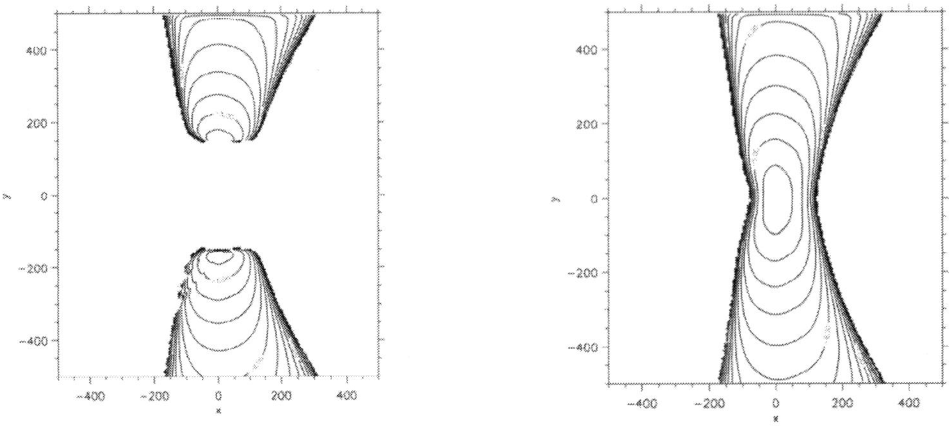

Figure 5. Model pickup proton density in 2 planes: $z = 0$ (left panel) and $z = 160$ AU (right panel). Shown are contours of log n, where n is in units of cm^{-3}. Plot scales are ± 500 AU.

Still, it should be recognized that this is only the first step. Unlike the secondary ENA models assuming weak scattering, the spatial retention mechanism does not require the source ENAs to be nearly radially streaming. Since the ribbon protons are taken to scatter to isotropy, their source can come from anywhere.

4. Further modeling

Before proceeding with the effect of additional source particles for the model ribbon, we construct a 3D version of the previous model. In keeping with the simple investigative picture, we consider the heliopause to be a sphere and continue to neglect the effects of flow deflection and field-line draping by taking the ISM magnetic field to be straight and uniform. Since the interstellar magnetic field will not enter the heliosphere, we take the pickup protons in the ISM to be reflected at the heliopause surface. The ISM plasma flows parallel to the plane of the 2D model, so the 3D system is composed of more 2D planes, each defined by $\mathbf{V}_{ISM}$ and $\mathbf{B}_{ISM}$, stacked in the z direction. The plasma properties in these planes are independent, so the model equations (4) - (6) can be solved separately within each plane, with only the spherically-symmetric sources and the effects of the heliopause boundary giving z-dependence of the solution.

We now wish to add the contribution of 1 keV ENAs from the inner heliosheath (IHS), defined as the spherical shell between the solar wind termination shock, taken to be at 100 AU, and the heliopause at 150 AU. To estimate the outward flux of heliosheath ENAs, we suppose the IHS to be uniformly filled with ENA-emitting material, which radiates 1 keV ENAs at some constant rate, isotropically in the rest frame of the material. We assume that the "global distributed flux" of 1 keV ENAs measured by IBEX [3] comes entirely from the IHS region. The spectrum of inward-streaming ENAs detected at IBEX is seen to vary with energy as $E^{-1.8}$ near 1 keV [24]. We assume that the IHS material is still flowing outward, with an average radial speed of 75 km s^{-1}. We then set the magnitude of the emitted flux to match these IBEX measurements, which we take to be ~ 80 (cm^2 sr s keV)$^{-1}$. A Compton-Getting transformation from the inward radial flux of the outward flowing emitters, then gives an outward ENA flux of 270 (cm^2 sr s keV)$^{-1}$ at 1 keV.

Adding the contribution of these ENAs to the secondary solar wind flux taken in [13] yields a higher density of pickup protons in the spatial region of the ribbon. We take the same parameter values for the ISM as in that paper. In Figure 5, we show the pickup proton density from the model equations for two values of z. Figure 5a shows the density contours in the $\mathbf{V}_{ISM}$-$\mathbf{B}_{ISM}$ plane containing the Sun (where the heliosphere is considered empty of these ribbon particles), and Figure 5b shows the solution in the plane 160 AU above (or below) that plane. The line-of-sight integrated ENA flux from these pickup protons, as would be seen at IBEX, is shown in Figure 6 as a function of

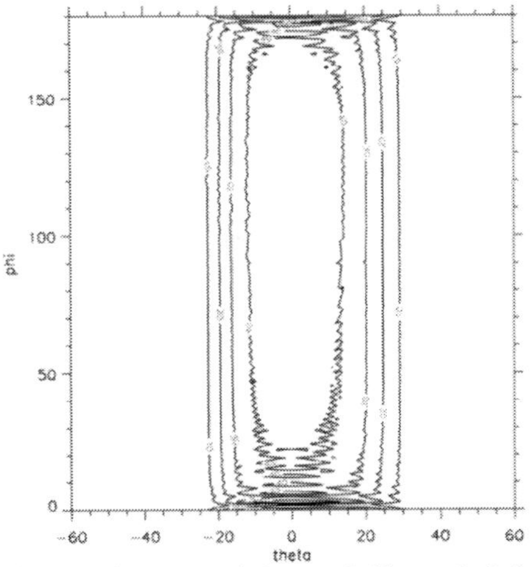

Figure 6. Model inward line-of-sight ENA flux as seen by IBEX as a function of angle from the upwind direction.

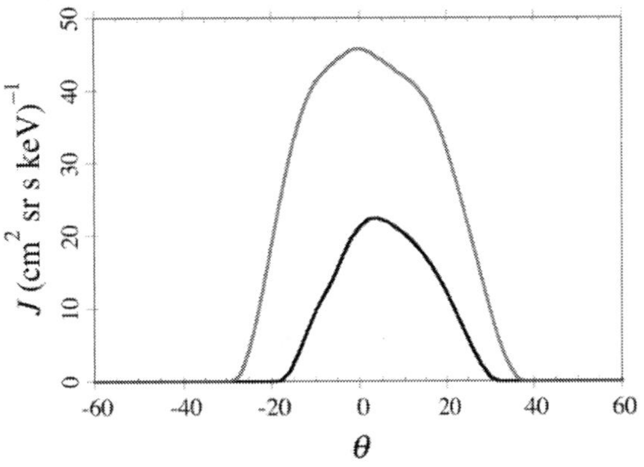

Figure 7. Model inward ENA flux in the upwind plane, as a function of angle across the ribbon. Upper red curve from this paper, lower black curve from [13].

angle on the sky. Here the wavy contours are caused by the limited grid resolution in z of our computation. The ENA flux across the ribbon structure in the upwind plane is given in Figure 7. The smaller black curve is the solution from [13], which only included the secondary solar wind source for the pickup protons. The new results here, including the IHS source, is shown by the larger, red curve. We see that the peak flux has approximately doubled, to 45.8 (cm^2 sr s keV)$^{-1}$, but still falls short of the observed values. The more spatially distributed IHS source also leads to a broadened width of the structure.

Clearly, there is more work to be done and there are many aspects of the simple model discussed here that can be improved.

5. Summary

We have described our recent work on the spatial retention model for the IBEX ribbon, first suggested by Schwadron & McComas [12]. We extended the dominant turbulence formalism of pickup proton isotropization, developed by Isenberg [15] to explain the heating of the core protons in the outer solar wind observed by Voyager 2. This extended mechanism was applied to the conditions in the nearby interstellar medium to generate a ribbon structure of ENAs emitted from a confined spatial volume. Our first work took the solar wind secondary ENAs as the source for the ribbon pickup protons, but this model only obtained a predicted flux of 22.4 (cm^2 sr s keV)$^{-1}$, as compared to the measured value of ~ 120 (cm^2 sr s keV)$^{-1}$. In this paper, we present new results, expanding the model to 3D and adding an estimated contribution of ENAs from the inner heliosheath. We find that the predicted ENA flux has improved, to 45.8 (cm^2 sr s keV)$^{-1}$, but this value is still insufficient.

However, the model thus far is still extremely simplified. We will be exploring the effects of adding more accurate sources, modifying the isotropization and spatial diffusion behavior, and including draping of the ISM magnetic field around the heliopause.

Acknowledgements

The author is grateful for valuable conversations with M. A. Lee, E. Möbius, N. A. Schwadron, B. J. Vasquez, and E. J. Zirnstein. This work was supported in part by NASA grant NNX13AF97G.

References

[1] McComas D J, et al. 2009 Global observations of the interstellar interaction from the Interstellar Boundary Explorer (IBEX) *Science* **326** 959

[2] McComas D J, et al. 2010 Evolving outer heliosphere: Large-scale stability and time variations observed by the Interstellar Boundary Explorer *J. Geophys. Res.* **115** A09113
[3] Schwadron N A, et al. 2011 Separation of the *Interstellar Boundary Explorer* ribbon from globally distributed energetic neutral atom flux *Astrophys. J.* **731** 56
[4] Heerikhuisen J, et al. 2010 Pick-up ions in the outer heliosheath: A possible mechanism for the Interstellar Boundary EXplorer ribbon *Astrophys. J.* **708** L126
[5] Möbius E, Liu K, Funsten H O, Gary S P and Winske D 2013 Analytic model of the *IBEX* ribbon with neutral solar wind based on ion pickup beyond the heliopause *Astrophys. J.* **766** 129
[6] Chalov S V, et al. 2010 Scatter-free pickup ions beyond the heliopause as a model for the Interstellar Boundary EXplorer ribbon *Astrophys. J.* **716** L99
[7] Wu C S and Davidson R C 1972 Electromagnetic instabilities produced by neutral-particle ionization in interplanetary space *J. Geophys. Res.* **77** 5399
[8] Lee M A and Ip W-H 1987 Hydromagnetic wave excitation by ionized interstellar hydrogen and helium in the solar wind *J. Geophys. Res.* **92** 11,041
[9] Florinski V, Zank G P, Heerikhuisen J, Hu Q and Khazanov I 2010 Stability of a pickup ion ring-beam population in the outer heliosheath: Implications for the IBEX ribbon *Astrophys. J.* **719** 1097
[10] Liu K, Möbius E, Gary S P and Winske D 2012 Pickup proton instabilities and scattering in the distant solar wind and the outer heliosheath: Hybrid simulations *J. Geophys. Res.* **117** A10102
[11] McComas D J, et al. 2012 The first three years of *IBEX* observations and our evolving heliosphere *Astrophys. J. Suppl.* **203** 1
[12] Schwadron N A and McComas D J 2013 Spatial retention of ions producing the IBEX ribbon *Astrophys. J.* **764** 92
[13] Isenberg P A 2014 Spatial confinement of the *IBEX* ribbon: a dominant turbulence mechanism *Astrophys. J.* **787** 76
[14] Isenberg P A, Smith C W and Matthaeus W H 2003 Turbulent heating of the distant solar wind by interstellar pickup protons *Astrophys. J.* **592** 564
[15] Isenberg P A 2005 Turbulence-driven solar wind heating and energization of pickup protons in the outer heliosphere *Astrophys. J.* **623** 502
[16] Smith C W, Isenberg P A, Matthaeus W H and Richardson J D 2006 Turbulent heating of the solar wind by newborn interstellar pickup protons *Astrophys. J.* **638** 508
[17] Isenberg P A, Smith C W, Matthaeus W H and Richardson J D 2010 Turbulent heating of the distant solar wind by interstellar pickup protons in a decelerating flow *Astrophys. J.* **719** 716
[18] Isenberg P A, Oughton S, Smith C W and Matthaeus W H 2010 Solar wind turbulent heating by interstellar pickup protons: 2-component model *Pickup Ions Throughout the Heliosphere and Beyond*, ed J A le Roux et al., (Melville, NY: AIP) p. 180
[19] Oughton S, Matthaeus W H, Smith C W, Breech B and Isenberg P A 2011 Transport of solar wind fluctuations: A two-component model *J. Geophys. Res.* **116** A08105
[20] Galeev A A and Sagdeev R Z 1988 Alfvén waves in a space plasma and its role in the solar wind interaction with comets *Astrophys. Space Sci.* **144** 427
[21] Johnstone A D, Huddleston D E and Coates A J 1991 The spectrum and energy density of solar wind turbulence of cometary origin *Cometary Plasma Processes*, ed A D Johnstone, (Washington, DC: AGU) p. 259
[22] Smith C W, et al. 2001 Heating of the low-latitude solar wind by dissipation of turbulent magnetic fluctuations *J. Geophys. Res.* **106** 8253
[23] Breech B, et al. 2008 Turbulence transport throughout the heliosphere *J. Geophys. Res.* **113** A08105
[24] Desai M I, et al. 2014 Energetic neutral atoms measured by the *Interstellar Boundary Explorer* (*IBEX*): Evidence for mulitple heliosheath populations *Astrophys. J.* **780** 98

13th Annual International Astrophysics Conference: Voyager, IBEX, and the Interstellar Medium IOP Publishing
Journal of Physics: Conference Series **577** (2015) 012015 doi:10.1088/1742-6596/577/1/012015

Effects of Large-Scale Simple Velocity Shear on a Fluctuating Interplanetary Magnetic Field

J. R. Jokipii, Jozsef Kóta and Joe Giacalone

Departments of Planetary Sciences and Astronomy, University of Arizona, Tucson, AZ, USA

E-mail: `jokipii@lpl.arizona.edu`

Abstract. The effects of large-scale simple velocity shear on a magnetic field in a low-β MHD fluid are examined in the limit that the magnetic field does not significantly affect the fluid flow. The large-scale fluid shear stretches the magnetic field and can cause significant field amplification. Application to the solar wind suggests that the latitudinal shear observed may contribute significantly to the observed periods of underwound or radial magnetic field.

1. Introduction and Background

Shear is commonly present in astrophysical fluids and affects the ambient magnetic field, which often can be regarded as being frozen in to the ambient thermal fluid. A particularly simple example of shear occurs when a quasi-steady flow in one direction (say, x or r) varies in a transverse direction (say, z or θ). Such a flow is observed in the solar wind near sunspot minimum, where the mainly radial wind flow changes from $\approx 300 - 400$ km/sec at low heliographic latitudes to ≈ 700 km/sec at high latitudes.

Clearly, in such a shear flow, one expects that any component of the magnetic field in the transverse direction, normal to the plane of the velocity shear, will be significantly affected, whereas the initial magnetic-field component parallel to the flow direction will initially, be unaffected. In this paper, we examine the effects of such such shear, neglecting the magnetic forces. We find that in this case, any initial transverse magnetic field results in amplification of the field parallel to the flow. Eventually, this may result in a situation where the effects of the magnetic field may not be neglected. We do not consider this limit in this paper.

2. Basic Considerations

We consider the effects of a prescribed simple shear on a magnetic field by considering a system where the magnetic field magnitude is small enough that it has negligible effect on the flow.

Consider a simple MHD flow and magnetic field system which illustrates the effects of shear. Let there be a steady fluid flow in the x direction which varies in the z direction, $\mathbf{U}(z) = U_x(z)\hat{\mathbf{e}}_x$. Then the magnetic-field evolution equation becomes:

$$\begin{aligned} \frac{\partial B_x}{\partial t} + U_x\frac{\partial B_x}{\partial x} &= B_z\frac{\partial U_x}{\partial z} \\ \frac{\partial By}{\partial t} + U_x\frac{\partial B_y}{\partial x} &= 0 \\ \frac{\partial Bz}{\partial t} + U_x\frac{\partial B_z}{\partial x} &= 0 \end{aligned} \quad (1)$$

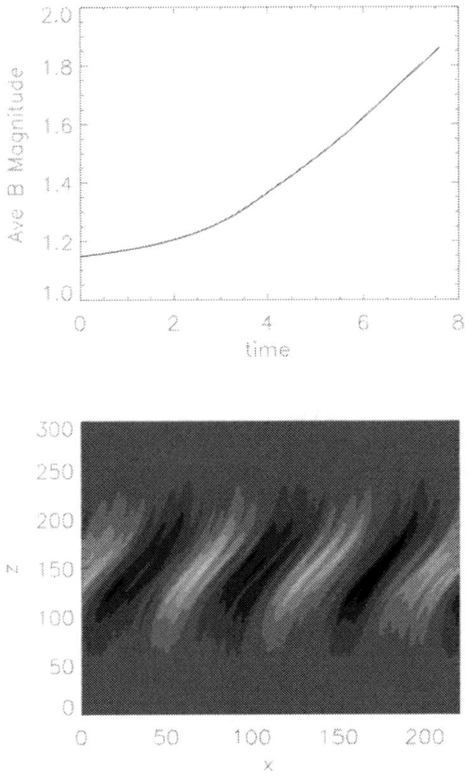

Figure 1. The solution in equation (3), where b_z is initially a simple cosine, showing the increase in the magnitude of the magnetic field with time (top) and contours of magnetic-field magnitude in the x,z plane.

These equations are straightforward to solve. The y and z components of the magnetic field are simply advected with the flow, whereas the x component in the direction of the flow vector changes because of the shear term $\partial U_x/\partial z$. Let there be an initial magnetic field, at t = 0,

$$\mathbf{B}(x, z, t_0) = B_0\hat{\mathbf{e}}_\mathbf{x} + b_z(x, t_0)\hat{\mathbf{e}}_\mathbf{z}. \tag{2}$$

Then, at later times the solution is

$$\mathbf{B}(x, z, t) = \left[B_0 + (t - t_0)\frac{\partial U_x}{\partial z} b_z(x - U_x(z)t, t_0)\right] \hat{\mathbf{e}}_\mathbf{x} + b_z(x - U_x t, t_0)\hat{\mathbf{e}}_\mathbf{z} \tag{3}$$

On the other hand, if the fluid at some plane $x = x_0$ is moved transversely in the z-direction at a varying velocity $U_z(t)$, the solution for $x > x_0$ is:

$$\mathbf{B}(x, z, t) = \left[B_0 + \frac{(x - x_0)}{U_x(z)}\frac{\partial U_x}{\partial z} b_z(z, x, t - (x - x_0)/U_x(z))\right] \hat{\mathbf{e}}_\mathbf{x} + b_z(x, (t - t0)/U_x(z))\hat{\mathbf{e}}_\mathbf{z}. \tag{4}$$

This case corresponds to a time-varying transverse fluid flow at the boundary $x = x_0$ which produces a time-varying transverse magnetic field which is then advected to $x > x_0$.

In both cases we see that the y component of the magnetic remains zero and the z component is advected in the x direction. The x component varies with time or position. In the first example of a specified initial magnetic field at zero time, the magnitude of the component of the magnetic field in the direction of the flow velocity increases without limit as the time increases to large values. Of course, this cannot happen because of the increasingly important dynamical effect of the magnetic field.

The behavior of the magnetic field for case 1 is illustrated in figure 1 for an initial magnetic field which is uniform and in the x direction and for a flow $U_x(z) = cos(z)$. The magnetic field grows so that at some point, the neglect of the magnetic-field stresses becomes inappropriate, and a more-complete analysis would be necessary.

The conclusions derived above in cartesian geometry also hold for a radial flow from the surface of a condensed object such as the Sun, where the radial flow speed varies with direction. If r is the radial direction and θ is the polar angle, the relevant magnetic-field evolution equations for latitudinal shear become:

$$\left[\frac{\partial}{\partial t} + U_r \frac{\partial}{\partial r}\right] rB_\theta = 0 \tag{5}$$

$$\left[\frac{\partial}{\partial t} + U_r \frac{\partial}{\partial r}\right] rB_\phi = 0 \tag{6}$$

$$\left[\frac{\partial}{\partial t} + U_r \frac{\partial}{\partial r}\right] r^2 B_r = rB_\theta \frac{\partial U_r}{\partial \theta} + \frac{rB_\phi}{sin\theta}\frac{\partial U_r}{\partial \phi} \tag{7}$$

Defining the quantities $S_r = r^2 B_r$ and $S_\theta = rB_\theta$, we obtain equations for S_r and S_θ which are very similar to the equations obtained above for the cartesian case. For the axisymmetric case, where θ is the polar angle, we find

$$\frac{\partial S_\theta}{\partial t} + U_r(\theta)\frac{\partial S_\theta}{\partial r} = 0 \tag{8}$$

$$\frac{\partial S_r}{\partial t} + U_r(\theta)\frac{\partial S_r}{\partial r} = \frac{\partial U_r}{\partial \theta} S_\theta \tag{9}$$

If we have a radial field B_0 and a transverse velocity $U_t(t)$ at $r = r_0$, the solution for $r > r_0$ can be written

$$S_\theta = B_0 r_0 \frac{U_t\left(t - \frac{(r-r_0)}{U_r(\theta)}\right)}{U_r(\theta)} \tag{10}$$

$$S_r = B_0 r_0^2 + \frac{(r - r_0)}{U_r(\theta)}\frac{\partial U_r}{\partial \theta} S_\theta \tag{11}$$

3. Velocity Shear in the Solar Wind and Underwound Magnetic Field

Observations of the solar wind over the past four decades (eg, Coles and Maagoe, 1972) have established that the solar wind at high heliographic latitude, and during the several years around sunspot minimum, has speed much larger than that in at low latitudes. The Ulysses observations during the first orbit of Sun showed (McComas, etal, 2000) that above a latitude of about 33° the solar wind speed was greater than 750 km/sec, whereas, at lower latitudes below about 17°, the speed varied more but was typically approximately 400 km sec. This implies a significant latitudinal gradient of approximately 1000 km/sec/radian.

Using these values in the above equations results in a significant enhancement of the radial component of the interplanetary magnetic field above that obtained for the Parker spiral corresponding to the solar-wind velocity at a given point.

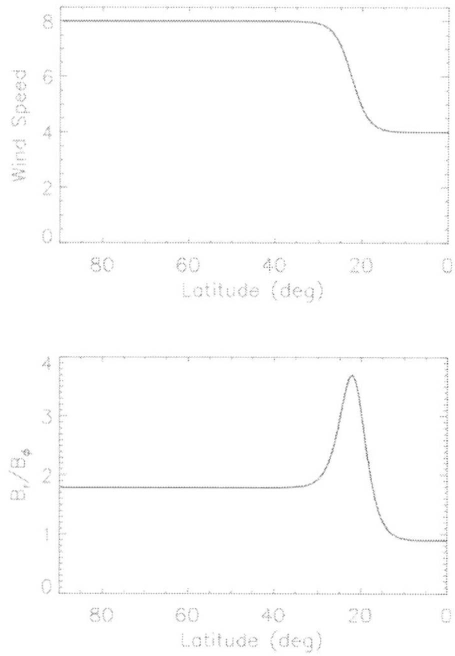

Figure 2. Illusration of a solution to equation (7) for plausible solar-wind parameters, showing a model latitudinal variation of the radial solar-wind velocity and the resulting ratio of the radial and azimuthal components of the magnetic field. In this case the shear was centered at approximately 20° latitude and the transverse velocity was 0.5 km/sec.

Schwadron(2002), Murphy etal (2002), Gosling and Skoug (2002), Schwadron and McComas (2005) and Lario and roelof (2010) have reported analyses of observations which show periods of time occur in which, for several hours to a few days, what is termed an "underwound" interplanetary magnetic field occur. That is, in contrast to the expected ratio of azimuthal to radial field components of order unity at on AU, the ratio was significantly larger. This would occur if the radial magnetic field were enhanced. Corresponding "overwound" periods are not reported. Murphy et al(2002) and Gosling and Skoug (2002) and Schwadron and McComas (2005) proposed models which essentially called upon temporal changes in the solar-wind velocity or rarefaction regions to change the field angle. These may indeed be operative.

We propose, here, that these underwound magnetic field periods may also be in part the result of the latitudinal shear discussed above. Schwadron (2002) and Schwadron and McComas (2003) discussed models which also involved latitudinal velocity shear, although their approach was different from that suggested here.

Because of the random walk or meandering of the interplanetary magnetic field any latitudinal component of the magnetic field will be acted upon by the latitudinal shear in the wind to increase the radial component of the magnetic field, as discussed above.

$$\frac{\partial U_r}{\partial \theta} \approx 1000 \frac{km}{sec\ radian}. \tag{12}$$

Consider a latitudinal interplanetary magnetic field produced at the Sun by a supergranulation velocity of approximately .5 km/sec. This produces a latitudinal meandering magnetic field in the solar wind. Jokipii, et al, 1995, demonstrated that this field could account for the magnetic field observed by Ulysses. This transversed field will be acted upon by the latitudinal shear in the solar wind as discussed above, and result in an amplified radial magnetic field. Figure (2) illustrates the ratio of radial to azimuthal field at 1 AU for reasonable parameters. Note in particular the region of underwound magnetic field.

4. Discussion and Conclusions

We have discussed the effects of a prescribed simple velocity shear on a magnetic field which is weak enough not to affect the flow. The shear is such that in cartesian coordinates, $\mathbf{U} = U_x(z)\hat{\mathbf{e}}_\mathbf{x}$ or in spherical coordinates $\mathbf{U} = U_r(\theta)\hat{\mathbf{e}}_\mathbf{r}$. In each case, the effect of the shear is to increase the magnitude of the magnetic field in the direction of the flow velocity if there is an initial magnetic field component in the direction of the velocity change.

Application of this to the observed solar-wind, where the flow is observed to be nearly radial but with a variation in latitude results in an enhanced radial magnetic field. In the range of latitudes where there is latitudinal shear, the radial component of the magnetic field is significantly enhanced over the case with no shear. It is suggested that this effect may contribute to the observed periods of "underwound" interplanetary magnetic field. These changes to the magnetic field will also have consequences for energetic-particle transport, as discussed previously by Intriligator, etal (2001).

5. Acknowledgements

This work was supported in part by NASA. I acknowledge helpful comments by a referee which significantly improved this paper.

References

Coles, W. A. and S. Maagoe, *J. Geophys. Res.*, **77**, 5622, 1972.
Gosling, J. T. and R. M. Skoug, *J. Geophys. Res.*, **107**, No A10, 1327, 2002.
Intriligator, D. S., et al, J. Geophys. Res., **106**, 10,625, 2001.
Jokipii, J. R., et al, Geophys. Res. Lett, **22**, 3385, 1995.
Lario, D. and E. C. Roelof, AIP Conference Proceedings 1216, page 639, 2010.
McComas, D. J., et al, J. Geophys. Res, bf 105, 10,419, 2000.
Murphy, N., Smith, E. J., and Schwadron, N. A., Geophys. Res. Lett., **29**, No 22, 2066, 2002.
Schwadron, N. A., Geophys. Res. Lett., **29**, No 14, 1863, 2002.
Schwadron, N. A. and D. J. McComas, Geophys. Res. Letters, 30, 1587, 2003.
Schwadron, N. A. and D. J. McComas, Geophys. Res. Letters, 32, 3112, 2005.

Journal of Physics: Conference Series **577** (2015) 012016 doi:10.1088/1742-6596/577/1/012016

3D maps of the local interstellar medium: searching for the imprints of past events

R. Lallement

GEPI/Observatoire de Paris, CNRS UMR8111, University Paris-Diderot, 5 Pl. J. Janssen, 92195 Meudon, France

E-mail: `rosine.lallement@obspm.fr`

Abstract.

Inversion of interstellar (IS) gas or dust absorbing columns measured along the path to stars distributed in distance and direction allows reconstructing the distribution of interstellar matter (ISM) in three dimensions. A low resolution IS dust map based on reddening measurements towards 23,000 nearby stars is used to illustrate the potential of the more detailed maps that are expected within the next several years. The map reveals the location of the main IS cloud complexes up to distances on the order of 600 to 1200 pc depending on directions. Owing to target selection biases towards weakly reddened, brighter stars, the map is especially revealing in terms of regions devoid of IS matter. It traces the *Local Bubble* and its neighboring cavities, including a conspicuous, giant, $\geq$1000 pc long cavity in the third quadrant located beyond the so-called βCMa *tunnel*. This cavity is bordered by the main constituents of the Gould belt, the well-known and still unexplained rotating and expanding ring of clouds and young stars, inclined by $\sim 20°$ to the galactic plane. Comparing the dust distribution with X-ray emission maps and IS gas observations shows that the giant cavity contains a large fraction of warm, fully ionized and dust-poor gas in addition to million K, X-ray bright gas. This set of structures must reflect the main events that occurred in the past; today however even the formation of the Gould belt is still a matter of controversy.

It has been suggested recently that the Cretaceus-Tertiary (KT) mass extinction is potentially due to a gamma-ray burst (GRB) that occurred in the massive globular cluster (GC) 47 Tuc during its close encounter with the Sun $\sim$70 Myrs ago. Such a hypothesis is based on computations of the cluster and Sun trajectories and the frequency of short GRBs in GC's. Given the mass, speed and size of 47 Tuc, wherever it crossed the Galactic plane it must have produced at the crossing site significant dynamical effects on the disk stars and IS clouds, and triggered star formation. On the other hand, a burst must have produced huge ionization and radiation pressure effects on the ISM. Therefore, identifying (or not) the corresponding imprints should provide additional clues to the extinction source and the ISM history. Interestingly, first-order estimates suggest that the Gould belt dynamics and age could match the expected consequences of the cluster crossing, and that the giant ionized, dust-free cavity could be related to gas ionization and dust evaporation by an intense flux of hard radiation such as produced by a GRB. Moreover, dust-gas decoupling during the crossing and after the burst could have produce a highly inhomogeneous dust to gas ratio, potentially explaining the high variability and pattern of the D/H ratio in the gas phase. Future Gaia data should confirm or dismiss this hypothesis.

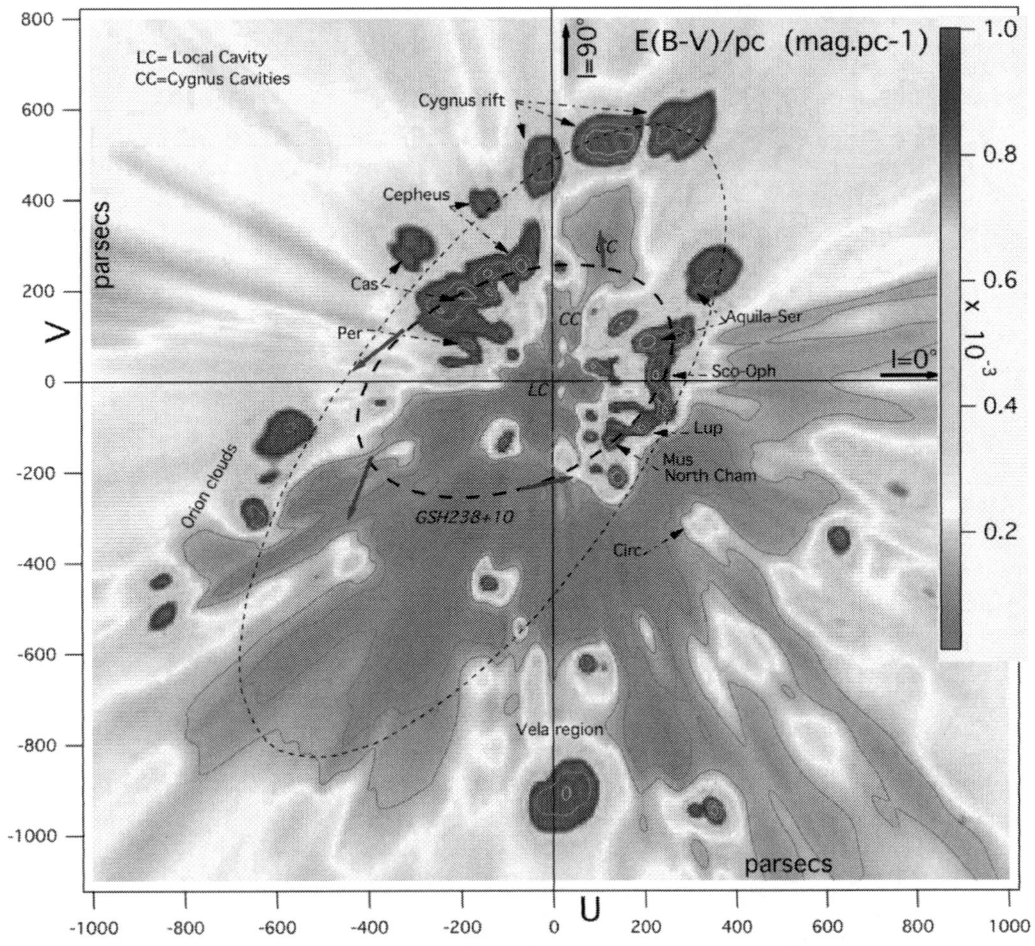

Figure 1. Differential color excess in the Galactic plane, derived by inversion of line-of-sight data (adapted from Lallement et al, 2014a). The Sun is at (0,0) and the Galactic center is to the right. Cavities (in red) are probably filled with hot (1 MK) gas or warm, tenuous ionized gas. Also shown is the Gould belt modeled by Perrot and Grenier (smaller ellipse) with four vectors (in blue) illustrating its expansion and rotation. The large ellipse represents the global structure discussed in the text made of the giant cavity in the third quadrant and the Gould belt.

1. 3D maps of the local ISM

Extremely detailed 2D, multi-wavelength observations of the Galactic ISM have been produced, as well as spectral cubes (2D+spectrum) in the radio domain. Still, we lack information on the distance to the observed interstellar structures. The missing information can be obtained by means of inversion of a large series of distance-limited data, i.e. from absorption measurements towards Galactic target stars. Absorption data can be reddening/extinction by dust, gaseous lines, or diffuse bands. The next decade should see significant new information about 3D distributions of gas and dust, owing, first, to the ESA Gaia mission, which will provide precise distances to a billion stars and, second, to the massive photometric/spectroscopic stellar surveys in progress or preparation that benefit from new multiplex techniques and large detectors.

An example of IS absorption data inversion is shown in Figure 1. A merged catalog of color excess measurements towards 23,000 stars within 2,000 pc has been inverted by means of a Bayesian technique devised by Vergely (2001), to produce a 3D distribution of differential reddening (Lallement et al, 2014). Figure 1 is a horizontal cut in the computed 3D distribution, here along the Galactic plane. Due to the limited dataset, a spatial correlation length must be imposed during the inversion process, resulting in a very poor resolution of the map, on the order of 15 pc, which means that structures of this size of smaller, as e.g. the local and tenuous clouds surrounding the Sun do not appear or are smeared. Instead, what is clearly mapped is the series of well known dense clouds (Scorpius-Ophiucchus, Aquila Rift, Cygnus, Cepheus, etc...) that are indicated in the Figure and are part of the so-called Gould belt (GB). The GB is an elliptical chain of dense clouds and regions of star formation and supernovae (SNRs) production that is contained in a flat disk inclined by $\sim 20°$ to the galactic plane (see e.g. Grenier (2004) for a recent review of the belt properties). The 3D maps show that close to the Plane, in the first (0-90°), second (90-180°) and fourth (270-360/0°) quadrants the belt is an arc-shaped structure that surrounds the Local Cavity (LC, see below) and two other empty cavities (here called the Cygnus cavities: CC on the map). The Orion and Vela clouds, on the other hand, form extensions of this arc at large distance in the third (180-270°) quadrant.

The photometric catalogs being limited in brightness, those stars that are weakly reddened also suffer less extinction and are preferentially observed. This results in biases towards low-extinction targets and in turn in a better mapping of the cavities compared to the clouds themselves. Indeed, the maps show quite well a series of regions devoid of dust or cavities, the $\sim$100 pc wide Local Bubble or Local Cavity (LC) around the Sun (Cox, 1998), and several cavities that are connected to the LC by tunnels including the Cygnus cavities. The Galactic Plane map in Fig 1 also reveals at larger distance in the third quadrant (between galactic longitudes 180 and 270 °) a conspicuous, very wide region devoid of dust that extends up to at least one kpc. This huge cavity is connected to the Local Bubble along its most elongated part in the third quadrant, the *region of bizarre emptiness* (Don Cox, 1998), also called the βCMa tunnel (Welsh, 1991, Gry et al, 1995). A cloud at $\sim$ 200 pc and longitude $\sim$240° forms a limited barrier between the two. We identify this huge cavity as the actual 3D counterpart to the giant radio shell GSH238+08+10 observed by Heiles (1998). A number of aspects of this cavity deserve further investigation. First, its axis is the same as that of the arc-shaped Gould belt's most compact fraction, as if the cavity and the belt were part of an ensemble. Second, the giant star-forming regions in Orion and Vela are both at the periphery of this super-cavity. Third, it is also aligned with an ionization gradient axis derived from white dwarf observations (shown in Fig 1), an axis directed towards l=$\sim$240° (Wolff et al, 1999). Fourth, this cavity must be largely filled with fully ionized warm gas, and only partially by the million-K X-ray emitting gas found in the cavities blown by stellar winds and SNRs, according to the comparisons between ROSAT diffuse X-ray emission maps and the 3D distribution (Puspitarini et al, 2014). Finally, a large quantity of the dust in this cavity has been evaporated, and there is evidence that the dust to gas ratio is generally weak. Figure 2 illustrates some comparisons between IS ionized calcium columns measured in absorption towards nearby stars and the corresponding columns of dust obtained through integration within the 3D distribution between the Sun and the target star. The third quadrant cavity is characterized by CaII to dust ratios that are much lower than in other regions, with differences that can reach three orders of magnitude. Enrichment in CaII is generally linked to dust evaporation, and high CaII to dust ratios are well known signs of such evaporative processes through the combination of both the dust decrease and the CaII increase. We note that CaII columns for stars in this cavity reach 10^{12}-10^{13} cm^{-2}, corresponding to 4.5 10^{19}-10^{20} H cm^{-2} (using local cloud CaII to H ratios) and to ionized cloud total lengths on the order of 100 to 1000 pc for typical warm gas densities on the order of 0.02-0.05 cm^{-3}. This confirms a large filling factor of the ionized gas. Finally, although much less marked, the NaI/dust ratio is also higher in the cavity (Fig 2). Since NaI is predominant in the denser clouds that are embedded in the cavity, this suggests that not only the ionized gas but also the denser, more neutral gas is dust-poor (see Puspitarini et al, 2014).

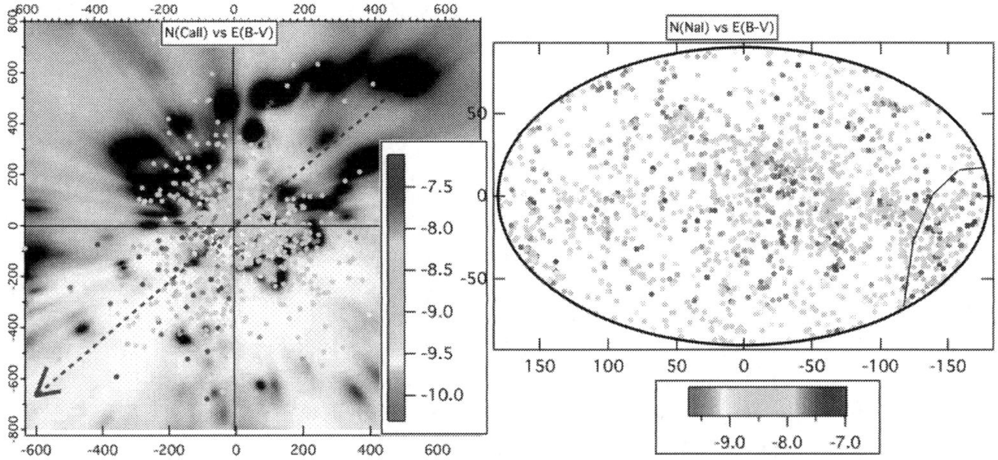

Figure 2. Left: Stars close to the Galactic plane for which we have ionized calcium (CaII) absorption measurements (filled circles), superimposed on a subset of the inverted differential color excess distribution (from Puspitarini et al. 2014). The color refers to the ratio between the CaII total column towards the star and the integrated color excess along the same line-of-sight. Stars located in the third quadrant are characterized by a particularly high CaII/E(B-V) ratio, showing that ionized, dust-poor gas is dominant. Right: Full-sky location of neutral sodium target stars. The color refers to the ratio between the NaI column and the integrated color excess. The third quadrant, especially below the GP, is characterized by high NaI/E(B-V) ratios, revealing again dust-poor gas.

2. The Gould belt and the local star properties

The 3D maps of the multi-phase ISM contain the imprints of its past history, a history that is mainly governed by the star births, lives and deaths by means of their radiation, winds and ejecta, but is also strongly influenced by dynamical events such as waves and tidal effects through their effects on the star formation rate. Hydrodynamical models of the multi-phase gas under the action of winds and supernovae (SNRs) are successful in reproducing most properties and shapes of multi-phase structures such as the Local Cavity, nearby bubbles and clouds (De Avillez & Breitschwerdt, 2009). Following such models the giant 3rd quadrant cavity that appears in the new maps could be the result of a major series of SNe, and the Vela, Orion and Gould belt structures could represent the next generation of star-forming regions along the compressed and subsequently cooled boundaries of the super-bubble. The large quantity of dust-free, ionized gas may result from the intense radiation field generated by the SNe that formed the giant cavity.

What is more difficult to reproduce is the Gould belt itself, which has the distinctive feature of associating a peculiar dynamics of its clouds and young stars with its inclination about the Plane. Over the last 10 years, no consensus has emerged about its origin (and on the contrary some skepticism reappeared about the mere existence of a particular event, see Elias et al, 2009). To recapitulate, two types of scenarios have been modeled and compared to data, the oblique impact of a massive external cloud or a series of explosive events, potentially initiated by a major one, such as a GRB. The former scenario naturally explains the inclination, while the latter is better in that it accounts for the expanding and rotating belt. Developing further the collision scenario, Olano (2001) modeled a 400 pc wide supercloud (SC) of 2 10^7 solar masses, initially moving ballistically outside the Plane and slowly rotating, then interacting with a major spiral arm $\sim$100 Myrs ago. During the braking phase the old ($\sim$500 Myrs or more) generation of stars from the SC decoupled from most of the gas, the central part of the SC partly collapsed to form the GB structures while retaining a global rotation, and external parts with lower angular momentum expanded and formed the Local Arm. During the last $\sim$100 Myrs young generations of stars formed and/or exploded in the compressed GB clouds, forming high pressure cavities and initiating a global expansion of the Belt. Globally, this scenario explains the Sirius supercluster, the Local Arm and the inclined and rotating Gould belt. Of particular interest here, this scenario predicts the formation of a large cavity left behind the braked supercloud, a global pattern that is interestingly very similar to the 3rd quadrant cavity observed in the prolongation of the Gould belt clouds, as discussed in Lallement et al. (2014a,b). The SC scenario has also been suggested by Lallement et al, (2014) as a source of enhanced D/H variability in the IS gas, following the pioneering work of Linsky et al (2006). If the SC were characterized by a long absence of star formation prior to the collision, deuterium-rich dust may have gradually formed in it through preferential adsorption of D on dust grains (Draine, 2003). During the braking phase dust-gas decoupling and cloud mixing may have created strong spatial inhomogeneities in the dust to gas ratio, and strong spatial inhomogeneities in the D abundance in the dust grains. Subsequent evaporation of the dust during star formation in OB associations released variable amounts of D and enhanced the D/H variability in the gas phase by comparison with the variability produced by dust evaporation from regions having the same dust to gas ratio and the same initial abundance of D in the grains. This may also explain the observed apparent link between the D/H ratio and the LC/ belt geometry (Linsky et al, 2006).

Despite all the advantages of the Olano supercloud scenario, there remains one puzzling coincidence, namely that elemental abundances in the super-cloud gas and stars prior to the encounter must be identical to abundances in the disk, given the observed absence of inhomogeneities in chemical abundance in the local gaseous ISM on one hand (e.g. O/H is about constant, see Andre et al (2003), and carbon shows little variability, see Sofia et al, 2004) and the absence of abundance variabilities among the nearby stars (e.g. Nieva et al 2012) whatever their kinematics. The second Gould belt formation scenario, an exploding event (or a series of explosions) with an initial asymmetry has been investigated in detail by Perrot and Grenier (2003) (herafter PG03). By following the evolution of the resulting expanding and rotating ring under the action of the galactic gravitational potential, differential rotation and ISM pressure gradients, PG03 have adjusted the parameters of the initial exploding event to reproduce the locations of the main clouds and OB associations of the belt. After careful examination of a large number of parameters and hypotheses they concluded that is impossible to reconcile the age of the Gould belt inferred from its dynamical structure, on the order of 25 Myrs, with the range of ages of its stars, on the order of 30 to 80 Myrs, in the frame of such models. Moreover, they found that the actual radial velocities of a number of young structures are in disagreement with the model.

Recently, the new and massive stellar surveys (SDSS/SEGUE, RAVE, LAMOST/LEGUE) have revealed spatially dependent vertical and radial motions of the stars within the first 1-2 kpc (see Carlin et al (2013) for a synthetic analysis). The stellar motion pattern is not simple, in particular the articulation with the Gould belt is still unclear. Following previous work by Minchev et al (2009) on high velocity streams and larger timescale, Widrow et al (2014) have devised a model of disk-satellite interaction adapted to the solar neighborhood and conclude that the local disk may be in an unrelaxed mode after the relatively recent crossing of a massive object with a vertical speed larger than that of the disk stars. This work considered as crossing candidates a dwarf galaxy or a dark matter subhalo; however a massive cloud as modeled by Olano (2001) is not excluded. A massive globular cluster, as suggested in the next section, would also produce similar non-negligible effects.

3. Is the Gould belt due to the encounter with the massive globular cluster 47 Tuc?

Massive compact globular clusters (GCs) have been recently proposed as the main sources of short/hard gamma ray bursts (GRBs) associated with the merging of neutron star binaries (see e.g. Gehrels et al, 2009). Because GRBs within $\sim$ 500pc of the Sun are among the potential sources of mass extinctions through lethal ozone destruction and/or global cooling due to nitrogen dioxide formation (e.g. Melott and Thomas, 2011), estimates of their statistical rate of occurrence in the Sun's vicinity are important clues to the extinction events. For this purpose, Domainko et al (2013) have reconstructed the orbits of the main globular clusters within the last 550 Myr and their distances to the Sun, based on their 3D-space velocities observed today (Dinescu et al 1999) and the solar trajectory parameters (Dehnen and Binney 1998). Interestingly, their study shows that the GRB rate exhibits three peaks at $\sim$ -70, -180 and -340 Myrs, all three due to encounters with the massive cluster 47 Tuc. The most recent peak indeed corresponds to the time period of the well known KT extinction. The range of distances of closest approach is very large due to uncertainties on the present distance and the proper motion of the cluster but values as small as 500-1000 pc are not excluded. 47 Tuc, presently located at 4.5 kpc from the Sun far below the Galactic plane (galactic coordinates (l, b) = (306°, -45°)), is one of the most massive and compact globular clusters, with an estimated mass of 1-2 million $M_{\odot}$ for a diameter of $\sim$ 40 pc.

Domainko et al (2013) considered primarily the orbitography of the clusters. However, the crossing of the Galactic disk by a GC as massive as 47 Tuc must have produced dynamical effects on the disk stars and gas, and the identification of such effects may reinforce the encounter/GRB scenario (but note that alternatively a close encounter with a GC may have occurred without any associated burst). Two types of question arise: first, does this encounter potentially explain the ordered stellar motions observed today within 1-2 kpc, the Gould belt star and gas properties and the enhanced star formation? And, second, if indeed a GRB occurred within 47 Tuc at a time it was close enough to the Sun, did it influence the formation of the cavities, ionization properties and structures that are seen today in the 3D maps? The answer to those questions evidently requires extensive modeling and is far beyond the scope of this article, however this problem illustrates well the potential use of three-dimensional maps. In what follows we very briefly discuss some general aspects of the encounter and burst.

The first and simplest one, is related to the energy scales. The stellar surface density being on the order of 50 sol.mass pc-2 in the Sun vicinity (Holmberg &Flynn 2004), a GC mass of 1.5 sol. mass is equivalent to the mass of a cylindric portion of the disk with a 200 pc diameter. According to Dinescu et al (1999) and Domainko et al (2014) the relative velocity between the GC and the Sun for the orbits of close approach is on the order of 100 km.s^{-1}, with a vertical component on the order of 50 km.s^{-1}. Since the average vertical velocity of the stars is on the order of 15-20 km.s^{-1}, comparing momenta shows that a crossing will generate large perturbations, in agreement with several models of encounters with dwarf galaxies or dark matter subhalos (e.g. Widrow et al, 2014). The width of the impacted area is also far from negligible. For the above cluster speed and mass, the impact parameter that leads to strong (90°) scattering of a star close to the mid-plane is on the order of 600 pc, according to the free particle approximation formula (2) of Widrow et al (2014). This provides an order of magnitude estimate of the size of the region that is significantly impacted dynamically, an estimated size that is in good agreement with the Gould belt size. A posteriori, it is verified that the diameter of the cluster is much smaller than the spatial scale of the impacted area, justifying models neglecting its finite size. A third aspect is the timescale. The Gould belt stars comprise not only the brightest young ($\leq$ 20 Myr) massive objects of the main associations, but also stars up to $\sim$ 60-80 Myr old, as shown by Torra et al (2000), Guillout et al (1998) and from open cluster analyses (Bobylev 2006). If enhanced stellar formation associated with the belt has resulted from strong dynamical perturbations, then this range of GB star ages is in good agreement with a triggering event $\sim$ 70 Myr ago. A fourth aspect is the geometry of the crossing. 47 Tuc has a large velocity component in the direction perpendicular to the Plane (Dinescu et al 1999) that we roughly estimate on the order of 50 km.s^{-1}. Widrow et al (2014) have studied in detail the influence of the vertical velocity of a crossing object and the subsequent oscillations and show that for the solar neighbourhood such a velocity corresponds to the transition between a simple bending-mode and higher-order modes. This leaves room for the formation of a pronounced *warp* and may explain a structure similar to the Gould plane. The persistence of a significant inclination for the stellar groups is naturally explained by the oscillation time around the disk, on the order of the time elapsed since the crossing, implying that no strong dissipation has occured yet. In the case of the gas, Perrot and Grenier also find an oscillation time on the order of 50 Myr.

A GRB produces very strong ionization of the surrounding ISM. At variance with long GRBs that occur in dense molecular clouds and ionize large amounts of their surrounding dense gas (see e.g. Krongold & Prochaska, 2013), the radiation from the short GRBs is entirely available for ionization of gas that is much farther away. With a number of ionizing photons on the order of up to 10^{58} (here we assume that the total radiation of a short GRB is a factor of 1000 less intense than the one of a long GRB and use long GRB numbers from Krongold & Prochaska, 2013), the burst is able to quasi-instantaneously ionize column-densities on the order of 5 10^{20} cm^{-2} in two cones of less than 50° opening angle each. Such columns are similar to the columns of ionized gas in the large 3rd quadrant cavity as discussed above. The effect on the dust must be quite strong too. Such a GRB in mid- or low- density environments must be very effective in destroying the dust, especially the smaller grains, in a large region of space. Crude estimates show that 0.1 micron grains will be evaporated over hundreds of pc where the radiation is not fully attenuated by the interaction with the gas. It is therefore tempting to associate the third quadrant ionized and dust-poor cavity with such a radiation burst. In this case, the ionization gradient (arrow in Fig 2) points to an ionizing source that has most impacted the ISM that is today within the 3rd quadrant cavity. On the other hand, the Sun and the ISM have moved independently since the burst and extrapolating back both motions would require specific models. A crude approximation consists in extrapolating back the Sun location using its present motion in the Local Standard of Rest (LSR), assuming the constancy of this motion over the last 65 Myr, and that the ISM during this period has always moved along with the LSR. About the second assumption, it is verified today, according to a synthetic study of absorption line Doppler shifts in the spectra of stars in all directions and within $\sim$500 pc (Lallement et al, in prep.), showing that the global motion of the Sun with respect to the IS gas is very similar in direction and modulus to the motion of the Sun in the LSR. Using for this motion the most recent value of Schönrich et al. (2010) puts the Sun at a distance of $\sim$ 1200 pc and coordinates (l, b)= (228°, -23°) w.r.t. the present coordinate system. This location corresponds to the extremity of the giant, ionized, dust-free cavity in the third quadrant. Is this is a coincidence or is there a relation of causality between a GRB and the formation of the giant cavity? Evidently more dynamical data and more realistic models are required to answer with confidence. It should be mentioned, finally, that the close encounter with 47 Tuc may have destabilized Oort cloud objects and triggered the infall of comets and meteorites (Montmerle 2014, private comm.) and that those events may have produced the KT extinction without the need for any GRB.

4. Summary

Detailed three-dimensional maps of the ISM are expected to be built within the next several years thanks to the massive stellar spectroscopic/photometric surveys in progress or preparation and thanks to the future stellar parallax distance determinations by the ESA mission Gaia (Perryman et al. 2001, de Bruijne 2012). As a precursor of such 3D maps I have presented a recently computed distribution of the differential reddening by the IS dust within $\sim$ 1,000 pc from the Sun. The mapping is based on 23,000 color excess measurements and assigns a location in 3D space to the main dense clouds in the sun vicinity, in particular the structures associated with the Gould belt. The map reveals particularly well the cavities devoid of dust, in particular a giant cavity beyond the so-called βCMa tunnel, bordered by the Orion and Vela star forming regions. The comparison with IS absorption data show that this cavity contains a large fraction of fully ionized, dust-poor gas.

Among their many potential uses, such 3D maps may help to study the past history of the nearby ISM. I discussed an example of a search for the imprints of past events, following the recent suggestion that a close encounter between the Sun and the massive cluster 47 Tuc may have occurred 70 Myr ago, and that a gamma-ray burst happened at that time in the cluster, potentially at the origin of the KT mass extinction. Interestingly, first-order estimates suggest that the inclined Gould belt of young stars and gas may indeed be related to the dynamical impact of such a crossing event, and that the ionization and dust depletion pattern seen today may have been produced by a short GRB.

By means of an unprecedented set of astrometric, spectrometric, and photometric observations of Milky Way stars Gaia is expected to improve the knowledge of the MW dynamics and evolution. Among all the expected results is the increased accuracy of globular cluster orbital elements and of the local arm motion, allowing one to confirm or dismiss an encounter with a massive globular cluster such as 47 Tuc. Accurate star distances and color excess measurements should allow building much more accurate and detailed 3D distributions of the IS matter, allowing, in turn, identifying in more detail the potential effects of a GC encounter and those of a GRB.

Acknowledgements: This work has been partly carried out with support from the French Research Agency (ANR) through the STILISM project. I am indebted to B.R. Sandel who has been an efficient language editor.

References:

[1] de Avillez, M.A., & Breitschwerdt, D. 2009, Astrophys. J. Let. 697 L158
[2] André, M. K., Oliveira, C. M., Howk, J. C., et al. 2003, Astrophys. J., 591, 1000
[3] Bobylev, V. V. 2006, Astronomy Letters, 32, 816
[4] de Bruijne, J. H. J. 2012, Astrophysics and Space Science, 341, 31
[5] de Zeeuw, P. T., Hoogerwerf, R., de Bruijne, J. H. J., Brown, A. G. A., & Blaauw, A. 1999, Astron. J., 117, 354
[6] Carlin, J. L., DeLaunay, J., Newberg, H. J., et al. 2013, Astrophys. J. Letters, 777, L5
[7] Cox, D. P. 1998, IAU Colloq. 166: The Local Bubble and Beyond, 506, 121
[8] Dehnen, W., & Binney, J. J. 1998, MNRAS, 298, 387
[9] Dinescu, D. I., Girard, T. M., & van Altena, W. F. 1999, Astron. J., 117, 1792
[10] Domainko W., Bailer-Jones, C.A.L., Feng F. 2013, MNRAS 432 (1) 258
[11] Draine, B. T. 2003, Annual Rev. Astron.. Astrophys., 41, 241
[12] Gehrels, N., Ramirez-Ruiz, E., & Fox, D. B. 2009, Annual Rev. Astron.. Astrophys., 47, 567
[13] Grenier, I. A. 2004, arXiv:astro-ph/0409096
[14] Gry, C., York, D. G., & Vidal-Madjar, A. 1985, Astrophys. J., 296, 593
[15] Guillout, P., Sterzik, M. F., Schmitt, J. H. M. M., Motch, C., & Neuhaeuser, R. 1998, Astron.. Astrophys., 337, 113
[16] Heiles, C., 1998, Astrophys. J., 498, 689

[17] Holmberg, J., & Flynn, C. 2004, MNRAS, 352, 440
[18] Krongold, Y., & Prochaska, J. X. 2013, Astrophys. J., 774, 115
[19] Lallement, R., Vergely, J. L., Valette B., Puspitarini L., Eyer L., Casagrande L. 2014a, Astron. & Astrophys. 561 A91
[20] Lallement R., Vergely J.L., Puspitarini L., Proc. Memorie SAIt, 2014b, in press
[21] Linsky, J. L., Draine, B. T., Moos, H. W., Jenkins E., Wood B. et al. 2006, Astrophys. J., 647, 1106
[22] Melott, A. L., & Thomas, B. C. 2011, Astrobiology, 11, 343
[23] Minchev, I., Quillen, A. C., Williams, M., et al. 2009, MNRAS, 396, L56
[24] Nieva, M.-F., & Przybilla, N. 2012, Astron. & Astrophys., 539, A143
[25] Olano, C. A. 2001, Astron. J., 121, 295
[26] Perrot, C. A., & Grenier, I. A. 2003, Astron. & Astrophys., 404, 519
[27] Pöppel, W. G. L., Bajaja, E., Arnal, E. M., & Morras, R. 2010, Astron. & Astrophys., 512, A83
[28] Perryman, M. A. C., de Boer, K. S., Gilmore, G., et al. 2001, Astron. & Astrophys., 369, 339
[29] Puspitarini, L., & Lallement, R., Vergely J.L., Snowden S.L. 2014, Astron. & Astrophys., in press
[30] Schönrich, R., Binney, J., & Dehnen, W. 2010, MNRAS, 403, 1829
[31] Sofia, U. J., Lauroesch, J. T., Meyer, D. M., & Cartledge, S. I. B. 2004, Astrophys. J., 605, 272
[32] Torra, J., Fernández, D., & Figueras, F. 2000, Astron. & Astrophys., 359, 82
[33] Vergely, J.-L., Freire Ferrero, R., Siebert, A., & Valette, B. 2001, Astron. & Astrophys., 366, 1016
[34] Welsh, B. Y. 1991, Astrophys. J., 373, 556
[35] Widrow, L. M., Barber, J., Chequers, M. H., & Cheng, E. 2014, MNRAS, 440, 1971
[36] Wolff, B., Koester, D., & Lallement, R. 1999, Astron. & Astrophys., 346, 969
[37] Wood, B. E., Linsky, J. L., Hébrard, G. 2004, Astrophys. J., 609, 838

13th Annual International Astrophysics Conference: Voyager, IBEX, and the Interstellar Medium IOP Publishing
Journal of Physics: Conference Series **577** (2015) 012017 doi:10.1088/1742-6596/577/1/012017

What fills the space between the partially ionized clouds in the local interstellar medium

Jeffrey Linsky[1] **and Seth Redfield**[2]
[1]JILA, University of Colorado and NIST, Boulder CO, USA 80309-0440
[2]Wesleyan University, Middletown CT, USA 06459-0001

E-mail: `jlinsky@jila.colorado.edu`, `sredfield@wesleyan.edu`

Abstract. The interstellar matter located between the warm clouds in the LISM and in the Local Cavity is now thought to be photoionized gas with temperatures in the range 10,000–20,000 K. While the hot stars ϵ CMa and β CMa are the primary photoionizing sources in the LISM, hot white dwarfs also contribute. We consider whether the Strömgren sphere gas produced by very local hot white dwarfs like Sirius B can be important in explaining the local intercloud gas. We find that the Strömgren sphere of Sirius can at least partially explain the intercloud gas in the lines of sight to several nearby stars. We also suggest that the partially ionized warm clouds like the Local Interstellar Cloud in which the Sun is located may be in part Strömgren sphere shells.

1. Properties of partially ionized warm clouds in the local interstellar medium

The Sun is surrounded by partially ionized warm gas extending out to a distance of about 15 pc in the local interstellar medium (LISM). We have learned about the properties of this gas from high-resolution spectra of absorption lines produced by interstellar gas along the lines of sight to nearby stars and from measurements of interstellar gas flowing into the heliosphere. [1] first reported that the radial velocities of interstellar absorption lines in the spectra of nearby stars are consistent with the interstellar gas flowing toward the Sun from the direction of the Scorpio-Centaurus Association. Subsequent investigations (e.g., [2, 3]) found that the interstellar gas flow has a number of velocity components with slightly different flow directions and speeds.

More recently, [4] identified 15 different velocity components of warm interstellar gas located within 15 pc of the Sun by analyzing interstellar absorption lines in HST spectra of 157 nearby stars. The flow vectors for each of these clouds predict accurate radial velocities for the interstellar gas in the directions of these clouds for all 34 new lines of sight observed by [5]. This success in predicting the radial velocities for these new sight lines indicates that these cloud vectors have accurate predictive power. Figure 1 shows the angular extent of the 15 clouds in Galactic coordinates. The Local Interstellar Cloud (LIC) is so-named because its angular extent covers nearly half of the sky, implying that the Sun is located just inside of the LIC or immediately outside. The decision as to which option is more likely to be valid requires the second type of data—measurements of interstellar gas flowing into the heliosphere.

Table 1 compares the parameters of the interstellar gas flowing into the heliosphere with the parameters of gas in the LIC obtained from absorption lines studied by [4]. The most recent measurements of the interstellar gas entering the heliosphere were obtained from IBEX satellite

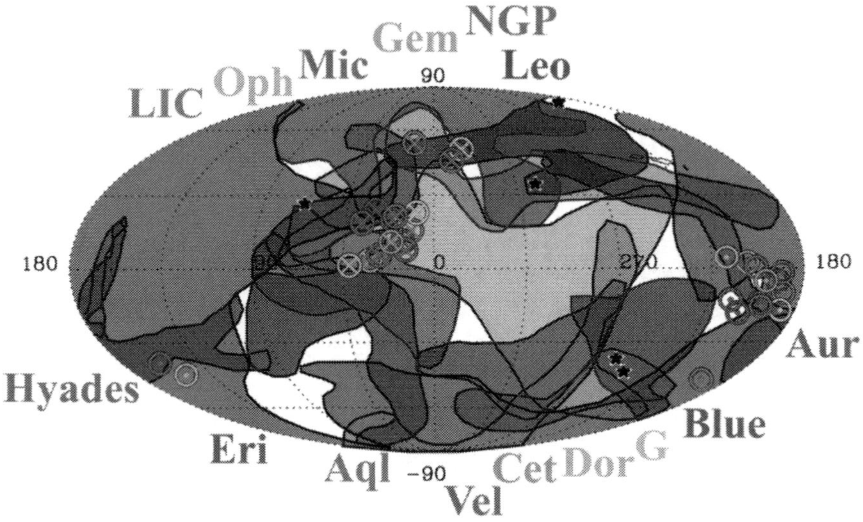

Figure 1. Morphologies of the 15 warm LISM clouds in Galactic coordinates with the Galactic Center direction in the center. The upwind heliocentric directions of the velocity vector for each cloud are indicated by the circled-cross symbols, and the downwind directions are indicated by the circled-dot symbols. The Sun is likely located just inside of the LIC (red) and the two closest clouds are the G cloud (brown) towards α Cen (1.32 pc) and the Blue cloud (dark blue) towards Sirius (2.64 pc). Figure from [4].

Table 1. Properties of the Local Interstellar CLoud.

Parameter	Redfield & Linsky 2008 [4]	McComas et al 2012 [6]
Flow speed (km s^{-1})	23.83 ± 0.90	23.3 ± 0.3
Galactic Longitude (deg)	187.0 ± 3.4	185.25 ± 0.24
Galactic Latitude (deg)	-13.5 ± 3.3	-12.03 ± 0.51
Temperature (K)	7500 ± 1300	6300 ± 390

observations of neutral hydrogen by [6, 7]. The flow speeds listed in Table 1 are for the inflow direction relative to the Sun's velocity through the LIC. The Galactic longitude and latitude are the direction of the upwind flows, and the gas temperatures were obtained from line widths. More precise measurements will be forthcoming from new IBEX data, but the present excellent agreement between the two sets of data provide strong evidence that the heliosphere is located just inside of the LIC. [8] computed a model for the LIC with neutral hydrogen number density $n_{\rm HI} = 0.19$–0.20 cm^{-3}, electron density $n_e = 0.007 \pm 0.01$ cm^{-3} and temperature $T = 6300$ K.

[4] find that the temperatures of other clouds lie in the range 5300–9900 K, and their neutral hydrogen column densities lie in the range $\log N_{\rm HI}$ = 17.2–18.8. The values of $n_{\rm HI}$ and n_e are uncertain although the clouds are likely partially ionized like the LIC.

The absence of interstellar absorption in the direction of the Sun's motion implies that the Sun will leave the LIC in less than 3000 years. [9] recently proposed that the inflow direction has changed over the last 40 years, suggesting that the heliosphere's environment is changing in our lifetime. However, the data upon which this result is based are not conclusive, and more data are needed before reaching a definitive conclusion. What will be the properties of the interstellar gas that the Sun will be entering, and what are the properties of the gas between the warm clouds and in the Local Cavity that envelopes these warm gas clouds?

2. Previous models for the intercloud gas

The classic papers of [10, 11, 12] predict that the interstellar medium should consist of three components: cold neutral and molecular gas ($T \leq 50$ K), warm neutral or ionized gas ($T \sim 8000$ K), and million-degree low-density ionized plasma. These papers assumed that the three components coexist in pressure and thermal equalibrium, which we now find is highly unlikely. For example, numerical simulations by [13], which include supernova explosions and realistic thermal processes, predict a very wide range of densities and temperatures in the ISM but no pressure equilibrium and no identifiable thermal phases. Observations are needed to determine which of these two very different models, or perhaps neither one, is realistic.

The nearby warm partially ionized gas clouds have properties roughly consistent with the warm component predicted by the classical models. Cold dense molecular clouds are observed typically by CO and HI 21-cm emission in many parts of the Galaxy. The nearest cold gas with a temperature of 15–30 K is the Leo cloud located at a distance between 11.3 and 24.3 pc from the Sun [14].

Our ideas concerning the properties of the gas surrounding the warm LISM clouds have undergone a radical change in the last 20 years. The gas between the clouds and extending out to roughly 100 pc from the Sun in what is now called the Local Cavity was originally assumed to be hot (roughly 10^6 K), fully ionized, and low density (roughly 0.005 cm^{-3}). This conclusion was based upon the predictions of the classical models and observations of diffuse soft X-ray emission consistent with the properties of the hot gas. This picture is no longer viable as X-ray emission from charge-exchange reactions between the solar wind ions and inflowing interstellar neutral hydrogen can explain most or perhaps all of the observed diffuse soft X-ray emission, except for the Galactic pole regions [15, 16, 17, 18]. Also, the intercloud and Local Cavity gas must have a much cooler temperature ($T \ll 10^6$ K) or very low emission measure as indicated by X-ray shadowing experiments (e.g., [2011]) and by extreme ultraviolet spectroscopy [19]. Since interstellar O VI absorption has not been detected within 58 pc of the Sun [20], the intercloud gas must be cooler than 300,000 K, yet still be mostly ionized so as to not show neutral hydrogen absorption.

Given the problems of explaining the intercloud (filler) gas with hot plasma, [21] proposed that the Local Cavity is an old supernova remnant with photoionized gas at a temperature of about 20,000 K. The likely photoionizing sources are primarily the hot stars ϵ CMa and β CMa that [22] found are the brightest extreme-ultraviolet sources observed by the Extreme Ultraviolet Explorer (EUVE) satellite, but several hot white dwarf stars also contribute to the ionizing radiation field.

3. Could the Strömgren spheres of hot white dwarfs explain the intercloud gas?

We consider here the effects of hot white dwarfs on their surrounding interstellar medium to test whether their ionizing radiation may play an important role in explaining the properties of the intercloud gas. In a classic paper, [23] showed that the EUV radiation ($\lambda < 912$ K) from a hot star completely ionizes hydrogen in its surrounding volume (an H II region) out to a distance now called the Strömgren radius where the build up of neutral hydrogen opacity absorbs the photoionizing radiation, producing a narrow partially ionized shell with neutral hydrogen gas outside. In this paper, Strömgren developed a simple model by assuming that the hot star is located in a constant density environment in which flows are ignored and time-independent photoionization of hydrogen is balanced by recombination. This model has been extended to include dust opacity, clumpiness, diffuse radiative transfer, dynamics, and the location of the star in a pre-existing cavity (e.g., [24, 25]). Here, we test whether the simple Strömgren sphere model can explain the gas-filling lines of sight to the closest stars using our LISM model. Following the suggestion of [26], we consider whether this filler gas can be understood simply as Strömgren spheres surrounding the many strong sources of extreme ultraviolet (EUV) radiation in the LISM, specifically hot white dwarfs and OB stars.

Table 2 lists the stars within 16 pc of the Sun for which absorption by neutral gas in nearby clouds has been identified by [4, 5]. We list the distances, Δd(neutral), along the lines of sight (LOS) through the identified clouds, assuming that the neutral hydrogen density is the same as that measured for the LIC ($n_{HI} = 0.195$ cm^{-3}) by [8] and that the remaining path length [Δd(ionized)] is filled by the presumably ionized filler gas.

Table 3 lists the nearby ($d < 20$ pc) sources of ionizing radiation (mostly hot white dwarfs) and the sizes of their surrounding volumes of Strömgren sphere-ionized gas. [26] computed the radii (R) of these H II regions assuming that the Strömgren spheres have stopped expanding, and that $n_e = 0.03$ cm^{-3}. We have recomputed the radii by assuming a gas-pressure balance with the LIC, $(n_{HI} + n_e + n_p + n_{He})T = 2710$ K cm^{-3}, where $n_{HI} = 0.195 \pm 0.005$ cm^{-3}, $n_e = n_p = 0.07 \pm 0.002$ cm^{-3} [8], $n_{He}/n_H = 0.1$, and $T = 7500$ K [4]. Table 3 lists $R(T)$ for various assumed temperatures computed from the values of $R(n_e = 0.03)$ in [26] scaled to higher densities using $R \sim (n_e/0.03)^{-2/3}$, which is based on equation 9-9 in [27].

Also listed in Table 3 are five more distant ionizing sources that dominate the EUV spectrum observed by the Extreme Ultraviolet Explorer (EUVE) spacecraft: ϵ CMa, whose emission dominates the 504–912 Å spectral region, β CMa, and the hot white dwarfs HZ 43, G191-B2B, and Feige 24 [8, 22, 28]. We compute their Strömgren sphere radii as a function of ionized-gas temperature scaling from the radii listed in [28] for $n_e = 0.04$ cm^{-3} and equation 9-9 in [27]. Following on the EUVE results of [22, 28] found that hot white dwarfs contibute only a small fraction of the total ionization in the Local Cavity, but we propose that for the short lines of sight to nearby stars, the Strömgren spheres of individual white dwarfs can provide much of the ionized filler gas between the partially ionized warm clouds.

We list here some rough estimates of the amount of ionized-filler gas in the LOS to nearby stars and whether nearby hot white dwarfs or other hot stars can supply the ionizing radiation to fill these sightlines with ionized gas. The computed Strömgren sphere radii are approximate given possible errors in the pressure balance and thermal equilibrium assumptions.

- **The LOS to Sirius:** The LIC and Blue are two partially ionized clouds in this LOS. Since the Sun is located inside of the LIC and Sirius is surrounded by the Strömgren sphere produced by EUV radiation from Sirius A and B, the LOS from the Sun to Sirius must consist first of LIC gas, then the Blue cloud, and then Strömgren sphere gas filling the remaining 1.92 pc. This path length of ionized gas is consistent with a Sirius A+B Strömgren sphere radius corresponding to a gas temperature of 10,000 K (see Table 3).
- **The LOS to α Cen:** Absorption by only the G cloud is seen in this LOS over a distance of 0.70 pc, if we assume that n_{HI} is the same as in the LIC. The remaining 0.62 pc could be filled with ionized gas from the Sirius A+B Strömgren sphere which cuts across the midpoint of the LOS from the Sun to α Cen at a distance of 2.21 pc. Since α Cen has an astrosphere, the LOS from the Sun to this star should consist first of a very thin layer of

Journal of Physics: Conference Series **577** (2015) 012017 doi:10.1088/1742-6596/577/1/012017

Table 2. Composition of Gas Along Selected Lines of Sight

Star	l	b	d(pc)	Cloud	log[N(HI)]	Δd(neutral)	Δd(ionized)
α Cen[a]	315.7	−0.9	1.32	G	17.6	0.70	0.62
Sirius A	227.2	−8.9	2.64	LIC	17.4	0.44	
				Blue	17.2	0.28	
				Sum		0.72	1.92
ϵ Eri[a]	5.8	−48.1	3.22	LIC	17.8	1.10	2.12
61 Cyg[a]	82.3	−5.8	3.49	Aql	17.8	1.10	
				Eri	17.8	1.10	
				Sum		2.20	1.29
Procyon	213.7	+13.0	3.51	LIC	17.9	1.56	
				Aur	17.6	0.70	
				Sum		2.26	1.25
ϵ Ind[a]	336.2	−48.0	3.63	LIC	16.6	0.14	3.49
61 Vir[a]	311.9	+44.1	8.53	NGP	(18.0)	(1.7)	(6.8)
β Com	43.5	+85.4	9.15	NGP	(18.0)	(1.7)	(7.4)
π^1 UMa	150.6	+35.7	14.6	LIC	17.85	2.30	12.3
τ Boo	358.9	+73.9	15.6	NGP	(18.0)	(1.7)	13.9
51 Peg	90.1	−34.7	15.6	Eri	(17.9)	(1.6)	
				Hya	(17.4)	(0.4)	
				Sum		(2.0)	(13.6)
χ Her[b]	67.7	+50.3	15.9	NGP	(18.0)	(1.7)	14.2

[a] Astrosphere detected by Wood et al 2005 [29] indicates that a cloud with neutral hydrogen must surround the star.
[b]No astrosphere detected by Wood et al 2005 [29].

LIC gas (showing no detectable absorption at the LIC velocity), then ionized gas from the Sirius A+B Strömgren sphere, and then the G cloud.

- **The LOS to ϵ Eri:** Spectra of ϵ Eri show absorption only at the LIC velocity, but the star has an astrosphere and, therefore, must be located in a partially neutral cloud. The LIC fills 1.10 pc along this LOS leaving 2.12 pc to be filled with ionized gas. Since Sirius is located 4.88 pc from ϵ Eri and 3.40 pc from the midpoint of the LOS from the Sun to ϵ Eri, we do not expect that Strömgren sphere gas from Sirius A+B can fill the missing 2.12 pc along the Sun-ϵ Eri LOS. Another hot white dwarf, 40 Eri B, is located 6.11 pc from ϵ Eri, so it is unlikely that its Strömgren sphere comes close to the Sun-ϵ Eri LOS. Instead, ϵ CMa, the largest source of ionizing radiation in the solar neighborhood [22] is the likely source for this ionized gas. If the assumptions of our model (thermal pressure equilibrium with no other sources of pressure) are appropriate, then the temperature of the ionized gas along this LOS would be about 40,000 K.
- **The LOS to Procyon:** Absorption by the LIC and Aur clouds leave 1.25 pc of path length to be filled with ionized gas. Ionizing radiation from Sirius A+B is the likely source of this ionized gas, since the separation of Sirius and Procyon is only 1.61 pc. This would indicate a Strömgren gas temperature of about 7,500 K.
- **The LOS to π^1 UMa, V368 Cep, MN UMa, δ Dra, 47 Cas, and ι Cep:** Spectra of these stars centered near l=130°, b=+30° with distances 14.4–35.4 pc all show absorption

Table 3. Hot White Dwarfs and their Strömgren Sphere Radii (pc)

WD	l	b	d(pc)	R(10,000K)	R(15,000K)	R(20,000K)	R($n_e = 0.03$)
Sirius B	227.2	-9.9	2.64	1.56	2.04	2.48	4.12
Sirius A+B	227.2	-9.9	2.64	1.93	2.51	3.05	5.08
40 Eri B	200.8	-38.0	5.04	0.44	0.58	0.70	1.17
GJ3753	123.3	+62.0	14.1	0.91	1.18	1.44	2.39
GJ433.1	284.9	+27.7	14.8	0.88	1.14	1.39	2.31
UZ Sex	245.3	+46.3	18.2	0.63	0.82	0.99	1.65
GD 50	188.9	-40.1	29.	1.88	2.47	2.99	4.11[a]
G191-B2B	155.9	+7.1	59.	8.18	10.72	12.98	17.86[a]
HZ 43	54.1	+84.2	68.	4.70	6.16	7.47	10.27[a]
Feige 24	166.0	-50.3	74.	9.55	12.51	15.16	20.85[a]
ϵ CMa	239.8	-11.3	124.	31.7	41.5	50.3	69.2[a]
β CMa	226.1	-14.3	151.	12.6	16.5	20.0	27.5[a]
n_e (cm^{-3})				0.129	0.0860	0.0645	0.03
$(n_e/0.03)^{-2/3}$				0.379	0.495	0.601	1.00
$(n_e/0.04)^{-2/3}$				0.458	0.600	0.727	1.00

[a] R($n_e = 0.04$) (pc).

only at the LIC velocity with no evidence for any other neutral gas in the LOS. Since the LIC lies in the immediate vicinity of the Sun, the remainder of these lines of sight must by ionized gas. GJ3753 (14.1 pc) and especially the hot white dwarf G191-B2B (59.9 pc) may be responsible for much of the ionizing radiation from this general direction.

- **The LOS to 61 Vir, β Com, τ Boo, and χ Her:** Spectra of these stars, all located at high Galactic latitudes (b> 44°), show absorption only by the NGP cloud with no evidence for absorption by the LIC or any other cloud. Since the closest star, 61 Vir (8.53 pc), has a detected astrosphere, the NGP cloud must be located near and in front of these stars. This leaves about 6.8 pc of the LOS to 61 Vir and similar path lengths toward the other stars to be filled with ionized gas. The high Galactic latitude white dwarfs especially HZ 43 but also GJ3753, GJ433.1, and UZ Sex could provide part of this ionizing radiation in addition to ϵ CMa and β CMa.

These examples support our proposal that ionized gas in the Strömgren spheres of hot white dwarfs is a component of the gas filling the lines of sight to nearby stars between the partially ionized warm gas clouds. A more detailed description of the morphology of the local ISM requires a three-dimensional model of the LISM clouds and a recalculation of the Strömgren spheres sizes based on modern models of hot white dwarfs.

Finally, we note that the thickness of the partially ionized outer shell of a Strömgren sphere is $\delta = (n_{HI}\sigma)^{-1}$, where $\sigma \approx 10^{-17}$ cm^2 [27] is the hydrogen-ionization cross section for EUV photons near 912 Å. Thus $\delta \approx 0.2$ pc. Depending on their distance from the Sun, filamentary warm clouds like the Mic Cloud could have roughly this thickness. The distance across the LIC is about 2.7 pc for log N(HI)=18.2 [4]. Thus the outer 0.2 pc of the LIC could be a Stŕomgren sphere shell and many of the other nearby partially ionized clouds could be fragments of Strömgren sphere shells.

Figure 2 shows the column density of neutral hydrogen computed from the geometric center

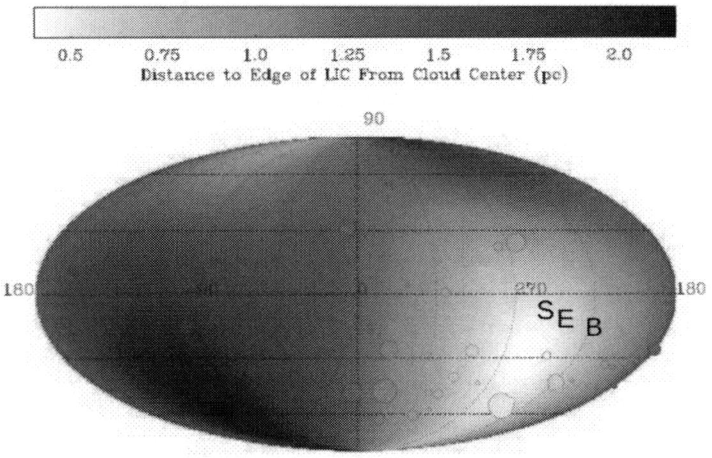

Figure 2. Column density of neutral hydrogen through the LIC as seen from its geometric center. The neutral hydrogen column density scale is above the figure. The symbols are S for the Galactic coordinates of the hot white dwarf Sirius B, E for ϵ CMa, and B for β CMa. Figure from Redfield et al. (in prep.).

of the LIC to its edge by Redfield et al (in prep.) based on the measured values of interstellar N(H I) with LIC radial velicities toward nearby stars. The region with Galactic longitude $225° \leq l \leq 300°$ and Galactic latitude $-90° \leq b \leq +30°$ shows very low H I column density corresponding to a skin depth of < 0.5 pc. The Galactic coordinates of Sirius, ϵ CMa and β CMa as seen from the center of the LIC are in this region. This is consistent with these three sources of ionizing radiation shaping the morphology of the LIC. Also, the short distance from the edge to the center of the LIC in this direction supports our proposal that the warm clouds are in part shells of Strömgren spheres.

4. Conclusions

The gas located between the warm LISM clouds and in the Local Cavity, which was previously assumed to be similar to the hot component in the classical interstellar medium models, is now thought to be photoionized gas with temperatures in the range 10,000–20,000 K. While the hot stars ϵ CMa and β CMa are the primary photoionizing sources in the LISM, hot white dwarfs also contribute. We consider whether the Strömgren sphere gas produced by very local hot white dwarfs like Sirius B can be important in explaining the local intercloud gas. We find that the Strömgren sphere of Sirius can at least partially explain the intercloud gas in the lines of sight to α Cen, Sirius, Procyon, ϵ Eri and other nearby stars. We also suggest that the partially ionized warm clouds like the LIC may be in part Strömgren sphere shells.

This work is a pilot study based on the LISM model of [4, 26] calculations for the Strömgren spheres of nearby hot white dwarfs. Redfield et al (in prep.) are now computing a new three-dimensional LISM model including the [5] sightlines. When this is available, we will reexamine the extent to which Strömgren sphere gas may explain the properties of the filler gas between the LISM warm clouds using the EUV luminosities of hot white dwarfs computed by [28].

5. Acknowledgements

We acknowledge support through the NASA HST Grant GO-11568 from the Space Telescope Science Institute, which is operated by the Association of Universities for Research in Astronomy, Inc. for NASA, under contract NAS 5-26555.

6. References

[1] Crutcher R M 1982, *Astrophys. J.* **254** 82
[2] Lallement R and Bertin P 1992 *Astron. Astrophys.* **266** 479
[3] Frisch P C, Grodnicki L and Welty D E 2002 *Astrophys. J.* **574** 834
[4] Redfield S and Linsky J L 2008 *Astrophys. J.* **673** 283
[5] Malamut C, Redfield S, Linsky, J L, Wood B E and Ayres T R 2014 *Astrophys. J.* **787** 75
[6] McComas D J, Alexashov D, Bzowski M, et al. *Science* **336** 1291
[7] Möbius E, Bochsler P, Bzowski M, et al. 2012 *Astrophys. J. Suppl.* **198** 11
[8] Slavin J D and Frisch P C 2008 *Astron. Astrophys.* **491** 53
[9] Frisch P C, Bzowski M, Livadiotis G, et al. 2013 *Science* **341** 1080
[10] Field G B, Goldsmith D W and Habing H J 1969 *Astrophys. J.* **155** L149
[11] McKee C F and Ostriker J P 1977 *Astrophys. J.* **218** 148
[12] Wolfire M, McKee C, Hollenbach D and Tielens A 1995 *Astrophys. J.* **453** 673
[13] Berghöfer T W and Breitschwerdt D 2002 *Astron. Astrophys.* **390** 299
[14] Peek J E G, Heiles C, Peek K M G, Meyer D and Lauroesch J T 2011 *Astrophys. J.* **735** 129
[15] Snowden S L, McCammon D, Burrows D N and Mendenhall J A 1994 *Astrophys. J.* **424** 714
[16] Cox D P 1998 *The Local Bubble and Beyond* ed D Breitschwerdt, M J Freyberg and J Trümper IAU Coll. 166, Lecture Notes in Physics **506** p. 121–131
[17] Koutroumpa D, Lallement R, Raymond J C and Kharchenko V 2009 *Astron. Astrophys.* **696** 1517
[18] Koutroumpa D 2012 *Astron. Nachr.* **333** 341
[19] Hurwitz M, Sasseen T P and Sirk M M 2005 *Astrophys. J.* **623** 911
[20] Barstow M A, Boyce D D, Welsh B Y, Lallement R, Barstow J K, Forbes A E, and Preval S 2010 *Astrophys. J.* **723** 1762
[21] Welsh B Y and Shelton R L 2009 *Astrophys Space Sci.* **323** 1
[22] Vallerga J V and Welsh B Y 1995 *Astrophys. J.* **444** 702
[23] Strömgren B 1939 *Astrophys. J.* **89** 526
[24] Yorke H W 1986 *Ann. Review Astron. Astrophys.* **24** 49
[25] McCullough P R 2000 *Publ. Astron. Soc. Pacific* **112** 1542
[26] Tat H T and Terzian Y 1999 *Publ. Astron. Soc. Pacific* **111** 1258
[27] Harwit M 1988 *Astrophysical Concepts* (New York: Springer-Verlag)
[28] Welsh B Y, Wheatley J, Dickinson N J and Barstow M A 2013 *Publ. Astron. Soc. Pacific* **125** 644
[29] Wood B E, Redfield S, Linsky J L, Müller H-R and Zank G P 2005 *Astrophys. J. Suppl.* **159** 118

Application of the theory of Large-Scale Quantization to the inner heliosheath

G Livadiotis
Southwest Research Institute, USA

E-mail: glivadiotis@swri.edu

Abstract. This paper addresses a remarkable feature of space plasmas that is related to their peculiar statistical behavior: The existence of a new quantization constant $\hbar_*$, similar to the standard Planck's quantization constant $\hbar$, but 12 orders of magnitude larger. The standard quantization through the Planck constant $\hbar$ may occur to individual particles, while the large-scale quantization through the new constant $\hbar_*$ applies only to clusters of correlated particles. In space plasmas, the value of $\hbar_*$ can be approximated by the division of the magnetosonic energy over the plasma frequency. Having the actual value of $\hbar_*$ been determined, it can be further applied in space plasma datasets to derive unknown plasma parameters when some key observable is missing, such as the magnetic field strength. This application is tested with solar wind data in the inner heliosphere, where the magnetic field is known, and then it is employed in the inner heliosheath plasma to estimate the unknown local magnetic field.

1. Introduction

Recent analyses of space plasmas revealed strong evidence for the existence of a new fundamental constant [1-7]. The new type of quantization constant $\hbar_*$ is similar to the Planck constant $\hbar$, but ~12 orders of magnitude larger. Planck's constant constitutes the smallest possible phase-space parcel for *individual and uncorrelated particles*, while the new quantization constant describes the smallest possible phase-space parcel for *collisionless particle systems characterized by local correlations*. The majority of space plasmas throughout the heliosphere are such systems.

The system's phase-space is represented by the $6N$-dimensional space of the position $\vec{x}$ and momentum $\vec{p}$ of the N particles of the system. The phase-space geometry invokes that there always exists a non-zero phase-space minimum parcel [8]; in other words, Classical Mechanics has the uncertainty principle built-in. While the existence of such non-zero phase-space cell is mathematically unavoidable, the exact value of this phase-space cell cannot be determined by mathematical theorems, but strictly from observations of the physical universe. The Planck constant discovered and determined from such observations that involve individual and uncorrelated particles. There is no reason for the phase-space cell of more complex systems to be strictly determined, again, by the Planck's constant, $\hbar$.

Imagine an intellectual phase-space version of Maxwell's demon that lives within the system and does not allow any type of physical structure of particles/quasi-particles to pass information through a phase-space smaller than a mega-cell given by $\hbar_c$, where $\hbar_c$ is larger than that of Planck's $\hbar$. This could be done, for example, by connecting various vicinal Planck's phase-space cells, i.e., by correlating the forces carrying information through these Planck's cells (Figure 1). Indeed, if there were no physical ways to pass information through a phase-space cell $\hbar_c > \hbar$, then, would the relevant physical laws alter and depend on $\hbar_c$ instead of $\hbar$? The first evidence that such a large-scale phase-space quantum may exist was shown in [1]. This mega-quantum characterizes systems with strong collective behavior and local correlations such as space plasmas.

The values of $\hbar_*$ were initially derived using four independent observational and theoretical analyses [1], that is, by examining: 1) a specific plasma with highly variant plasma parameters (Ulysses spacecraft measurements for solar wind); 2) a variety of 11 different space plasmas; 3) the limit of entropy $S \to 0$, where the classic Sackur-Tetrode equation of entropy that applies to particle systems with correlations (e.g., Earthy gases) is generalized to apply also to systems with correlated particles (i.e., collisionless plasmas); and, 4) explosive events and other plasma phenomena which can be related to the new type of quantization constant. The physical meaning of $\hbar_*$, as the smallest phase-space parcel, did not imply that this is some universal constant similar to the Planck's constant $\hbar$. Indeed, the values of $\hbar_*$ could be dependent on the plasma parameters and generally on the type of the plasma. Surprisingly, a similar value was found through all the measurement methods [1,6], supporting the constancy and universality of the value of $\hbar_* \sim 1.2 \cdot 10^{-22}$ Js.

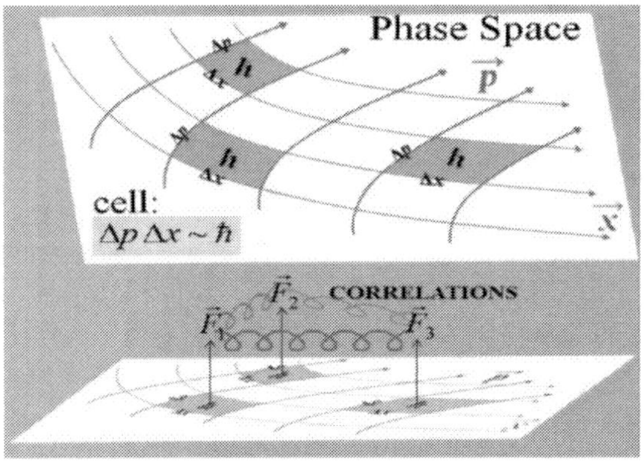

Figure 1. Sketch of phase-space, indicating the Planck's non-zero phase-space parcel, $\hbar$ (upper sketch), but this may change with the presence of correlations (lower sketch).

2. Statistical behavior and Large-Scale Quantization in space plasmas

Numerous independent developments in space plasma physics have revealed the peculiar statistical behavior of this category of plasmas: Particle populations in space plasmas reside in stationary states out of thermal equilibrium (e.g., [9] and refs therein). Thermal equilibrium is a special stationary state; systems at thermal equilibrium have their distribution functions of particle velocities stabilized into a Maxwell distribution, which is connected with the classical framework of Boltzmann-Gibbs statistical mechanics. However, Maxwell distributions are extremely rare in space plasmas. On the contrary, the vast majority of space plasmas resides out of thermal equilibrium, and is described by the kappa distributions (e.g., [10-14]). These non-Maxwellian distributions cannot embody the Boltzmann-Gibbs statistics, but instead, must be described within a generalized statistical framework (e.g., see [15,16]).

The origin of this vastly different statistical behavior between classical systems and space plasmas is the manifestation of correlations between the plasma particles. The stronger the correlation the further the plasma resides from thermal equilibrium [9]. On the other hand, the presence of correlations causes the phenomenon of the large-scale quantization.

The critical difference between the Planck constant $\hbar$ and the large-scale quantization constant $\hbar_*$, is that the latter applies – and it has been detected – in systems with local correlations between their particles. Local correlations are manifested by the presence of a correlation length between particles. For example, the Debye length in space plasmas gives a measure of the smallest correlation length, i.e., particles within such a length are always correlated, forming the Debye sphere [5]. The correlation length divides the system into an ensemble of clusters of correlated particles, e.g., Debye spheres (Figure 2). Particles within "correlation clusters" participate altogether to the large-scale quantization.

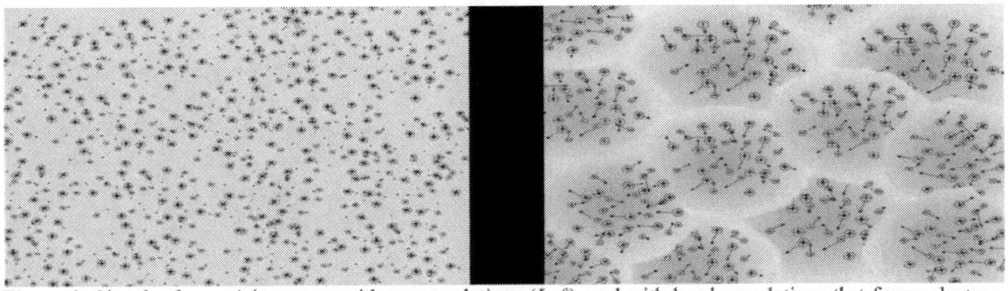

Figure 2. Sketch of a particle system with no correlations (*Left*), and with local correlations that forms clusters of correlated particles - the correlation clusters (*Right*). Large-scale quantization characterizes such systems with local correlations between their particles (e.g., space plasmas).

The scheme in Figure 3 demonstrates how the large-scale quantization gets related to the non-equilibrium statistical behavior of space plasmas and the competition between correlations and collisions. While correlations shift plasmas away from thermal equilibrium, collisions destroy correlations, recovering plasmas back at thermal equilibrium [1,9]. Thus, the large-scale quantization appears in collisionless plasmas, i.e., in plasmas where the correlation length is smaller than the mean free path – the average collision length.

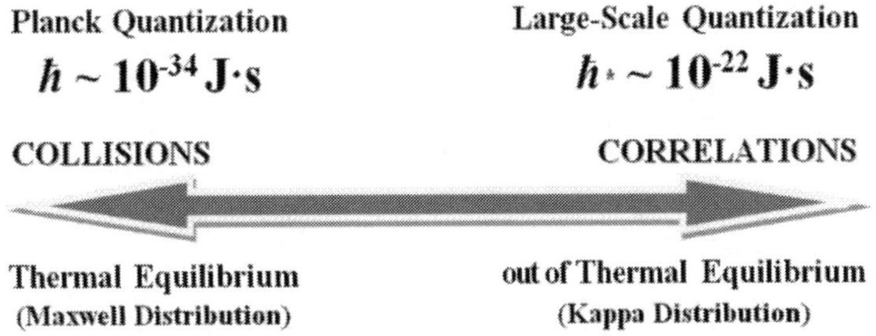

Figure 3. The non-equilibrium statistical behavior of space plasmas is related to the large-scale quantization. These physical phenomena are both caused by long-range interactions that induce local correlations. The particle correlations must be strong enough to defy particle collisions that lead the system back to thermal equilibrium.

3. Large-scale quantization and Quantum Mechanics

3.1. The uncertainty principle

Quantum Mechanics requires a minimum phase-space volume, and this phase-space "quantum" is typically recognized as the Planck constant. This limitation of phase-space follows the Heisenberg uncertainty principle [17]. In particular, individual particles, or particle systems with no local correlations, have a minimum energy level ε_p connected with the Planck constant, $\varepsilon_p \sim \hbar\omega$, while the lifetime associated with this lowest energy level is $t_p \sim 1/(2\omega)$; the frequency ω denotes the particle's basic oscillation mode. Therefore, for an arbitrary energy exchange that involves these particles, $\Delta\varepsilon \geq \varepsilon_p$, and for an arbitrary duration of this process, $\Delta t \geq t_p$, the uncertainty principle is given by

$$\Delta\varepsilon\,\Delta t \geq \varepsilon_p\, t_p = \tfrac{1}{2}\hbar \ . \tag{1}$$

As explained in Introduction, no mathematical or analytical argument can point to a particular value of the phase-space cell, $\hbar$. Structured systems, more complex than simply a group of uncorrelated individual particles, may exist and be characterized by a quantization constant different than that of the Planck constant. Down to the scale of Debye length, plasmas cannot be described through the approximation of a continuous fluid that applies for larger scales; instead, a new type of interpretation must be given that treats the smallest parcels of plasma as phase-space "quanta". This concept led to the development of the large-scale quantization that extends in scales of space and time larger than those of Planck's quantization [1,6]. It has been shown that there is a minimum energy $\varepsilon_c \sim \hbar_* \omega$ and a lifetime $t_c \sim 1/(2\omega)$ associated with the clusters of correlated particles, leading to a new type of quantum-cell and uncertainty principle,

$$\Delta\varepsilon\,\Delta t \geq \varepsilon_c\, t_c = \tfrac{1}{2}\hbar_* ,\ \hbar_* >> \hbar \ . \tag{2}$$

The inequality $\hbar_* \geq \hbar$ acknowledges that the Planck constant is still unique as the smallest possible quantization in the case where more than one quantization values exist.

3.2. The energy-frequency relation

Uncorrelated particles can transfer their energy in Planck's quanta of $\hbar\cdot\omega_{ph}$, that is the energy of a quasiparticle, e.g., a photon of frequency ω_{ph}. In a similar way, correlated particles can transfer their energy in larger quanta, $\sim\hbar_*\cdot\omega_{pl}$, where each quantum corresponds again to a quasiparticle; in plasmas, the transmitted quasiparticle is typically a plasmon of frequency ω_{pl}.

The smallest energy that can be transferred from an individual particle that is not subject to correlations into one quasiparticle, e.g., a photon, is $\varepsilon_p \sim \hbar\cdot\omega_{ph}$, while the smallest energy that can be transferred from a particle that is subject to local correlations into one quasiparticle, e.g., a plasmon, is $\varepsilon_c \sim \hbar_*\cdot\omega_{pl}$ (Figure 4) [1,6].

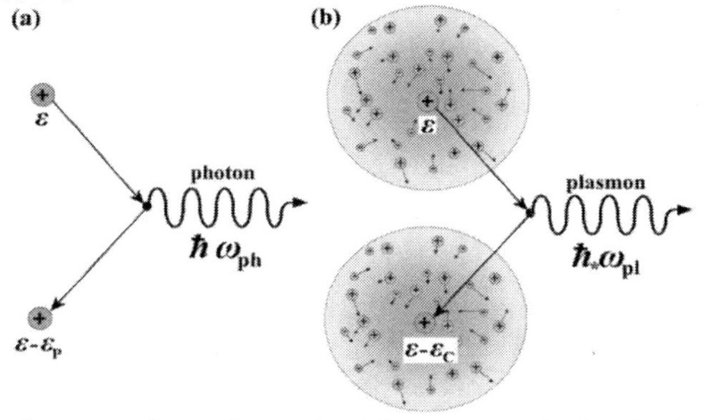

Figure 4. (a) The smallest energy that can be transferred from an uncorrelated particle is of the order of $\sim\hbar\cdot\omega$. The quasi-particle that carries this energy quantum may be a photon oscillating with frequency ω_{ph}. (b) For a correlated particle in plasma, the smallest energy that can be transferred is of the order of $\sim\hbar_*\cdot\omega$ (with $\hbar_*/\hbar\sim 10^{12}$). The quasi-particle is typically a *plasmon* oscillating with the plasma frequency ω_{pl}. (Taken from [6])

The transformation of the plasma particle's energy (e.g., kinetic, magnetic, etc) into the quasi-particle's transmitted energy constitutes one of the basic methods of determining measurements of the large-scale quantization constant, i.e., $\hbar_* \sim \varepsilon_c / \omega_{pl}$.

4. Application

4.1. The approximation of $\hbar_*$ in space plasmas

The amount of particle energy that can be transferred away from the correlation cluster (e.g., the Debye sphere - if the correlation length is given by the Debye length) via a quasiparticle can be estimated by the average particle energy. This sums the thermal and magnetic energy, leading to the magnetosonic energy $\varepsilon_c \sim \frac{1}{2}(m_i + m_e) U_{ms}^{\ 2}$ [6], where U_{ms} is the fast magnetosonic speed (m_i, m_e are the ion/electron masses). Also, the plasmon's frequency ω is typically given by the primary plasma frequency, ω_{pl}. As a consequence, the value of $\hbar_*$ can be approximated by $\hbar_* \sim \frac{1}{2}(m_i + m_e) U_{ms}^{\ 2} / \omega_{pl}$, or more analytically [5,6], by

$$\hbar_* \cong 0.0705 n^{-\frac{1}{2}} B^2 + 2.447 \gamma n^{-\frac{1}{2}} T_i (1+\alpha)/2 \ , \tag{3}$$

where the units are $\hbar_*[10^{-22}\text{J}\cdot\text{s}]$, $B[\text{nT} = 10\mu\text{G}]$, $n[\text{cm}^{-3}]$, $T_i[\text{MK}]$; γ is the average polytropic index $\gamma \equiv (\gamma_i + \alpha\gamma_e)/(1+\alpha)$, $\alpha = T_e/T_i$, ratio between electron/ion temperatures.

The actual value of $\hbar_*$ can be approximated by $\sim \frac{1}{2}(m_i + m_e) U_{ms}^{\ 2} / \omega_{pl}$, only when the plasma wave spectrum is dominated by the plasma frequency, ω_{pl}. Indeed, typically, space plasmas exhibit a dominant oscillation at the plasma frequency. (As an example, Figure 5 plots the wave spectrum of solar wind taken by STEREO A [18]). In general, more frequencies may be important in the particle's energy. Then, the energy ε_c is distributed over all these frequencies, i.e.,

$$\varepsilon_c = \hbar_* \bar{\omega} \ , \text{ where } \bar{\omega} \equiv \int \omega D(\omega) d\omega \ , \ \int D(\omega) d\omega = 1 \ , \tag{4}$$

where the integration is over all the wave spectrum.

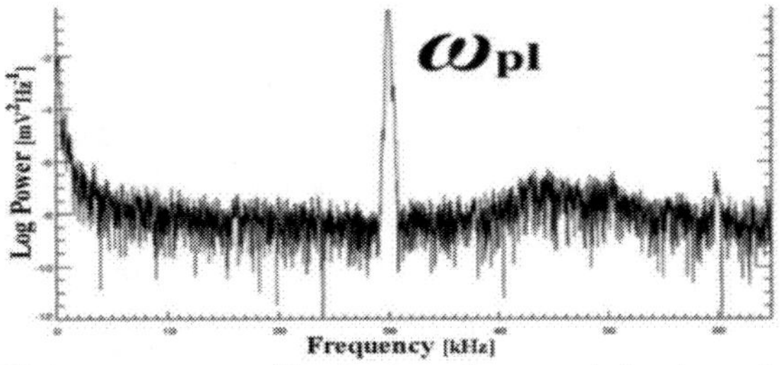

Figure 5. Typical STEREO A wave spectrum. The dominant frequency is the plasma frequency ω_{pl} [18].

The constancy of the approximation $\hbar_* \sim \frac{1}{2}(m_i + m_e) U_{ms}^{\ 2} / \omega_{pl}$ in space plasmas can be demonstrated with two examples of datasets, the ACE data for solar wind near Earth (~1AU), and Ulysses data for solar wind that is for heliocentric distances between Earth and Jupiter (1.1AU<r<5.5AU) (see Figure 6).

Journal of Physics: Conference Series **577** (2015) 012018 doi:10.1088/1742-6596/577/1/012018

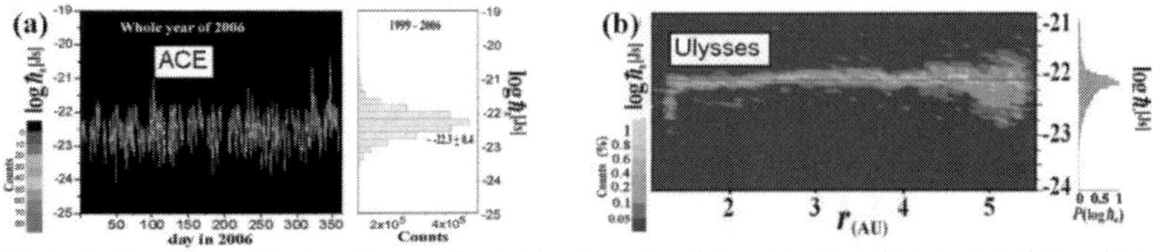

Figure 6. The value of $\hbar_*$ is well-approximated by the ratio of the magnetosonic energy and the plasma frequency, i.e., $\hbar_* \sim \varepsilon_C/\omega_{pl}$. Here this constancy is shown with values of $\hbar_*$ derived using Eq.(3) for (a) ACE [6] and (b) Ulysses datasets of solar wind [1].

4.2. Derivation of the magnetic field

We can exploit the constancy of $\hbar_*$ in space plasmas throughout the heliosphere, and use the approximation $\hbar_* \sim \varepsilon_C/\omega$ to derive unknown plasma parameters, e.g., magnetic field strength, density, ion and electron temperature [4,6]. Indeed, the dominant oscillation is given by the plasma frequency, $\omega \sim \omega_{pl}$, hence the large-scale quantization can be approximated by the relation $\hbar_* \sim \varepsilon_C/\omega_{pl}$, which is expanded in Eq.(3). This relation can be reversed, and expressed in terms of the unknown parameter, e.g., the magnetic field strength,

$$B \cong \sqrt{16.74 n^{\frac{3}{2}} - 34.7 \gamma\, n T_i (1+\alpha)/2} \ , \tag{5}$$

with the same units as in Eq.(3).

First, we use Ulysses measurements of the solar wind over a broad range of heliocentric distances 1.1AU<r<5.5AU. In Figure 7(a), the observational values of B are co-plotted with the modelled ones. Both sets of values follow similar distributions. The modelled B values appear with some larger variance around the statistical mode (most frequent value) at B~0.5-0.6nT.

After this verification, the method can be also applied in space plasmas with unknown magnetic field. Such an example is the inner heliosheath. There, the temperature, density [12], and polytropic index have been estimated, but the magnetic field strength is unknown. The plasma parameters have been estimated as radial averages for each sky direction. Figure 7(b) shows the results, that is a histogram of all the derived values of the magnetic strength. The mode of the distribution is ~2μG, a value that characterizes the whole inner heliosheath. The derived value is consistent with the in situ measurements taken by Voyager 1 and 2.

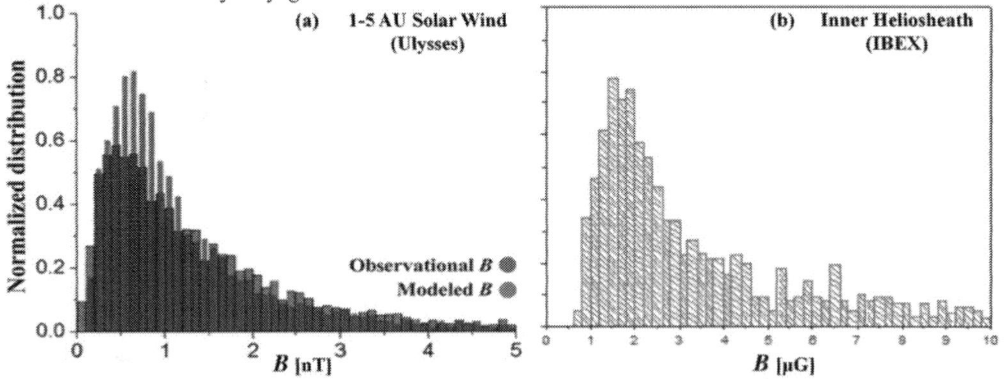

Figure 7. Estimation of the magnetic field strength values, (a) for solar wind using Ulysses measurements, and comparison with the respective observational values [6]; (b) for the inner heliosheath using IBEX measurements.

5. Conclusion

Recent publications revealed strong evidence that space plasmas are characterized by a new type of quantization, called the Large-Scale Quantization, where the new quantization constant $\hbar_*$ is similar to the standard Planck's quantization constant $\hbar$, but 12 orders of magnitude larger.

While the standard quantization through the Planck constant $\hbar$ characterizes interactions of individual particles, the large-scale quantization through the new constant $\hbar_*$ applies only to interactions involving clusters of correlated particles. Local correlations are realized when a correlation length is assigned between the system's particles. The Debye length in space plasmas constitutes the smallest possible correlation length. The correlation length divides the system into an ensemble of clusters of correlated particles. The particles within each of these "correlation clusters" participate altogether to this new type of quantization.

In space plasmas, the value of $\hbar_*$ can be approximated by the division of the magnetosonic energy over the plasma frequency. Having the value of $\hbar_*$ already been determined by various theoretical and observational methods, it can be further applied in space plasma datasets to derive unknown plasma parameters when some key observable is missing, such as the magnetic field strength. This application is tested with solar wind data in the inner heliosphere, where the magnetic field is known, and then is used in the inner heliosheath plasma to estimate the unknown local magnetic field.

References

[1] Livadiotis G and McComas D J 2013 Evidence of Large-Scale Quantization in Space Plasmas *Entropy* **15** 1116

[2] Witze A 2013 Space plasmas share a secret *Nature News*, doi:10.1038/nature.2013.13159

[3] Livadiotis G and McComas D J 2013 Large-Scale Quantization in plasmas *APS Bul. Plasma Phys* **58** GO6.00005

[4] Livadiotis G and McComas D J 2014 Large-Scale Quantization in space: Summary and applications *ASP Conf Ser* **484** 131

[5] Livadiotis G and McComas D J 2014 Electrostatic shielding in plasmas and the physical meaning of the Debye length *J Plasma Phys* **80** 341

[6] Livadiotis G and McComas D J 2014 Large-Scale phase-space quantization from local correlations in space plasmas *J Geophys Res* **119** 3247

[7] Livadiotis G 2014 Lagrangian temperature: Derivation and physical meaning for systems described by kappa distributions *Entropy* (In Press)

[8] Gromov M 1985 Pseudo holomorphic curves in symplectic manifolds *Invent math* **82** 307

[9] Livadiotis G and McComas D J 2013 Understanding Kappa Distributions: A Toolbox for Space Science and Astrophysics *Space Sci Rev* **75** 183

[10] Ogasawara K, et al. 2013 Characterizing the dayside magnetosheath using energetic neutral atoms: IBEX and THEMIS observations *J Geophys Res* **118** 3126

[11] Nicolaou G, Livadiotis G, and Moussas X 2014 Long-Term Variability of the Polytropic Index of Solar Wind Protons at 1 AU *Sol Phys* **289** 1371

[12] Livadiotis G, et al. 2011 First sky map of the inner heliosheath temperature using IBEX spectra *Astrophys J* **734** 1

[13] Livadiotis G, et al. 2012 Pick-up ion distributions and their influence on ENA spectral curvature *Astrophys J* **751** 64

[14] Livadiotis G, et al. 2013 Pressure of the proton plasma in the inner heliosheath *Astrophys J* **762** 134

[15] Livadiotis G 2009 Approach on Tsallis statistical interpretation of hydrogen-atom by adopting the generalized radial distribution function *J Math Chem* **45** 930

[16] Tsallis C 2009 *Introduction to Non-extensive Statistical Mechanics: Approaching a Complex World* (New York: Springer)

[17] Heisenberg W 1927 Über den anschaulichen inhalt der quantentheoretischen Kinematik und Mechanik *Z Phys* **33** 879

[18] Thejappa G, MacDowall R J, and Bergamo M 2012 In situ detection of strong Langmuir turbulence processes in solar type III radio bursts *J Geophys Res* **117** A08111

Journal of Physics: Conference Series **577** (2015) 012019 doi:10.1088/1742-6596/577/1/012019

Interstellar Gas Flow Vector and Temperature Determination over 5 Years of IBEX Observations

E. Möbius[1], M. Bzowski[2], S.A. Fuselier[3,4], D. Heirtzler[1], M. A. Kubiak[2], H. Kucharek[1], M.A. Lee[1], T. Leonard[1], D.J. McComas[3,4], N. Schwadron[1], J.M. Sokół[2], P. Wurz[5]

1 Space Science Center & Department of Physics, University of New Hampshire, Durham, NH 03824, USA
2 Space Research Centre, Polish Academy of Sciences, Warsaw, Poland
3 Southwest Research Institute, San Antonio, TX, USA
4 University of Texas, San Antonio, TX 78228, USA
5 Physikalisches Institut, Universität Bern, 3012 Bern, Switzerland

E-mail: Eberhard.moebius@unh.edu

Abstract. The Interstellar Boundary Explorer (IBEX) observes the interstellar neutral gas flow trajectories at their perihelion in Earth's orbit every year from December through early April, when the Earth's orbital motion is into the oncoming flow. These observations have defined a narrow region of possible, but very tightly coupled interstellar neutral flow parameters, with inflow speed, latitude, and temperature as well-defined functions of inflow longitude. The best-fit flow vector is different by ≈ 3° and lower by ≈ 3 km/s than obtained previously with Ulysses GAS, but the temperature is comparable. The possible coupled parameter space reaches to the previous flow vector, but only for a substantially higher temperature (by ≈ 2000 K). Along with recent pickup ion observations and including historical observations of the interstellar gas, these findings have led to a discussion, whether the interstellar gas flow into the solar system has been stable or variable over time. These intriguing possibilities call for more detailed analysis and a longer database. IBEX has accumulated observations over six interstellar flow seasons. We review key observations and refinements in the analysis, in particular, towards narrowing the uncertainties in the temperature determination. We also address ongoing attempts to optimize the flow vector determination through varying the IBEX spacecraft pointing and discuss related implications for the local interstellar cloud and its interaction with the heliosphere.

1. Introduction

The neutral gas component of the local interstellar cloud (LIC) that surrounds our solar system reaches the inner heliosphere as an interstellar neutral (ISN) wind due to the relative motion of the Sun and the interstellar material. On its way into the inner heliosphere the ISN wind composition and structure is altered on its passage through the heliospheric interface region, its flux is reduced by ionization losses, and its course is deflected by the Sun's gravitation and for H also by radiation pressure. This interaction of the surrounding interstellar medium with the heliosphere and its journey into the inner solar system has been described in numerous modeling studies, e.g. [1 – 5]. Among ISN constituents, He provides the most pristine sample of the surrounding LIC inside the heliosphere. Taking a sample of the LIC is important for a variety of reasons. In particular, the state of the LIC defines the boundary of the heliosphere, it is the only accessible sample of cosmic bulk gaseous material other than from the solar system, and the Sun is currently located apparently at the edge of an interstellar cloud, with the prospect of exiting relatively soon on astronomical scales [6, 7] and a longer term history of dramatically changing interstellar conditions [8].

Local measurements of the ISN flow inside the heliosphere started with UV backscatter observations of H [9] and He [10], followed by pickup ion detection [11, 12], and finally neutral gas imaging of the flow itself [13, 14]. A concerted attempt to consolidate the measurements with these different optical and in-situ methods resulted in a set of consensus parameters for the LIC [15] and the realization that neutral gas imaging [16] should provide the most detailed and accurate information about the distribution function of the neutral gas in the LIC.

It came as a surprise that the first quantitative analysis of the ISN flow observations with the Interstellar Boundary Explorer (IBEX) [17, 18] appeared to suggest a different set of parameters, whose range included a different flow vector at its center with about the same temperature as found with Ulysses GAS [16] or at the edge of the range a similar flow vector, but with much higher temperature. Together with finding a larger interstellar magnetic field than before from the IBEX ENA observations [19], one potential consequence is the realization that the heliosphere likely has no fast bow shock [20]. The search for reconciliation of the different results, among others, has led to the proposal that the interstellar flow direction may have been changing over the course of the space age, supported by a compilation of 40 years of published interstellar flow direction results [21]. Stimulated by these intriguing possibilities, re-analysis efforts of the Ulysses GAS observations, including formerly unanalyzed data from the last fast latitude passage in 2007, are ongoing [22-24].

Recently, [25] have opposed the idea of a changing ISN flow by questioning the validity and interpretations of several of the results used in [21], albeit with several invalid arguments, which are addressed in [26]. In this paper, we will concentrate on the IBEX observations, firm up the temperature determination, and discuss the constraints that provide the IBEX flow vector values. The latter two have been called in question by [25] and yet are key to the IBEX analysis. We will also put the determination of the ISN flow vector into perspective with the interstellar magnetic field direction, derived among others from IBEX Ribbon observations [27-29]. Interstellar species, such as H and O, which are affected substantially by the heliospheric interface, should be deflected in a plane that is defined by the ISN inflow vector and the interstellar magnetic field direction [30-33]. The paper will close with a preview to IBEX observations with a modified spin axis orientation that should allow a more accurate determination of the ISN flow vector.

2. Instrumentation and Observations

IBEX was launched into a highly elliptical Earth orbit in October 2008 to observe energetic neutral atoms (ENAs) with minimum interference from the near Earth environment [34]. The satellite carries two single-pixel high-sensitivity ENA cameras, IBEX-Hi [35] and IBEX-Lo [36], whose fields-of-view (FoV) point perpendicular to the spin axis and whose combined energy range is 10 – 6000 eV with overlap between 300 and 2000 eV.

IBEX-Lo has been optimized to observe interstellar H, He, and O flow distributions. Incoming neutral atoms pass a collimator, which determines the sensor FoV of 6.5° FWHM with regular resolution and 3.2° FWHM in high-resolution mode. The collimator and a set of repeller electrodes are biased to reject most of the ambient electrons and ions. After passing the collimator the neutral atoms are converted into negative ions on a highly polished conversion surface, which is covered with a vapor-deposited diamond-like carbon layer. The negative ions are then electrostatically collected into an electrostatic analyzer, where they are selected for energy, followed by post-acceleration in a potential of 16 kV (7 kV after summer 2012) and mass analysis in a triple-time-of-flight (TOF) section [36]. The triple TOF analysis also provides for superior rejection of noise and background contributions [37]. While H and O are detected and identified through the ready generation of converted H^- and O^- ions, noble gases, such as He and Ne, do not produce stable negative ions [38]. Therefore, these species are analyzed through negative ion sputter products of the carbon and an ever-present water layer, which are produced with a combination of H^-, C^-, and O^- in characteristic proportions for each species and energy that was calibrated before flight. In particular, interstellar neutral (ISN) He produces a well characterized specific signature in IBEX-Lo [14, 18].

For each detected atom, energy step, TOF values, and a time tag with 4.2 ms resolution are registered in the form of individual events. The sensor also accumulates rates for each TOF and single detector. The events and rates are passed to a central electronics unit (CEU), which culls electron events due to their short TOF and accumulates histograms at 6° resolution in spin angle for atoms identified as H and O, separate for each of the 8 energy steps. Events with full information are buffered in the CEU and formatted into the telemetry. If the number of events exceeds the telemetry allocation, which occurs during the passage of the peak ISN flow, a sample of events is taken so that the full angle range is covered and preference is given to events with valid triple TOF detection over those with single TOF signals. Absolute rates for the angular distributions can be found through normalization to the H and O histograms [36].

Because IBEX-Lo provides energy bands with a resolution $\Delta E/E = 0.7$ and the energy information is almost completely lost for He in the sputter process, the detailed velocity distribution of the ISN flow is recovered from the observed angular distributions. IBEX is a roughly Sun-pointing, spinning satellite with the spin axis re-oriented towards the Sun after each 7–8 day orbit during the first three years and each half orbit arc of 4.6 days after raising the perigee into a lunar synchronous orbit in 2011 [39]. Complete full-sky ENA maps are obtained with a resolution comparable to the sensor FoV over a period of six months. As a consequence, IBEX samples the heliospheric and interstellar ENA distributions at 1 AU in a plane that is approximately perpendicular to the Earth-Sun line. This is equivalent to observing ENAs, which arrive from the heliospheric boundary or beyond, near the perihelia of their trajectories, regardless of their flow direction at infinity. This geometry allows significant simplification in the reconstruction of the flow distribution at infinity so that analytic approximations could be used in one of the analysis methods [40, 18]. However, the geometry also couples IBEX observables, and this coupling leads to a basic degeneracy in the observed parameters.

IBEX observes the ISN flow each year from late December through early April, when the Earth moves into the oncoming flow thus boosting the atom energy in the observer frame, with the passage of the He flow peak at ecliptic longitude $\lambda_{Peak} = 130.6 \pm 0.7°$ [18] around January 31. While the flow could, in principle, be also observed in the fall, when the Earth is overtaking the flow [41], the neutral atom energy in the spacecraft frame is too low, and no usable signal has been detected thus far.

3. Drivers for the Interstellar Flow Parameters

To understand the interstellar neutral gas flow vector and temperature results obtained by IBEX and their implications it needs to be reemphasized 1) which parameters are tightly coupled through the IBEX observables, 2) where the remaining degeneracy comes from, and 3) how it can be broken, although with an appreciably larger uncertainty. The end result of the coupling between the different LIC neutral gas parameters (flow longitude $\lambda_{ISM\infty}$, latitude $\beta_{ISM\infty}$, speed $V_{ISM\infty}$, and temperature $T_{ISM\infty}$) is a narrow "tube" of possible solutions in the 4-dimensional space of possible parameters. The three parameters ($V_{ISM\infty}$, $\beta_{ISM\infty}$, and $T_{ISM\infty}$) are related to $\lambda_{ISM\infty}$ (as independent variable) through a well-defined functional dependence, which because of its extension in $\lambda_{ISM\infty}$ constitutes a degeneracy due to the IBEX observation of the ISN flow over a limited portion of the Earth's orbit [17, 18, 20]. The "tube" of allowable values in velocity reaches to the velocity vector derived by [16] with Ulysses GAS, with a slightly differing combination in $\lambda_{ISM\infty}$, $\beta_{ISM\infty}$, and $V_{ISM\infty}$, but a noticeably different optimum $\lambda_{ISM\infty}$ and $V_{ISM\infty}$. As noted in the original papers, the IBEX vector could still be compatible with the Ulysses velocity within mutual uncertainties, but a substantially higher ($\approx$30%) ISN temperature is associated with the Ulysses velocity vector.

As discussed in [18] and [40], the determination of the ISN inflow vector is tied to the observation of the flux maximum (the ISN bulk flow vector at 1 AU) perpendicular to the Earth-Sun line with the Sun-pointing IBEX spacecraft (shown in Figure 1 for a sample ISN trajectory). Now $V_{ISM\infty}$ and $\lambda_{ISM\infty}$

are uniquely connected via the hyperbola equation for the true anomaly θ_∞ of the trajectory, which is related to bulk flow peak location λ_{Peak} at 1 AU and flow direction $\lambda_{\text{ISM}\infty}$ at infinity:

$$V_{ISM\infty} = \sqrt{\frac{GM_S}{r_E} \cdot \left(\frac{-1}{\cos(\theta_\infty)} - 1 \right)} \quad \text{and} \quad \theta_\infty = \lambda_{\text{ISM}\infty} + 180^\circ - \lambda_{\text{Peak}} \tag{1}$$

The location of the flow maximum λ_{Peak} was determined independently with the same assumptions on ambient conditions [17, 18] to $\leq\pm0.7^\circ$, which constrains the functional relationship $V_{\text{ISM}\infty}(\lambda_{\text{ISM}\infty})$ in (1) to within $\leq\pm0.7^\circ$ in $\lambda_{\text{ISM}\infty}$ and $\leq\pm0.5$ km/s in $V_{\text{ISM}\infty}$. However, a large range of values along the function satisfies the observed ISN flow maximum at 1 AU equally well.

Simultaneously, functions for the ISN flow direction in latitude $\beta_{\text{ISM}\infty}(\lambda_{\text{ISM}\infty})$ and the temperature $T_{\text{ISM}\infty}(\lambda_{\text{ISM}\infty})$ can be obtained from the ISN flow peak location in latitude and its angular width. Because the statistical uncertainties of the peak location and angular width of the ISN flow are typically 2% and 1%, respectively, the combined uncertainties of the two relations are mostly controlled by the uncertainties in relation (1) and potential systematic effects, which we address for the derived temperature below. The observed angular distribution of the ISN flow constitutes a local Mach cone of width $v_{\text{Th}}/V_{\text{ISM\_1AU}}$, with the bulk flow $V_{\text{ISM\_1AU}}$ and thermal speed v_{Th} at 1 AU, thus fixing the ratio $T_{\text{ISM}\infty}/V^2_{\text{ISM\_1AU}}$. In turn, $T_{\text{ISM}\infty}$ is tied to the adopted ISN speed $V_{\text{ISM}\infty}$ and thus to $\lambda_{\text{ISM}\infty}$.

As shown in Figure 1, the flow latitude Ψ_{Peak} at 1 AU in the rest frame is connected to the ISN flow latitude $\beta_{\text{ISM}\infty}$ at infinity through the true anomaly of the bulk flow as

$$\tan(\psi_{Peak}) = \frac{\tan(\beta_{ISM\infty})}{|\sin(\theta_\infty)|} = \frac{\tan(\beta_{ISM\infty})}{|\sin(\lambda_{ISM\infty} + 180^o - \lambda_{Peak})|} \tag{2}$$

and thus again to the ISN inflow longitude $\lambda_{\text{ISM}\infty}$. These relations constitute a degeneracy in the determination of $V_{\text{ISM}\infty}$, $\beta_{\text{ISM}\infty}$, and $T_{\text{ISM}\infty}$ as a function of $\lambda_{\text{ISM}\infty}$. However, relation (2) also provides the means to break the degeneracy with IBEX observations. When latitude distributions are combined for different longitude λ_{Obs} positions with IBEX as observer, then, λ_{Obs} replaces λ_{Peak} in Equation (2). With λ_{Obs} now as a variable, the resulting relation $\Psi_{\text{Peak}}(\lambda_{\text{Obs}})$ depends solely on $\lambda_{\text{ISM}\infty}$. The comparison of ISN flow distributions observed over entire observation seasons with the predicted relation (2) as a function of $\lambda_{\text{ISM}\infty}$ and respective χ^2-minimization in [18] resulted in the reported optimum inflow longitude $\lambda_{\text{ISM}\infty} = 79^\circ +3.0^\circ(-3.5^\circ)$. Independently, [17] obtained a similar result through numerical modeling. The uncertainties in $\lambda_{\text{ISM}\infty}$ are the basis for the bounding values of the resulting parameter range from the IBEX observations used in [20].

To illustrate the basis for the inflow direction in the IBEX analysis, the right panel in Figure 1 shows the peak location of simulated ISN flow distributions in latitude Ψ_{Peak} for the usable observation

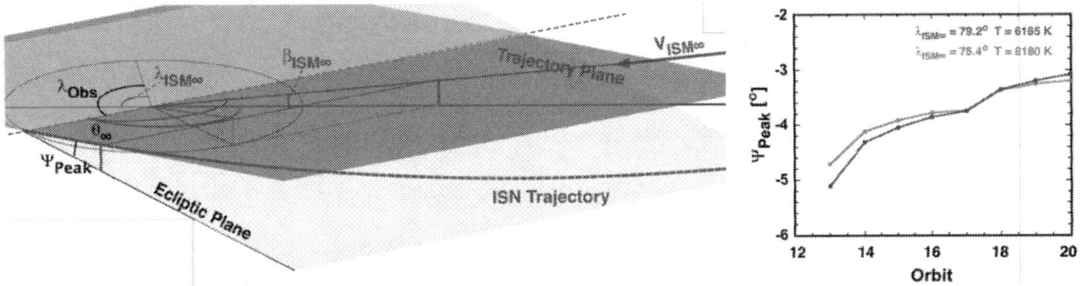

Fig. 1: ISN flow trajectory in its orbit plane (blue), which intersects the ecliptic plane (yellow) along the observer-Sun line and is inclined by the angle Ψ_{Peak} of the observed flow peak (shown in the inertial frame). The ISN flow vector at infinity $V_{\text{ISM}\infty}$ is inclined by the inflow latitude $\beta_{\text{ISM}\infty}$. The local peak flow angle Ψ_{Peak} varies with the true anomaly θ_∞ of the trajectory and thus with observer location λ_{Obs} and inflow longitude at infinity $\lambda_{\text{ISM}\infty}$. The right panel shows the simulated variation $\Psi_{\text{Peak}}(\lambda_{\text{Obs}})$ as averages over the 2009 IBEX orbits for $\lambda_{\text{ISM}\infty} = 75.4^\circ$ and 79.2°.

Journal of Physics: Conference Series **577** (2015) 012019 doi:10.1088/1742-6596/577/1/012019

periods during the 2009 ISN season orbits. Shown are the results for two ISN flow parameters sets on the center line of the parameter tube as defined in equations (1) and (2), using the flow longitude $\lambda_{ISM\infty}$ of [16] (red) and that of [17, 18] (blue). The visible difference between the two curves with different $\lambda_{ISM\infty}$ along the IBEX tube is a steeper variation of Ψ_{Peak} as a function of orbit (longitude) towards earlier orbits, for the $\lambda_{ISM\infty}$ value stated in [17, 18] over that in [16]. The optimization used in [18] is solely sensitive to the shape of the curves in Figure 1. The shape of these curves also dominates the χ^2-minimization for simulated flow distributions in [17], but this method is also sensitive to the overall shape of the observed angular distributions.

4. Temperature Determination

The temperature required by the IBEX relations for the velocity vector of [16] is much higher than reported there. The authors of [19] argue that the width of the angular distribution, which now solely influences the derived temperature, could be brought into agreement with the Ulysses temperature value by assuming unknown instrumental effects in IBEX. We have determined $T_{ISM\infty}$ from the width of the latitude distribution in the analytic method [18]. The value obtained through χ^2-minimization by [17] is also dominated by the latitudinal variation of the fluxes. The left two panels of Figure 2 show angular distributions in latitude for the two peak orbits in 2009 and 2010 that have been mainly used to determine the He temperature in [18]. We show 2048-spin accumulations here for two time periods that are selected for almost identical ecliptic longitude close to the ISN flow maximum, as constrained by the ISN good times according to [18] for these orbits (Table 1 in [18]). Identical longitude is very important for comparable widths as can be seen in the right panel, which shows the width obtained from every second 512-spin period during ISN good times of orbit 64. The width decreases by ≈0.7° over the course of 7 days or 7° in longitude. Given this variation, the angular widths obtained in 2009 and 2010 are in remarkable agreement. To match the ISN flow maximum longitude (130.6±0.7°) in the temperature determination, [18] combined these orbits with the previous ones. The data used in Figure 2 are made available as 512-spin (≈2 hour) time periods for the ISN flow good times, as in Table 1 of [18], under IBEX Data Release 6 (http://ibex.swri.edu/researchers/publicdata.shtml#dr6). The angular distributions are provided in 6° resolution, with both the sum of direct and histogram events listed for full correction of the telemetry limitations, similar to Data Release 3 with longer time intervals.

As discussed in [18] the peak height and thus the width of these distributions as shown in Figure 2 could be affected by particle rate dependent suppression of the data flow of individual events through the CEU. In such a case, the peak of an angular distribution would be reduced the most, thus increasing the apparent width of the original angular distribution. To account for such potential, at the time

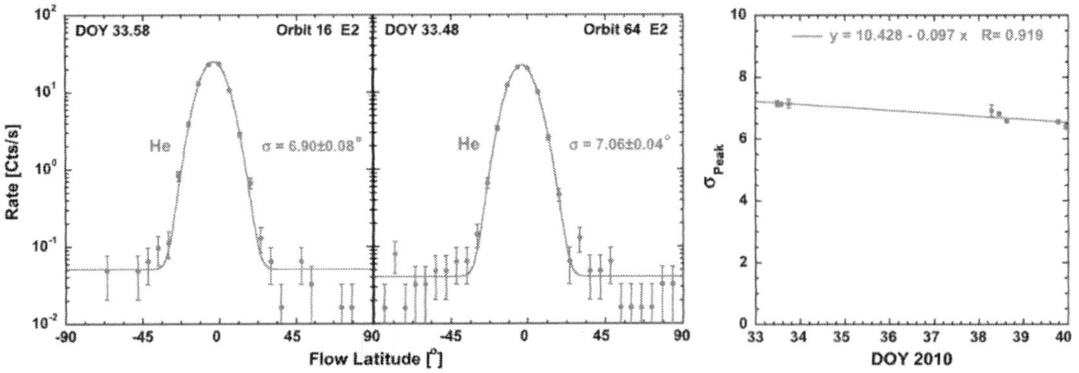

Fig. 2: Left: Latitude distributions for two 2048-spin periods in orbits 16 and 64 from ISN flow good times [18] that are only 0.3° in ecliptic longitude apart (orbit 64 is 0.3 days earlier or 0.3° closer to 0°) with Gaussian fit and statistical error bars. Right: Derived width after subtracting the collimator function as a function of time over orbit 64 taken over every second 512-spin period during the ISN flow good times. The angular width decreases as a function of time and thus ecliptic longitude.

unknown, effects [18] estimated an upper limit in terms of a maximum CEU event processing time $\tau \leq$ 5 ms. They modeled the processing analytically similar to a dead time and compared observations of the ISN flow at IBEX-Lo energy steps 2 and 4, the latter with a rate lower by a factor of 5 and thus much reduced interface traffic. In [19] the authors claim that an instrumental dead time $\tau \approx 7$ ms would allow reconciliation of the angular distributions observed with Ulysses and IBEX, although [18] indicated that $\tau \leq 5$ ms is an upper limit and [20] reported that the effect on the width and thus the temperature was much smaller based on follow-up testing and simulation, as discussed in more detail below.

The telemetry limitation for high-flux neutral atom sources, such as the ISN flow and the Earth's magnetosphere, was anticipated in the IBEX mission planning, and therefore H and O energy-angle histograms with 6° resolution are accumulated onboard IBEX to normalize the reported event rates. Because the width of the IBEX-Lo FoV is already 6.5° FWHM and to eliminate any potentially unanticipated effects on the data flow by the limited telemetry at angular resolution <6°, [17] and [18] restricted the angular resolution to 6°, thus using the histograms (accumulated in the CEU) for full correction of any telemetry limitations and their effects.

In addition, the hardware interface between IBEX-Lo and the CEU has a known transfer time of 1.354 ms for each event, which increases for the first 1° of each 6° bin to 2.188 ms because IBEX-Lo housekeeping and rates are also transmitted. Both transmission times are much shorter than the previously estimated maximum processing time. The response of the interface to a large variety of event rate combinations was verified with a hardware simulator and computed in Monte Carlo simulations. We found that the majority of the events transmitted by the interface (typically 300–400 s^{-1}) originate from an unanticipated substantial electron flux through the sensor mentioned by [18]. These electron events, along with other events that fall outside the TOF criteria for either H or the combination of C and O, are all fed into the CEU for selection. While electrons are readily identified based on their short or missing TOF signals and completely eliminated from ENAs, they still load the interface with a base traffic. However, because the ISN flow distributions contribute only a small fraction ($\lesssim$20%) to the total interface traffic, the effect on the shape of these distributions is very small. A change of the sensor mode in mid 2012 excludes the majority of these events that do not have any valid TOF signal by explicitly requiring the presence of at least the TOF between the two carbon foils. After that mode change the interface load has no noticeable effects on the data flow anymore

Figure 3, which shows two consecutive angular distributions of the ISN flow taken early February 2013 in the new IBEX-Lo mode (TOF2 required, left panel) and in the original normal mode (no TOF required, right panel), demonstrates the overall data flow reduction of the peak rate that is ≈20% reduced in the old mode, but no noticeable broadening of this distribution in comparison with the other distribution. Contrary to expectation, the width in the old mode, (σ after subtraction of the collimator function) appears even slightly smaller (by 0.32°), 0.13° of which may be attributed to the fact that this angular distribution is taken ≈1.3 days later when the ISN flow distribution is already slightly narrower (c.f. Figure 2). Also, simulations have demonstrated that in the old mode the potential increase in width of the observed angular distributions is limited to at most 1%, or to an apparent increase in the

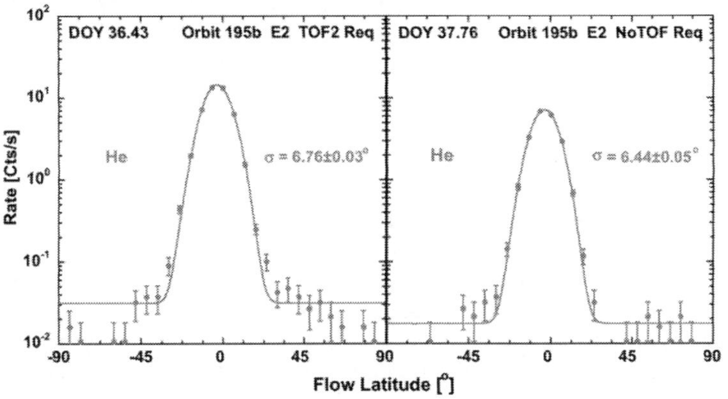

Fig. 3: Latitudinal distribution for the entire time period when IBEX-Lo was returned to the mode used 2009 to mid 2012 (right) together with an equally long time period in the new mode that removes all events with no valid TOF signal from the traffic into the CEU (left). Although added traffic tends to reduce the event rate (by ≈20%) the angular width is slightly smaller, 0.13° of which may be attributed to the 1.3 days later time.

Journal of Physics: Conference Series **577** (2015) 012019 doi:10.1088/1742-6596/577/1/012019

deduced temperature of <2%, as fully supported by the comparison in Figure 3. Therefore, any systematic instrumental effects on the temperature determination from IBEX observations larger than 2% can be excluded for the observations analyzed thus far.

In addition, we have tested whether broad non-uniform background, for example, from a secondary neutral helium component underneath the ISN flow distribution could lead to a broader combined distribution. The higher temperature observed by IBEX would require the addition of a component with at least 15% of the ISN flow peak height, which would be visible in large non-thermal wings that are not observed. Therefore, such an effect is also excluded.

Figure 4 shows the inflow speed and temperatures of He and the O and Ne combination obtained with IBEX [17, 18, 20] as a function of $\lambda_{ISM\infty}$ along the resulting narrow tube in LIC parameter space. T_{He} is now shown with only a 2% reduction over the width of the measured distribution, reflecting the maximum effect on the data flow, i.e. much less than the combined error bar. The lower reduction reflects the current understanding of the effect of the electron TOF events on the overall ISN signal. The temperatures of the different LIC species are in agreement within their respective error bars. This agreement provides another argument against substantial instrumental effects in the temperature because these would not likely lead to the observed isothermal result. The 1-σ range in $\lambda_{ISM\infty}$ as obtained in [18] and [17] along the parameter tube is indicated by open triangles, and the center value is indicated by the open squares. For comparison, the LIC parameters in [16] are also shown with their error bars. While the $V_{ISM\infty}(\lambda_{ISM\infty})$ combination falls at the edge of the IBEX range as reported before, the Ulysses temperature is substantially lower.

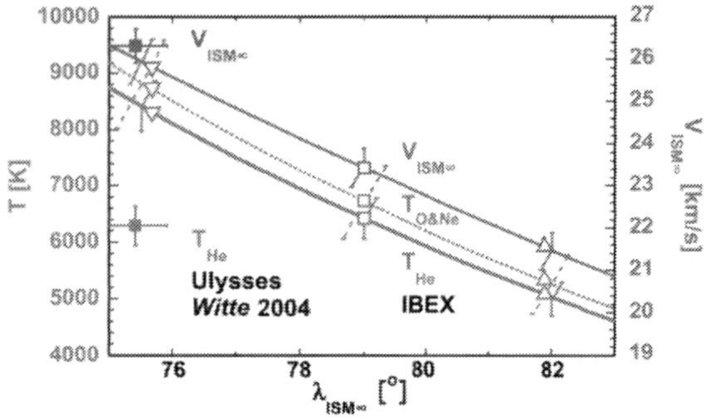

Fig. 4: ISN inflow speed $V_{ISM\infty}$ and temperatures T_{He} and $T_{O\&Ne}$ as a function of flow longitude $\lambda_{ISM\infty}$ obtained with IBEX [17, 18, 20] with error bars perpendicular to the parameter tube. The ranges along the tube are indicated by open triangles and the center value by open squares. Ulysses GAS values (solid squares) from [16] are shown in J2000 coordinates with the respective error bars.

5. Other Consequences and Future Perspectives

As has been demonstrated in the previous sections and in accordance with earlier results in [17, 18, 20], the IBEX ISN observations lead to a narrow tube of parameters for the LIC velocity vector and temperature, for which $V_{ISM\infty}$, $\beta_{ISM\infty}$, and T_{ISM} can be expressed as unique functions of $\lambda_{ISM\infty}$. Driven mostly by the observed variation of the flow peak in latitude as a function of observer longitude, a best fitting ISN flow longitude $\lambda_{ISM\infty} = 79°$ is found [20], although the range in longitude along the tube reaches all the way to the previously favored value $\lambda_{ISM\infty} = 75.4°$ (in J2000 coordinates [15, 16]). Interestingly, the flow longitude favored by the IBEX observations results in a LIC temperature T_{ISM} = 6300±360 K as also reported in [15, 16], while lower longitudes are tied to increasing temperatures, with T_{ISM} = 8300±450 K for $\lambda_{ISM\infty} = 75.4°$ as a consequence of the local Mach cone measurement. We have excluded instrumental effects and the influence of non-uniform backgrounds as potential reasons for observing substantially higher temperatures with IBEX than with Ulysses GAS. In addition, the IBEX angular distributions that are used to determine the temperature from their width are obtained with much superior counting statistics, and the ISN O+Ne temperature is the same within mutual error bars. Therefore, the relation found for the LIC temperature with IBEX appears very solid.

When putting these results into context, several intriguing possibilities arise. Firstly, the vector favored by the IBEX results appears to agree better with the LIC velocity vector in [42], who argued based on the local ISN velocity from [15, 16] and their cloud velocities that the heliosphere might be in a transition region between the LIC and the G-cloud. Secondly, together with a trend found over the entire space age, the possibility of a change in the interstellar flow vector over time, as proposed by [21] cannot be simply dismissed, as suggested in [25]. In such a scenario, the Ulysses GAS results by [16] and the IBEX observations would be consistent with sampling the same LIC temperature, but different velocity vectors, or encountering different portions of turbulent eddies at the edge of the LIC, where astronomical studies place the solar system [6, 7]. Alternatively, and this may be born out in more recent analyses of GAS observations [23, 24] and more extended IBEX observations, the local LIC flow vector could consolidate to values that are in agreement, but with higher LIC temperatures than previously found with GAS. LIC temperatures determined by absorption line measurements show values between ≈5800 and ≈10,000 K [26]. Because these represent a variety of line-of-sight averages, the local observation in the heliosphere could point to a locally higher temperature, perhaps, at the edge of the LIC.

An accurate determination of the pristine LIC flow vector is important also for another reason. Already [17] pointed out that a relatively subtle change in inflow longitude by ≈3° over the previous value has drastic consequences when taken in context with the inflow direction of ISN H as reported by [30, 43], which is deflected from the pristine flow direction due to the draping of the interstellar magnetic field B_{ISM} over the heliosphere [31-33]. According to global heliospheric modeling [32], the direction in the sky, into which the H flow is deflected, is determined by the orientation of the B_{ISM}-$V_{ISM\infty}$ plane. The relatively substantial change of the B_{ISM}-$V_{ISM\infty}$ plane orientation in the sky caused by a 3° change in $\lambda_{ISM\infty}$ is illustrated in the left panel of Figure 5. The most accurate determination of the B_{ISM} direction can be expected from the location of the IBEX ribbon, which has been interpreted within most models [27, 28, 44] as showing the locus in the sky where $\vec{B}_{ISM} \cdot \vec{r} = 0$ holds, preferably with solar wind involvement [45-47]. The most accurate determination of the ribbon center, which according to these models represents the up- or down-field direction in the sky, is from [29]. As discussed by [48] in terms of a secondary ENA model [45, 47], higher energy ribbon ENAs would probe larger distances from the Sun and thus should represent progressively more pristine interstellar conditions. The ribbon center as the B_{ISM} direction has been validated recently based on the streaming direction inferred from TeV cosmic ray anisotropies [49]. Shown in the right panel of Figure 5 are the ribbon centers for different ENA energies, along with the nose direction and the orientation of the B_{ISM}-$V_{ISM\infty}$ plane for the LIC velocity vectors from Ulysses GAS and IBEX. The Ulysses GAS result is more closely aligned with the expected B_{ISM} direction at the lower ENA energies, while the IBEX result agrees better with the direction from higher energies. Therefore, an accurate determination of the LIC inflow vector will be of utmost importance for probing the outer heliosheath where B_{ISM} controls the flow pattern around the heliosphere.

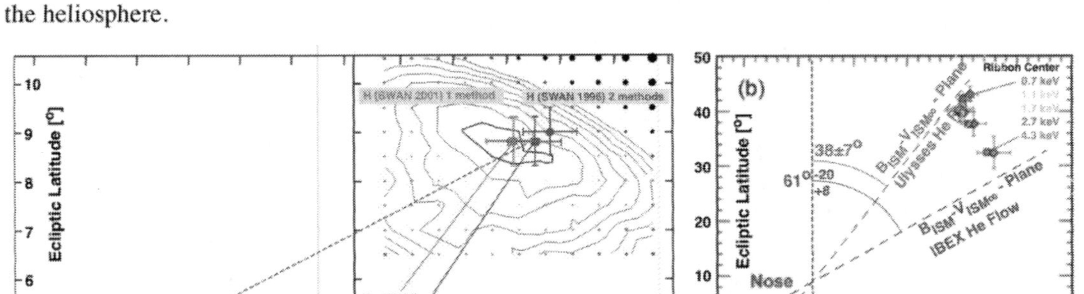

Fig. 5: Left Panel: Flow direction of ISN H from SOHO SWAN [43] and of He from Ulysses GAS (red [16]) and IBEX (blue [17, 18] along with projected B_{ISM}-$V_{ISM\infty}$ - planes (adapted from [43]). Right Panel: IBEX ribbon center locations in the sky for different ENA energies [29] along with the heliospheric nose and projected B_{ISM}-$V_{ISM\infty}$ - planes based on Ulysses and IBEX observations.

To provide improved leverage to IBEX observations in breaking the inherent degeneracy along the narrow LIC parameter tube, the IBEX team has started with a variation in the pointing of the IBEX spin axis to substantial negative latitudes, as a means to uniquely identify where in the four-dimensional parameter space tube the ISN flow actually lies. Figure 6 shows how the observed ISN flow peak in latitude at a given observer longitude varies as a function of inflow longitude for different IBEX spin axis orientations. Generally, a spin axis pointing below the ecliptic (more negative values in e_z) results in a steeper functional behavior and thus provides stronger leverage on the determination of the ISN inflow longitude. In 2014, the observations were made with a spin axis orientation of ≈-5°, interspersed with an axis orientation of ≈0°. The new observations are currently under evaluation.

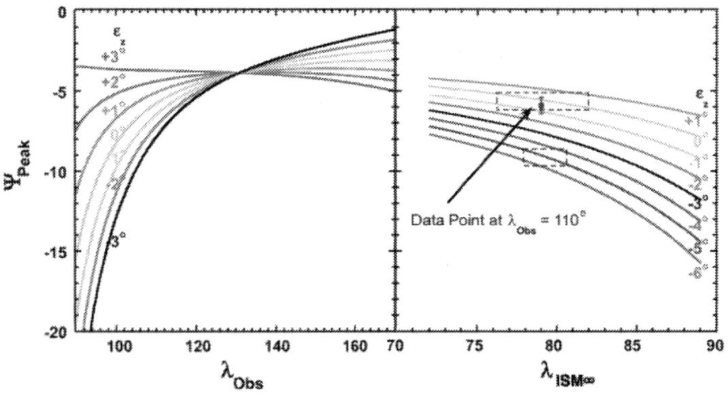

Fig. 6: ISN Flow peak location in latitude Ψ_{Peak} as a function of observer longitude λ_{Obs} (left) and as a function of $\lambda_{ISM\infty}$ for λ_{Obs} = 110° (right) with different spin axis orientations ε_z. The ability to discern $\lambda_{ISM\infty}$ by exploiting the variation of Ψ_{Peak} with λ_{Obs} is improved with progressively steeper functional dependence for more negative spin axis pointing. Included in the right panel are a typical data point with error bar and resulting error boxes in $\lambda_{ISM\infty}$ for ε_z = 0° (2012&13) and -5° (2014).

6. Conclusions

IBEX observations of the ISN He flow during early winter each year provide a versatile tool to determine the physical parameters of the LIC, with the potential for continuing improvement with multiple data sets and varying observation strategies. Multiple year observations have confirmed the narrow tube of LIC parameters $V_{ISM\infty}$, $\beta_{ISM\infty}$, and $T_{ISM}(\lambda_{ISM\infty})$ with small error bars perpendicular to these functional dependencies and a larger uncertainty around the center value of $\lambda_{ISM\infty} = 79^{+3}_{-3.5}{}^{o}$ (c.f. Fig. 4). In particular, the T_{ISM} values have been solidified in a detailed analysis of the IBEX data path, which has shown that the observed angular distributions may only be widened by ≤1% (or T_{He} by ≤2%) in the worst-case scenario. Therefore, the He temperature found from IBEX observations is ≈30% higher than that reported in [16] when adopting the same ISN flow vector. Recent re-analyses of the Ulysses GAS observations point to an underestimate of the temperature in the earlier analysis [23, 24].

We have combined the Ulysses GAS and the IBEX ISN flow vectors with the determination of the H ISN flow from SOHO SWAN observations as shown in [43] and compared the resulting B_{ISM}-$V_{ISM\infty}$ planes with the B_{ISM} direction as derived from the IBEX ribbon center, following models that connect the ribbon location to the locus where $\vec{B}_{ISM} \cdot \vec{r} = 0$ is satisfied. Then the plane derived with the Ulysses direction is nearly in line with the ribbon center at 0.7 keV, while the IBEX direction agrees better with the ribbon center at the highest energy 4.2 keV, as derived in [29]. In terms of a secondary ENA model for the ribbon, the center at higher energies would come closer to a B_{ISM} direction at larger distances [48], i.e. supposedly under more pristine (undraped) conditions. This presents one more reason to derive the ISN flow vector with higher precision. To achieve this goal, the IBEX spin axis orientation has been turned to larger negative latitude, which promises a better sensitivity of the observations to the inflow longitude, which is still most uncertain in the IBEX analysis.

Acknowledgments: We thank all the men and women who made IBEX a reality. This work was carried out as a part of the IBEX mission, funded as a part of NASA's Explorers Program. M.B., M. K., and J. S. are supported by grant 2012-06-M-ST9-00455 of the Polish National Science Center.

References

[1] Fahr, H., Kausch, T. & Scherer, H. 2000, *A&A* **357**, 268.
[2] Zank, G.P., & Müller, H.R., 2003, *JGR* **108**, 1240.
[3] Alexashov, D. & Izmodenov, V. 2005, *A&A* **439**, 1171.
[4] Müller, H., Florinski, V., Heerikhuisen, J., et al. 2008, *A&A* **491**, 43.
[5] Zank, G., Pogorelov N., Heerikhuisen J., et al. 2009, *Space Sci. Rev.* **146**, 295.
[6] Cheng, K, & Bruhweiler, F., 1990, *ApJ* **364**, 573.
[7] Slavin, J., & Frisch, P., 2002, *ApJ* **565**, 364.
[8] Frisch, P. C., 2006, *Solar Journey*, vol. 338, Dordrecht: Springer.
[9] Bertaux, J., & Blamont, J., 1971, *A&A* **11**, 200.
[10] Weller, C., & Meier, R., 1974, *ApJ* **193**, 471.
[11] Möbius, E., Hovestadt, D., & Klecker, B., 1985, *Nature* **318**, 426.
[12] Gloeckler, G., Geiss, J., Balsiger, H., et al., 1992, *A&A* **92**, 267.
[13] Witte, M., Banaszkiewicz, M., & Rosenbauer, M., 1996, *Space Sci. Rev.* **78**, 289.
[14] Möbius, E., Bochsler, P., Bzowski, M., et al. 2009a, *Science* **326**, 969.
[15] Möbius, E., Bzowski, M., Chalov, S., et al. 2004, *A&A* **426**, 897.
[16] Witte, M., 2004, *A&A* **426**, 835.
[17] Bzowski, M., Kubiak, M., Möbius, E., & et al., 2012, *ApJS* **198**, art ID 12.
[18] Möbius, E., P. Bochsler, M. Bzowski, et al., 2012, *ApJS* **198**, art ID 11.
[19] Schwadron, N. A., Allegrini, F., Bzowski, M., et al. 2011, *ApJ* **731**, 56.
[20] McComas, D.J., Alexashov, D., Bzowski, M., et al., 2012, *Science* **336**, 1291.
[21] Frisch, P. C., Bzowski, M., Livadiotis, G., et al., 2013, *Science* **341**, 1080.
[22] Katushkina, O., A., Izmodenov, V.V., Wood, B.E., McMullin, D.R., *ApJ*, in press, 2014.
[23] Bzowski, M., et al., 2014, *ApJ*, subm..
[24] Wood, B., Mueller, H.-R., & Witte, M., 2014, *ApJ*, subm.
[25] Lallement, R., & Bertaux, J.L., 2014, *A&A* **565**, id.A41..
[26] Frisch, P.C., Bzowski, M., Drews, C., et al., 2014, *ApJ*, subm..
[27] McComas, D. J., Allegrini, F., Bochsler, P., et al. 2009a, *Science* **326**, 959.
[28] Schwadron N A, Bzowski M, Crew G B, et al., 2009, *Science* **326,** 966.
[29] Funsten, H., R. DeMajistre, P. C. Frisch, et al., *ApJ* **776**, Art ID 30, 2013.
[30] Lallement, R., Quemerais, E., Bertaux, J. L., et al., 2005, *Science* **307**, 1447.
[31] Izmodenov V.V., Alexashov D.B., Myasnikov A.V., 2005, *A&A* **437**, L35.
[32] Pogorelov, N. V., Heerikhuisen, J., Mitchell, J. J., et al., 2009, *ApJ. Lett.* **695**, 31.
[33] Opher, M., Stone, E. C. , Gombosi, T., 2007, *Science* **316**, 875.
[34] McComas, D. J., Allegrini, F., Bochsler, P., et al. 2009b, *Space Sci. Rev.* **146**, 11.
[35] Funsten, H. O., Allegrini, F., Bochsler, P., et al. 2009, *Space Sci. Rev.* **146**, 75.
[36] Fuselier, S., Bochsler, P., Chornay, D., et al., 2009, *Space Sci. Rev.* **146**, 117.
[37] Wurz, P. S.A. Fuselier, E. Möbius, et al., 2009, *Space Sci. Rev.* **146**, 173.
[38] Hotop, H., & Lineberger, W.C., 1984, *J. Phys. Chem. Ref. Data* **14**, 731.
[39] McComas, D.J., Carrico, J.P., Hautamaki, B., et al., 2011, *Space Weather* **9**, CiteID S11002.
[40] Lee, M., Kucharek, H., Möbius, E., et al., 2012, *ApJS* **198**, art ID 10.
[41] Möbius, E., Kucharek, H., Clark, G., et al. 2009b, *Space Sci. Rev.* **146**, 149.
[42] Redfield, S., & Linsky, J., 2008, *ApJ* **673**, 283.
[43] Lallement, R., Quemerais, E., Koutroumpa, D., et al., 2010, *AIP Conf. Proc.* **1216**, 555.
[44] McComas, D.J., Lewis, W.S., & Schwadron, N.S., 2014, *Rev. Geophys.* **52**, doi:10.1002/2013RG000438.
[45] Heerikhuisen, J., Pogorelov, N. V., Zank, G. P., et al., 2010, *ApJ Lett.* **708**, L126.
[46] Kucharek, H., Fuselier, S.A., Wurz, P., et al., 2013, *ApJ* **776**, art ID 109..
[47] Schwadron, N. A. & McComas, D. J., 2013, *ApJ* **764,** 92.
[48] Möbius, E., K. Liu, H. Funsten, S. P. Gary, D. Winske, 2013, *ApJ* **766**, art ID 129.
[49] Schwadron, N. A., Adams, F. C., Christian, E. R., et al., 2014, *Science* **343,** 988.

Journal of Physics: Conference Series **577** (2015) 012020 doi:10.1088/1742-6596/577/1/012020

Suprathermal ions in the solar wind from the *Voyager* spacecraft: Instrument modeling and background analysis

B M Randol and E R Christian

NASA Goddard Space Flight Center, Greenbelt, MD 20771, USA

E-mail: brentrandol@gmail.com

Abstract. Using publicly available data from the *Voyager* Low Energy Charged Particle (LECP) instruments, we investigate the form of the solar wind ion suprathermal tail in the outer heliosphere inside the termination shock. This tail has a commonly observed form in the inner heliosphere, that is, a power law with a particular spectral index. The *Voyager* spacecraft have taken data beyond 100 AU, farther than any other spacecraft. However, during extended periods of time, the data appears to be mostly background. We have developed a technique to self-consistently estimate the background seen by LECP due to cosmic rays using data from the *Voyager* cosmic ray instruments and a simple, semi-analytical model of the LECP instruments.

1. Introduction

The solar wind is a plasma consisting of electrons, protons, and other ions that flows out from the Sun at all times. During nominal conditions, or "quiet times," the core of the velocity distribution of solar wind protons is roughly Maxwellian, with thermal temperatures on the order of 10 eV. Outside this Maxwellian core, the velocity distribution exhibits power law behavior beginning at around several keV in the solar wind frame [1].

The spectral index of this power law, γ, is observed to have a common value of 1.5 when expressed as $j \propto E^{-\gamma}$, where j is the differential number flux and E is the kinetic energy [2, 3, 4]. Other analyses show different values [5], with an apparent dependence on solar cycle [6].

Many theories exist for the formation of power laws, typically falling into the two categories of generalized shock acceleration [7, 8, 9] and stochastic acceleration [10, 11, 12, 13, 14, 15], though magnetic reconnection can also produce power laws [16, 17]. These mechanisms can of course operate in concert, and there is some evidence of this for stochastic and shock acceleration [18]. Many theories try to specifically address the formation of the common spectral index [19, 20, 21, 22, 23, 17]. More data is needed to help determine which of these theories best predicts the real world.

2. Data

The *Voyager* spacecraft were launched in 1977 and have been traveling outward through the heliosphere ever since, collecting useful solar wind data. In this study of suprathermal ions, we analyze data recorded by the Low Energy Charged Particle (LECP) experiment [24]. LECP consists of two subsystems, the Low Energy Particle Telescope (LEPT) and the Low Energy

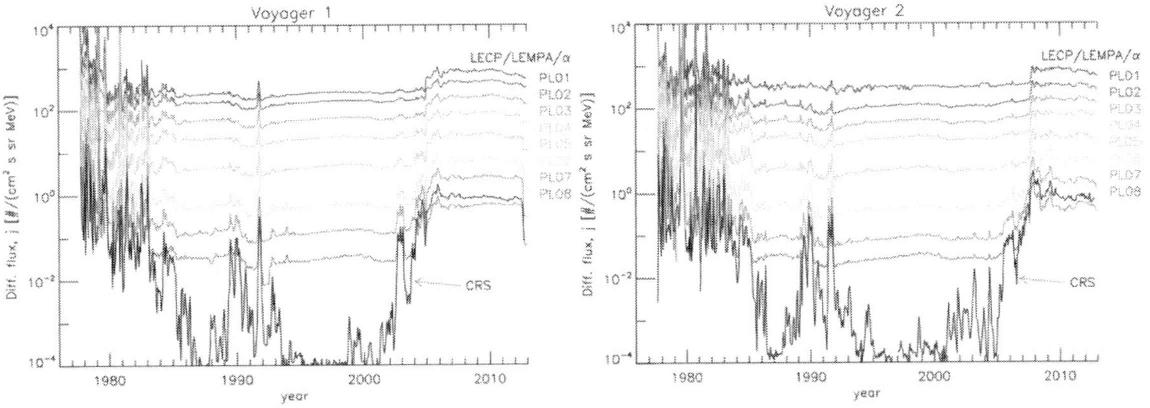

Figure 1. Measured 27-day, direction-averaged differential particle fluxes from the eight PL channels from the *Voyager 1* (left) and *Voyager 2* (right) LECP/LEMPA/α detector (denoted by different colored lines) and 27-day ~2.2 MeV differential fluxes from CRS (black lines).

Magnetospheric Particle Analyzer (LEMPA), each of which has several detectors. Here we focus on the LEMPA/α detector because of its lower low-energy threshold and less complicated priority scheme as compared to LEPT.

However, analyzing LECP/LEMPA/α data is challenging because it is sensitive to high-energy particles. Figure 1 shows the *Voyager* LECP/LEMPA/α measured differential fluxes in eight energy channels averaged over look-angle (refered to as PL01, PL02, etc.) and averaged over 27-day time intervals, along with differential fluxes measured by the *Voyager* Cosmic Ray System (CRS) [25]. The mean energy of the CRS energy channel (2.2 MeV) falls between the mean energies of the PL07 and PL08 channels (1.5 and 2.7 MeV, respectively). Hence the fluxes of the three channels should mostly track each other, and indeed they do during times of high fluxes. However, at other times, most noticeably in the late 1980s and mid-to-late 1990s, they differ by many orders of magnitude. In fact, during these times, the last channel of LECP is correlated strongly with the last CRS channel, which has a mean energy higher by nearly two orders of magnitude (170 MeV).

A strong correlation also exists between the LECP/LEMPA/α data and data from a channel of the LECP/LEMPA/β detector, referred to as EB05 [24]. Because the β detector faces the walls of the instrument, it is not sensitive to low-energy ions. However, both the α and β detectors are sensitive to high-energy particles that penetrate the walls of the instrument. Figure 2 shows the 27-day and look-angle averaged EB05 count rate vs. the 27-day PL01-PL08 differential fluxes. Notice that the correlation between the two data sets is nearly perfect when the differential flux in each channel is at its lowest, which agrees well with our comparison of the CRS and LECP/LEMPA/α data. This correlation has been noted before, and the background due to penetrating particles was estimated for a few specific time periods through an analysis of the LECP/LEMPA/α anisotropy data [26].

Here, we take a different approach: construct a mathematical model of the instrument's response to high-energy particles, fit semi-empirical functions to measured cosmic ray fluxes, and convolve the two to determine the contribution of cosmic rays to the measured flux of LECP.

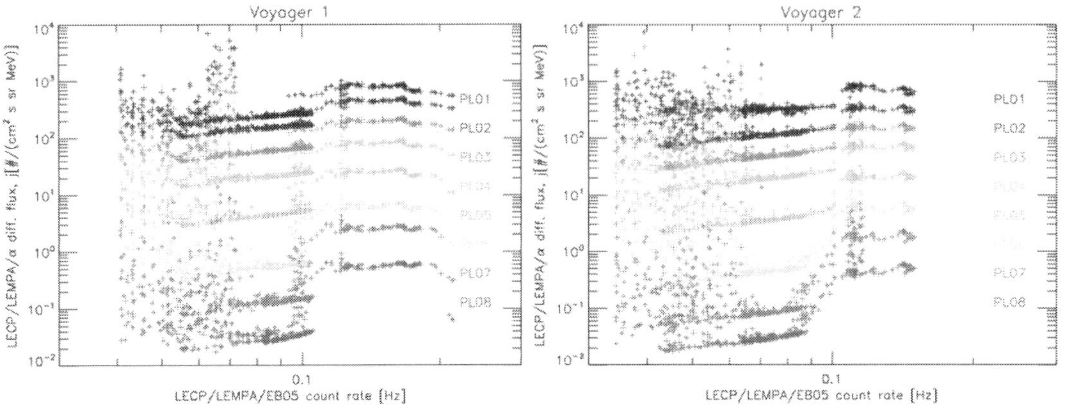

Figure 2. Measured 27-day differential particle fluxes from the eight PL channels (from the LECP/LEMPA/α detector), averaged over look-angle vs. EB05 count rate (from the LECP/LEMPA/β detector). The colors represent the different channels.

3. Analysis

To compare with data, we must compute j_i, the differential number flux at each energy channel, i. j_i is defined through the relation $j_i = C_i/G_i$, where G_i is further defined by the relation $G_i = G'(b_i' - a_i')$. G' is the calibrated geometric factor of each detector, and a_i' and b_i' are the calibrated lower and upper limits of each energy channel, respectively. We take the values of G', a_i', and b_i' from [27]. C_i, the count rate, is found generally through the following equation

$$C_i = \sum_s \int \epsilon_{is}(E, \mathbf{\Omega}, \mathbf{x}) j_s(E, \mathbf{\Omega}) d\mathbf{x} d\mathbf{\Omega} dE, \tag{1}$$

where $\epsilon_i(E, \mathbf{\Omega}, \mathbf{x})$ is the detector efficiency for a given energy channel i, as a function of incoming particle energy E, species s, incoming particle direction $\mathbf{\Omega}$, and incident particle position $\mathbf{x}$.

We use spherical coordinates ($d\mathbf{\Omega} = d\phi \sin\theta d\theta$) and make several assumptions to this very general equation for mathematical and conceptual simplicity:

(i) Axisymmetry, so as to make the ϕ integral trivial
(ii) Incoming flux is isotropic in the instrument frame of reference, such that $j(E, \mathbf{\Omega}) = j(E)$
(iii) Incoming flux is from protons.

Axisymmetry is needed for mathematical tractability in the current analysis but could potentially be relaxed in a future study. For the second assumption, the largest source of anisotropy is the shifting of particle distributions from the solar wind frame to the measured frame, an effect that decreases with increasing energy. For typical solar wind bulk energies (of order 1 keV), the effect on a power law spectrum with the common spectral index in the solar wind frame is completely negligible by 1 MeV. Also, the CRS instrument has multiple apertures that measure different look directions at the same energy. The similarities in differential flux for the different apertures demonstrate that isotropy is a good assumption at these energies ($\sim$2 and $\sim$5 MeV) for most time periods. Since we are focusing in this analysis on the effects of high-energy penetrating particles, we can safely assume isotropy. The last assumption of just protons has been investigated, and is a very good approximation for the times considered. We will re-examine the assumptions of isotropy and just protons in a future study.

Journal of Physics: Conference Series **577** (2015) 012020 doi:10.1088/1742-6596/577/1/012020

With the above simplifications, C_i becomes

$$C_i = \int R_i(E) j(E) dE, \tag{2}$$

where $R_i(E)$ is the energy response function, which we define as an integral of the efficiency over position and look-direction. We also assume that the size of the detector is small relative to the entire instrument, such that we can perform the position integral separately, obtaining a separable active area, $A(\theta)$. For our simple calculation, we assume $A(\theta) = A_f|\cos\theta| + A_s \sin\theta$, where A_f is the surface area of the face of the detector, and $A_s = 2\pi w l_0$, where l_0 is the thickness of the detector at normal incidence and $w = \sqrt{A_f/\pi}$ is the radius of the detector (assuming the shape of the detector is a circular cylinder). l_0 was determined to be 89.1 μm for *Voyager 2* and 96.5 μm for *Voyager 1* [27]. The entire active area, $A(\theta)$, is shown in Figure 3a (solid line), along with the contribution from just the side of the detector, $A_s \sin\theta$ (dotted line). Also shown is the active area, $A(\theta)$, multiplied by the angular weighting factor, $\sin\theta$ (dashed line).

An expression for the energy response function is then

$$R_i(E) = 2\pi \int_{\theta_i^a(E)}^{\theta_i^b(E)} \sin\theta \, A(\theta) d\theta, \tag{3}$$

where $\theta_i^a(E)$ and $\theta_i^b(E)$ are found through $\Delta^{\mathrm{Si}}(E, \theta_i^{\{a,b\}}) = \{a_i, b_i\}$. $\Delta^{\mathrm{Si}}(E,\theta)$ is the energy deposited in the detector, and a_i and b_i are the electronic thresholds, which differ from a_i' and b_i' at low energies due to the combined pulse-height defect. The electronic trigger thresholds for the first few channels are given in [27]. The rest we assume to be the same as the calibrated limits.

Particles lose energy as they pass through material, and there are several ways to estimate this process. One way is by taking the average energy lost, $\overline{\Delta}$, divided by the material's thickness, l, a ratio usually referred to as the stopping power. Theoretically, this mean value is derived from a probability distribution of energy loss, $P(\Delta)$, where Δ is an arbitrary value of energy loss. However, this energy loss distribution is in general not symmetric about its most probable value, Δ', with a tail extending to high Δ. For a given l, $P(\Delta)$ is more asymmetric at higher energies than at lower energies; for large thicknesses and/or lower energies, $P(\Delta)$ approaches a Gaussian [28]. Here, we use the most probable value as given by the Landau-Vavilov-Bichsel (LVB) equation, Equation (30.11) in [29], altering it slightly by limiting the total energy loss to E, the initial energy of the particle. With this modification Δ' behaves thusly (neglecting shielding for the moment): it equals E until reaching the energy at which particles begin to pass through the detector, decreases with E to its "minimum ionizing energy," and then increases more slowly to a constant level (see [29] and references therein). The LVB equation does not compare well with experiments at low energies due to its simplified stopping cross-section. However, we chose to use it for this analysis due to its simple analytical form. Other estimates of the differential energy loss that are more accurate at low energies are not analytical, such as GEANT4 [30], which we will consider using to find the instrument response in a future study.

The instrument consists of both the detector itself and all the shielding and structural material surrounding it. Figure 3b shows the thickness of each material used in our model as a function of polar angle, θ, measured relative to the boresight of the detector. For simplicity, we estimate the complex shielding and structure by a single 3 mm thick layer of W for $\theta > \theta'$, where $\theta' = 22.5°$ is the half-angle of the aperture of the instrument. This approximation is reasonable since W has a much larger stopping power than any other component. Thicknesses of the Si detector and the Al foil, which acts as a light barrier, are obtained from [27]. The thicknesses in Figure 3b depend on θ due to simple geometric effects: surfaces that face perpendicular to the boresight,

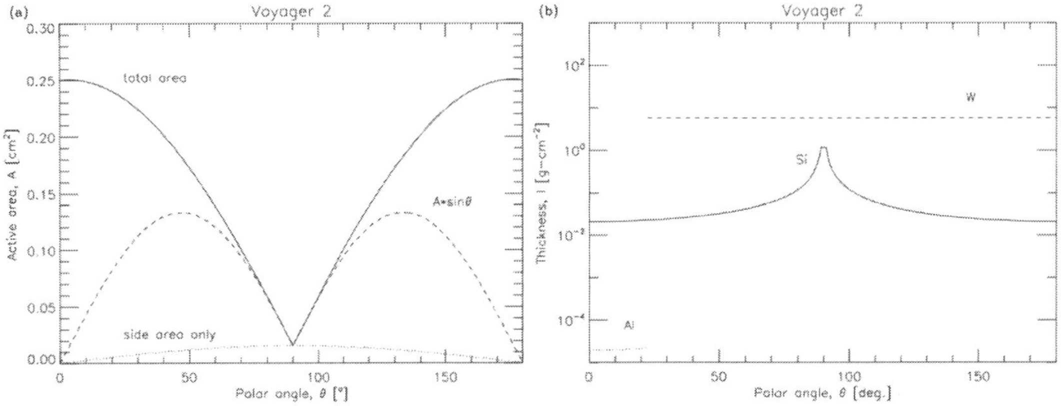

Figure 3. Active area (left) and thicknesses (right) of various materials of the *Voyager 2* LECP/LEMPA/α detector as a function of polar angle, θ, measured from the boreisght of the instrument. Calculations of the same quantities but for *Voyager 1* are very similar; see text for the differences.

like the face of the detector, vary as $|\cos\theta|^{-1}$, and parallel faces, in this case just the side of the detector, vary as $(\sin\theta)^{-1}$. Note that the detector is parallel-facing between the angles θ_p and $\pi - \theta_p$, where $\theta_p = \arctan(w/l_0)$. We neglect any thickness of the Al outside the aperture as well as any shielding by the body of the spacecraft itself and the other instruments, again for conceptual and mathematical simplicity (and because the model works well without this level of detail in the shielding).

Using our shielding and detector thickness estimates as a function of polar angle, θ, we first calculate the energy lost in the shielding (either Al or W), then if the particle has enough energy to have made it through the shielding, we calculate the energy deposited in the Si detector using the particle's remaining energy. Figure 4 shows $\Delta^{\mathrm{Si}}(E,\theta)$ as a color map. The shape of the map can be understood through the variation of shielding and detector thickness with θ. For $\theta < \theta'$, the only shielding is a thin layer of Al, so particles pass through it with almost no energy loss except at the very lowest energies. For $\theta > \theta'$, low-energy particles are slowed significantly by the shielding, causing a cut-off around 100 MeV, while at the same time depositing the most energy around $\theta = 90°$ because that's where the detector is thickest. At high energies (>1 GeV) particles at all angles are not slowed much by the shielding, so the variation is completely due to the detector thickness.

From our calculation of Δ^{Si}, we find $\theta_i^{a,b}$ and numerically calculate the integral in Equation 3. The result is shown in Figure 5, normalized by G, the ideal geometric factor of the instrument calculated with the following

$$G = 2\pi \int_0^{\theta'} \sin\theta\, A(\theta) d\theta. \tag{4}$$

For $A(\theta) \approx A_f \sin\theta\cos\theta$ and $\theta' = 22.5°$, G is approximately equal to 0.115 cm^2-sr. Note that the *Voyager* calibrated geometric factors, G', are slightly lower, 0.04 and 0.113 cm^2-sr for *Voyager 1* and *Voyager 2*, respectively (see [27] for more information). In Figure 5, different channels are represented by different colors as denoted in the figure (the color scheme is the same as in Figures 1 and 2). At low E, the normalized response functions are box-like, with lower and upper limits occurring roughly at a_i' and b_i', respectively, which are shown as dotted lines in the figure. At the lowest energies, the responses are reduced in width and shifted to higher E as compared to the calibrated limits, a_i' and b_i' due to energy losses in the Al layer. The LVB

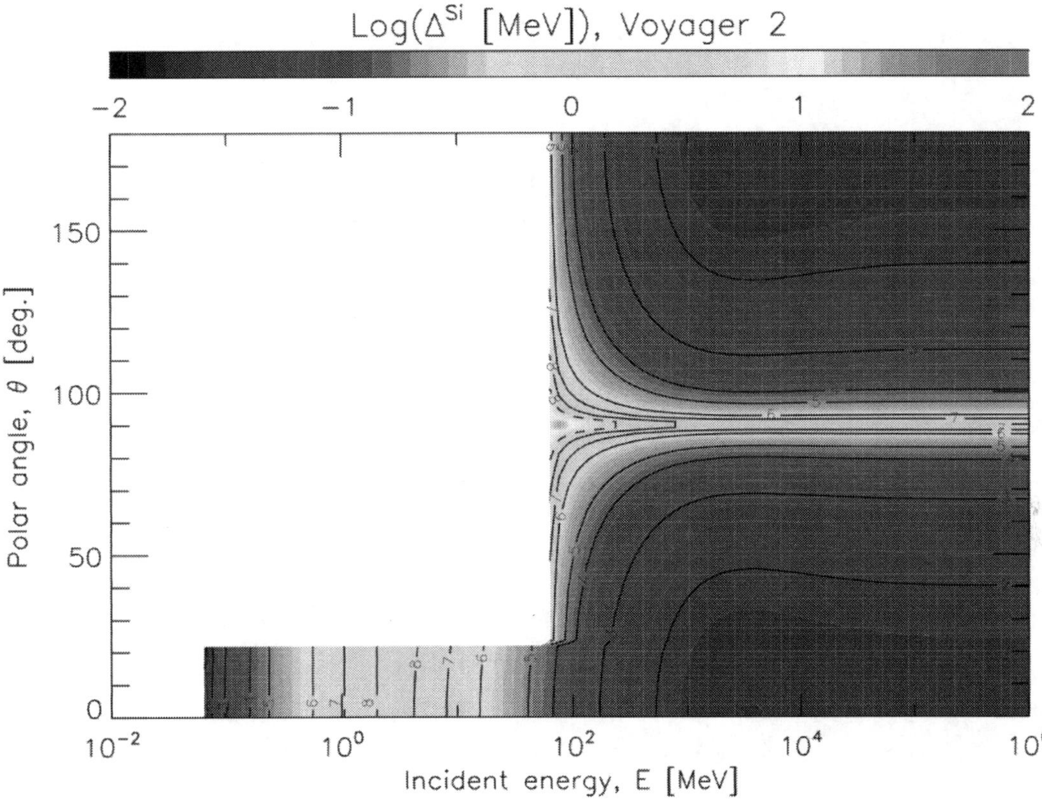

Figure 4. Energy deposited in the *Voyager 2* LECP/LEMPA/α detector as a function of incoming energy and polar angle for protons. Also shown are the channel thresholds (solid lines represent a_i and the dashed line represents b_8).

equation is not as accurate for very thin targets [29], and so energy loss in the Al is probably overestimated in this analysis. Beyond $E \approx b_8$, the response continues with channels in reversed order and with the same height. Then, depending on the channel, the response rises about a decade, goes through a local minimum, and finally levels off to a value inversely proportional to the channel number. The much larger response beyond ~100 MeV is due to the larger angular acceptance of the instrument at these energies.

The next step of the analysis is to estimate a model differential flux, j. Our chosen model spectrum consists of three components: a power law with an exponential roll-over due to suprathermal solar wind ions, a modulated power law due to galactic cosmic rays (GCRs), and another modulated power law due to anomalous cosmic rays (ACRs). For the suprathermal power law, we adopt a form equivalent to that given in [3], $j = c_0(E\ [\mathrm{MeV}])^{-\gamma_0} \exp(-(E/E_0)^{\alpha_0})$, where E is particle kinetic energy, γ_0 is the spectral index at low energies, c_0 is the normalization constant, E_0 is the rollover energy, and α_0 is the rollover sharpness index. For the GCR component, we use the "force-field" approximation [31] to modulate the spectrum: $j(E) = ((E + M)^2 - M^2)((E + M + |z|\Phi)^2 - M^2)^{-1} j_\infty(E + |z|\Phi)$, where $|z|$ is the absolute value of the particle's electric charge relative to the proton charge, Φ is the effective potential of modulation, and M is the particle's rest energy. Since we are just considering protons, z and M are 1 and 938 MeV, respectively. $j_\infty(E)$ is the spectrum at the boundary,

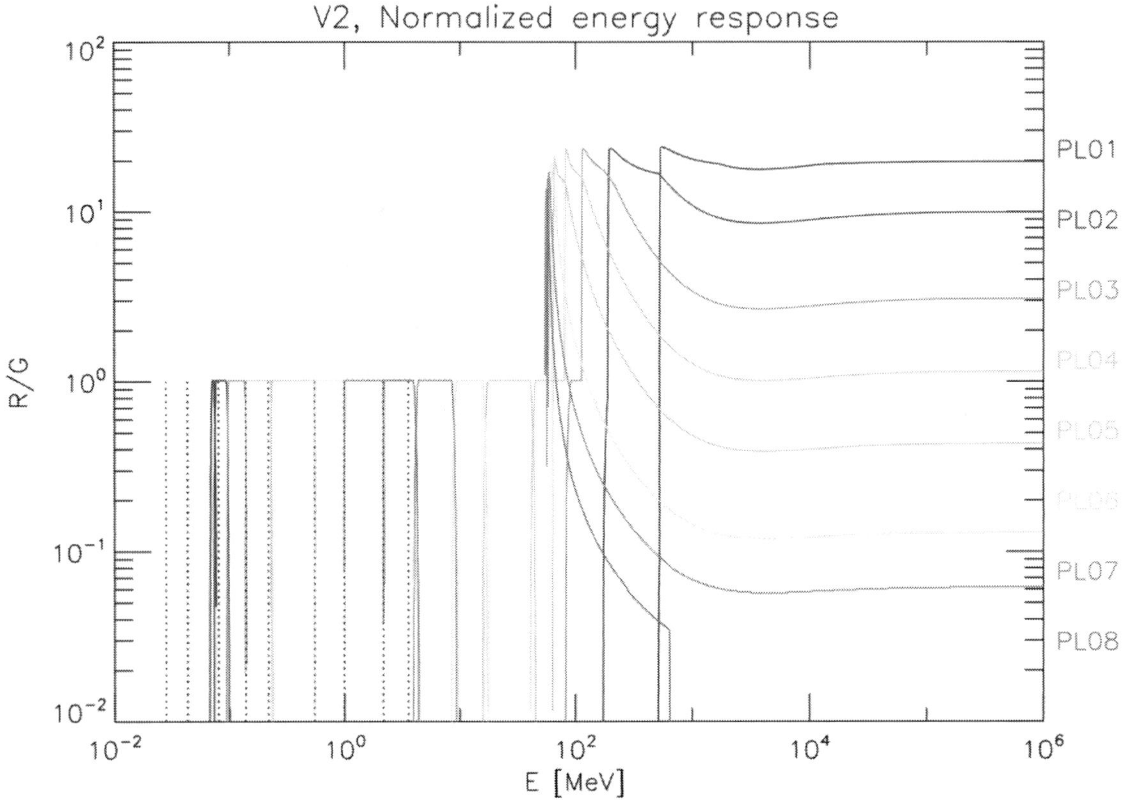

Figure 5. Response of the *Voyager 2* LECP/LEMPA/α detector, found through Equation 3. Color corresponds to energy channel. Dotted lines are calibrated energy limits, a'_1 through a'_8 and b'_8

the form of which for this analysis we assume to be a triple power law [31]: $j_\infty(E) = c_1(E\ [\mathrm{MeV}])^{-\gamma_1}(1 + c_2(E\ [\mathrm{MeV}])^{-\gamma_2} + c_3(E\ [\mathrm{MeV}])^{-\gamma_3})^{-1}$. For the ACR component, we use the diffusion-convection approximation to modulate the spectrum inside the termination shock [31], $j(E) = j_{TS}(E)\exp(-c_6\left((E+M)((E/M+1)^2-1)^{1/2}\right)^{-1})$, and a power law with an exponential-type rollover at high energies for the spectrum at the termination shock [31], $j_{TS}(E) = c_5(E\ [\mathrm{MeV}])^{-1}\exp(-c_4(E/62\ \mathrm{MeV})^{\gamma_4})$.

4. Results and discussion

Figure 6 shows as "x" symbols differential fluxes from *Voyager 1* LECP/LEMPA/α in viewing sector 2 (see e.g. [32] for viewing direction definitions) averaged over the entire year of 1999, plotted vs. logarithmic mean channel energy, $E_i = \sqrt{a'_i b'_i}$ The horizontal error bars show a'_i and b'_i. The open circles represent data from *Voyager 1* CRS averaged over the year 1999 [33]. Filled circles are year-averaged LEPT proton fluxes. Open diamonds represent CRS proton data as available on CDAWeb. The open squares represent data from instruments at 1 AU, as summarized in [34]. Although this data is taken at 1 AU and not during the same period, solar modulation has little effect at these energies. The dashed lines represent a fit of our model to the data, found with the parameters given in Table 1. The solid line represents j_i, our calculation of the spectra measured by LECP, given j, the model functions. Finally, the dotted line represents

Table 1. Parameters for the model fluxes. c_0, c_1, and c_5 have units of particles/(MeV-cm^2-sr-s), E_0 has units of MeV, and Φ has units of MV. All other constants are unitless.

	V1	V2
j_0	9.5×10^{-4}	9.5×10^{-4}
γ_0	1.5	1.5
E_0	2.6	2.6
Φ	2.0×10^{2}	2.0×10^{2}
c_1	5.3×10^{5}	5.3×10^{5}
γ_1	2.8	2.8
c_2	1.3×10^{4}	5.6×10^{4}
γ_2	1.0	1.3
c_3	2.8×10^{8}	2.8×10^{8}
γ_3	2.9	2.9
c_4	1.9	1.9
γ_4	2.5	2.5
c_5	1.0	1.0
c_6	8.0×10^{2}	8.0×10^{2}

the calculated flux of LECP for $E > 4.0$ MeV (not visible under solid line).

For illustrative purposes, we also show, as solid colored lines, the normalized response, R/G', multiplied by the input differential flux, j, for each energy channel of the detector (PL01-08), with the same color scheme as in Figures 1, 2, and 5. These lines represent the integrand of equation 2, with the calibrated geometric factor, G', factored out in order to show that at the lower energies, R/G' times j equals the incident model flux. Naturally, these curves rise at a certain energy, just as the response does, but then quickly fall, following the trend of the input model flux. Note that, although the lowest energy portions of the integrand are comparable to or higher than the higher energy portions, they contribute very little to the total expected flux in each channel. This is beacuse it is the area under each curve that matters, and the width of the lower energy portions is extremely small compared to the widths of the curves at higher energies (the horizontal axis is logarithmic).

Using a simple model for the instrument and a simplified formula for the energy deposition of protons in the shielding and detector, combined with a realistic model for the flux of cosmic rays, we have shown that the basic shape and overall level of flux from *Voyager* LECP/LEMPA/α is reproduced. Thus, we have shown, through our physics-based approach, that cosmic rays do indeed cause the background in LECP. However, the detailed shape is not matched exactly. To match the shape exactly, a more accurate but complex description of the instrument and a more accurate but non-analytical method of determing the energy deposition in the shielding and detector may be needed. With these improvements to the model, we hope to remove the cosmic ray background to estimate the flux of suprathermal particles in the heliosphere and beyond. In a future study, we plan to compare our method with existing methods, most notably an empirical method that uses the LEMPA/β detector data.

Acknowledgments

We thank the members of the *Voyager* and LECP teams, without which this work would not have been possible. We especially thank Rob Decker for helpful discussions. BMR was supported by an appointment to the NASA Postdoctoral Program at the Goddard Space Flight Center,

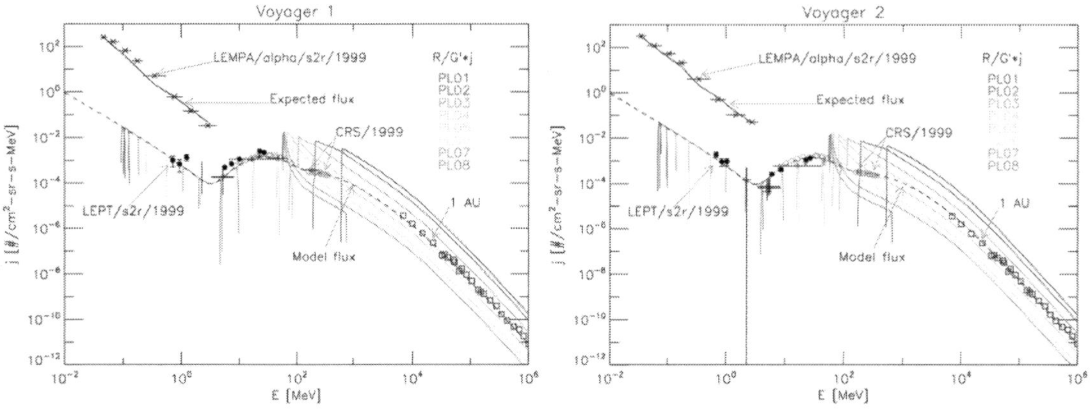

Figure 6. Differential flux vs. kinetic energy. "x" symbols are *Voyager 1* (left) and *Voyager 2* (right) LECP/LEMPA/α data from viewing sector 2 averaged over the entire year of 1999. Open circles are proton fluxes from *Voyager 1* CRS as presented in [33]. Open squares represent proton data taken at 1 AU, as presented in [34]. Filled circles are year-averaged LEPT proton fluxes. Open diamonds represent CRS proton data as available on CDAWeb. The dashed line is the model input flux. The solid black line is the model output flux. The solid colored lines represent the normalized response (R/G') multiplied by the incident flux for each channel, PL01-08. The dotted line (not visible under solid black line) is the model output flux corresponding to the input flux between 4 MeV and 1 TeV.

administered by Oak Ridge Associated Universities through a contract with NASA.

References

[1] Gosling J T, Asbridge J R, Bame S J, Feldman W C, Zwickl R D, Paschmann G, Sckopke N and Hynds R J 1981 *J. Geophys. Res.* **86** 547–554

[2] Gloeckler G, Fisk L A, Mason G M and Hill M E 2008 *Particle Acceleration and Transport in the Heliosphere and Beyond* American Institute of Physics Conference Proceedings vol 1039, ed G Li *et al* (Melville, NY: American Institute of Physics) pp 367–374

[3] Gloeckler G, Fisk L A, Mason G M, Roelof E C and Stone E C 2012 *Physics of the Heliosphere: A 10 Year Retrospective*, American Institute of Physics Conference Proceedings vol 1436, ed J Heerikhuisen *et al* (Melville, NY: American Institute of Physics) pp 136–143

[4] Hill M E and Hamilton D C 2010 *Pick-up ions throughout the Heliosphere and Beyond* American Institute of Physics Conference Series vol 1302, ed J le Roux *et al* (Melville, NY: American Institute of Physics) pp 58–63

[5] Mewaldt R A *et al* 2001 *Solar and Galactic Composition: A Joint SOHO/ACE Workshop* American Institute of Physics Conference Series vol 598, ed R F Wimmer-Schweingruber (Melville, NY: American Institute of Physics) pp 165–170

[6] Desai M I, Dayeh M A and Mason G M 2010 *Twelfth International Solar Wind Conference* American Institute of Physics Conference Proceedings vol 1216, ed M Maksimovic *et al* (Melville, NY: American Institute of Physics) pp 635–638

[7] Fisk L A and Lee M A 1980 *Astrophys. J.* **237** 620–626

[8] Giacalone J, Jokipii J R and Kóta J 2002 *Astrophys. J.* **573** 845–850

[9] Chalov S V 2008 *Astron. Lett.* **34** 646–651

[10] Isenberg P A 1987 *J. Geophys. Res.* **92** A2

[11] Bogdan T J, Lee M A and Schneider P 1991 *J. Geophys. Res.* **96** 161–178

[12] Schwadron N A, Fisk L A and Gloeckler G 1996 *Geophys. Res. Lett.* **23** 21

[13] le Roux J A and Ptuskin V S 1998 *J. Geophys. Res.* **103** 4799

[14] le Roux J A, Matthaeus W H and Zank G P 2001 *Geophys. Res. Lett.* **28** 3831–3834

[15] le Roux J A, Zank G P and Matthaeus W H 2002 *J. Geophys. Res.* **107** A7

[16] Kowal G, de Gouveia Dal Pino E M and Lazarian A 2011 *Astrophys. J.* **735** 102–111
[17] Drake J F, Swisdak M and Fermo R 2013 *Astrophys. J. Lett.* **763** L5
[18] Desai M I, Mason G M, Gold R E, Krimigis S M, Cohen C M S, Mewaldt R A, Mazur J E and Dwyer J R 2007 *Space Sci. Rev.* **130** 243–253
[19] Fisk L A and Gloeckler G 2006 *Astrophys. J. Lett.* **640** L79–L82
[20] Fisk L A and Gloeckler G 2008 *Astrophys. J.* **686** 1466–1473
[21] Jokipii J R and Lee M A 2010 *Astrophys. J.* **713** 475–483
[22] Zhang M 2010 *J. Geophys. Res.* **115** A12102
[23] Zhang M and Schlickeiser R 2012 *Astrophys. J.* **756** 129
[24] Krimigis S M, Bostrom C O, Armstrong T P, Axford W I, Fan C Y, Gloeckler G and Lanzerotti L J 1977 *Space Sci. Rev.* **21** 329–354
[25] Stone E C, Vogt R E, McDonald F B, Teegarden B J, Trainor J H, Jokipii J R and Webber W R 1977 *Space Sci. Rev.* **21** 355–376
[26] Zhang M 2005 *Astrophys. J.* **624** 1038–1048
[27] Krimigis S M, Carbary J F, Keath E P, Bostrom C O, Axford W I, Gloeckler G, Lanzerotti L J and Armstrong T P 1981 *J. Geophys. Res.* **86** 8227–8257
[28] Glazov L G and Sigmund P 2007 *Nucl. Instrum. Meth. B* **256** 50–56
[29] Beringer J *et al.* 2012 *Phys. Rev. D* **86** 010001
[30] Agostinelli S *et al.* 2003 *Nucl. Instrum. Meth. A* **506** 250–303
[31] Caballero-Lopez R A and Moraal H 2004 *J. Geophys. Res.* **109** A01101
[32] Decker R B, Krimigis S M, Roelof E C, Hill M E, Armstrong T P, Gloeckler G, Hamilton D C and Lanzerotti L J 2005 *Nature* **309** 2020–2024
[33] Cummings A C, Stone E C and Steenberg C D 2002 *Astrophys. J.* **578** 194–210
[34] Fulks G J 1975 *J. Geophys. Res.* **80** 1701–1714

Plasma and Flows in the Heliosheath

John D. Richardson[1] **and Robert B. Decker**[2]

[1] Kavli Center for Astrophysics and Space Sciences, Massachusetts Institute of Technology, Cambridge, MA 02139 USA

[2] Applied Physics Laboratory, Johns Hopkins University, Laurel MD USA

E-mail: `jdr@space.mit.edu`

Abstract. Voyager 2 (V2) is now 20 AU deep into the heliosheath; if the heliosheath width were similar to that in the Voyager 1 (V1) direction, then V2 is 2/3 of the way to the heliopause. We present recent V2 observations, compare with observations from V1, and compare with model predictions. The speed of the flows observed by V2 in the heliosheath are, on average, remarkably constant at 150 km/s. The flow angle has changed dramatically, however, and is now over 60° from radial, with more of the turning occurring in the RT than RN planes. These flows are very different than those at V1, where the speed was always below 100 km/s and decreased across the heliosheath. Models predict VR at V2 well, but the the flow angle predictions differ from the observations. The average density and temperature of the plasma decreased by a factor of two in 2008, recovered in 2011, and have not changed significantly since 2011.

1. Introduction

Voyager 2 has observed plasma in the heliosheath since the termination shock crossings at 84 AU in 2007 [1,2,3,4]. Since the V1 plasma instrument (PLS) is not working, these are the first observations of plasma in the heliosheath. V1 plasma flow speeds have been derived from V1 Low Energy Charged Particle experiment (LECP) and Cosmic Ray Subsystem (CRS) data [5,6,7,8,9], but plasma densities and temperatures cannot be determined. The exception is recent V1 plasma wave data from outside the boundary at 122 AU where low-energy particle intensities decreased and galactic cosmic ray intensities and the magnetic field increased [10,11,12,13]; this boundary may be the heliopause. These data show plasma waves at a frequency which implies a density of 0.06-0.08 cm^{-3} [14], consistent with V1 being in the local interstellar medium. The plasma emissions may be triggered by the arrival of merged interaction regions at V1. If this boundary were the heliopause V2, which crossed the termination shock 2.5 years after V1, could be within a few years of a heliopause crossing. This paper presents V2 heliosheath plasma data out to 104 AU, compares these data to the V1 observations and to V2 flows derived from LECP data, and discusses the implications of these results.

2. Data

The V2 PLS instrument observes ions and electron currents with energies/charge from 10-5950 eV/q every 192 seconds using four Faraday Cups [15]. The observed currents are fit with convected isotropic proton Maxwellian distributions to find the plasma velocity, density, and temperature. The cups have acceptance angle responses which are flat to 45°, then decrease to 0 at 60°. To determine the velocity, currents must be observed in at least three of the four

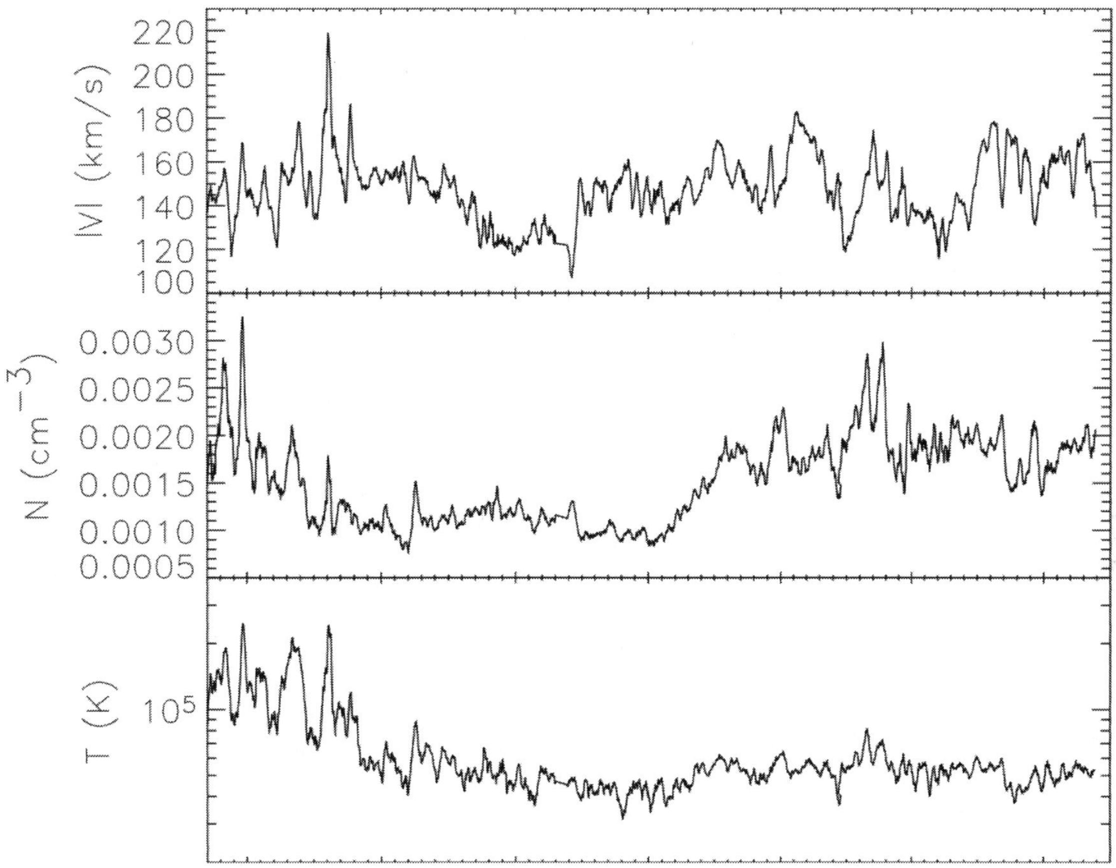

Figure 1. The plasma speeds, density, and temperature measured by the plasma instrument on V2.

Faraday cups; flow angles outside the cup acceptance angle and noise allow only about 15% of the spectra to be analyzed. When data are available in three cups, Maxwellian distributions fit the data well.

Figure 1 shows the plasma speed, density, and temperature in the heliosheath from the termination shock out to 105 AU. The speed has remained constant at about 150 km/s throughout the heliosheath, a major surprise [16]. It has fluctuated from 120 to 180 km/s, but on average has not changed contrary to model predictions of a speed decrease. These observed speeds are very different from those observed at V1, where the speed after the termination shock was only 100 km/s and the speed decreased monotonically with distance until it approached zero in 2010.

The density just after the termination shock was 0.002 cm-3, about twice that in the upstream solar wind. The density then decreased by a factor of 2 in 2008 and remained low until 2011,

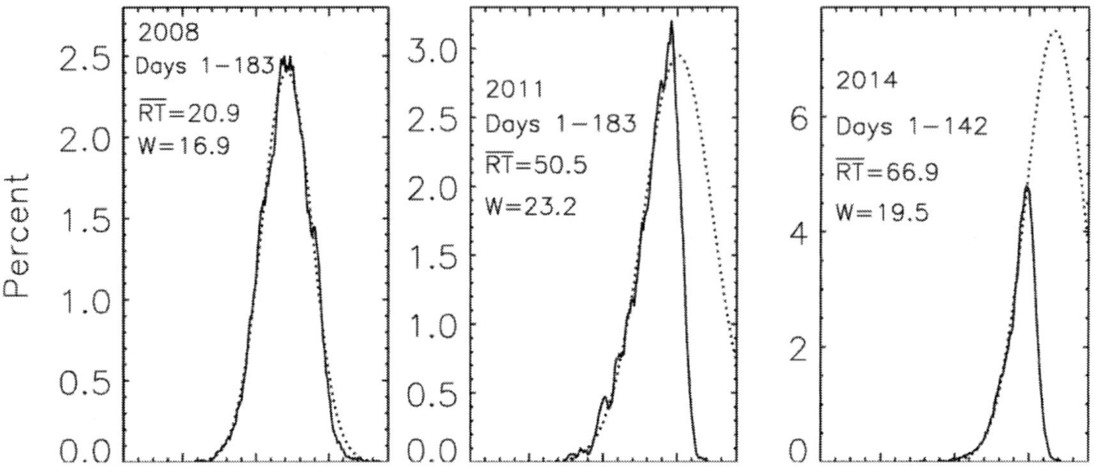

Figure 2. Distributions of the flow angles in the RT plane observed by V2 (solid line) and a Gaussian fit to the data with angles below 50° (dashed line). The years and days used for this analysis and the average RT angles and distribution widths from the fits are given in each panel.

when increased back to its pre-2008 level. This period of low density corresponds to solar minimum in the outer heliosphere so it seems likely this decrease is a solar cycle effect, with V2 observing more high-speed, low-density wind [17]. However, the speed does not change in this region as one would expect if V2 were in high-speed wind.

The average flow angle in the RT direction is about 60°. (The RTN system has R radially outward, T in the plane of the solar equator and positive in the direction of solar rotation, and N completes a right-handed system.) Since the response of the cups goes to zero at 60°, many spectra cannot be analyzed. The heliosheath is a highly fluctuating region, so flow angles are constantly changing and V2 rarely remains in flow at too high an angle to derive plasma parameters for more than a couple hours at a time. These large fluctuations in the heliosheath plasma give a wide distribution of plasma parameters which are well fit by Gaussian distributions [18]. In the case of the RT flow angles, we assume the distributions are Gaussian and fit the data below the instrument cutoff; the Gaussian fit parameters then give the average flow angle of the data. Figure 2 shows the Gaussian fits to the RT angles for days 1-183 in 2008 and 2011 and for days 1-142 in 2014. The cutoff in the observed distribution occurs at about 50°. In 2008 this effects only a few percent of the observations. In 2011 the peak in the RT angles is very close to the cutoff value. In 2014 the peak is well beyond the cutoff, but the Gaussian fit to the data below the peak gives an average flow angle of 66° in the RT plane.

The top panel of Figure 3 shows the three components of velocity. The VT component has

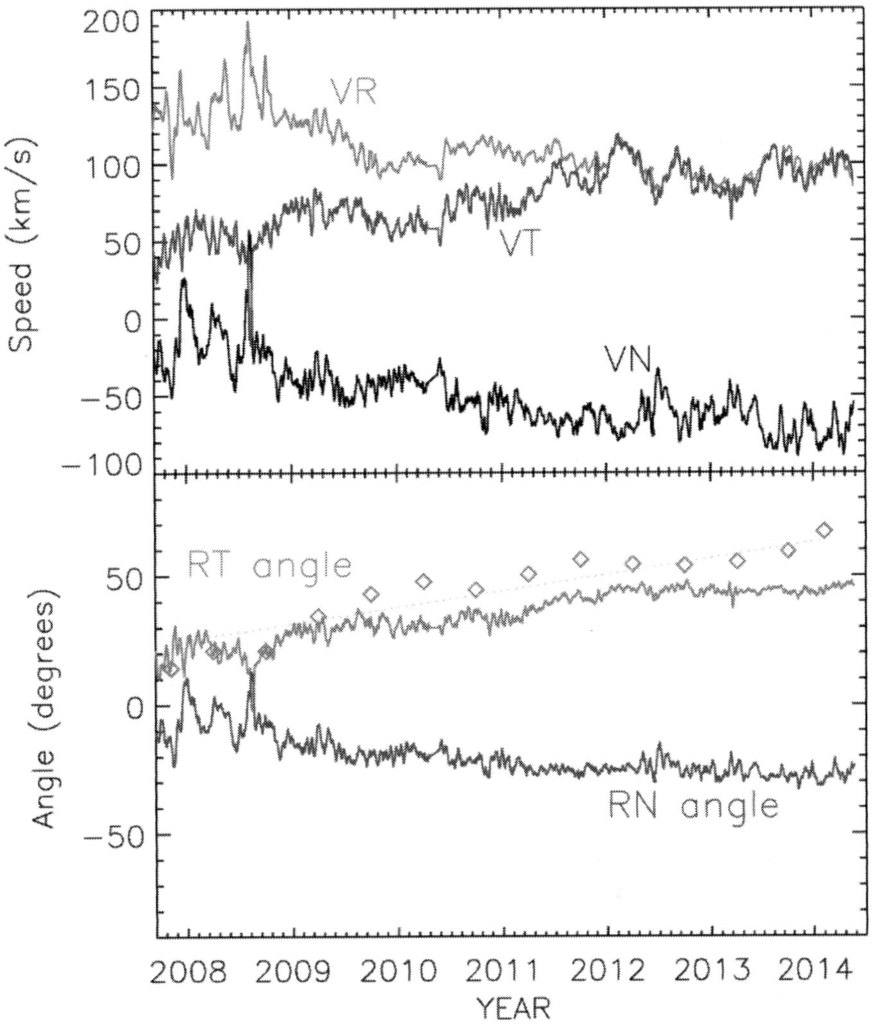

Figure 3. The top panel shows the three components of velocity observed by V2; VT is not corrected for the instrument cutoff. The bottom panel shows 11-day averages of the flow angles at V2. The blue diamonds show values of the RT angle corrected for the instrumental angle cutoff.

not been corrected for the instrument response. VR decreases across the heliosheath from 130 to about 90 km/s; this decrease is much less than that observed at V1. VT increases from 40 km/s to over 100 km/s, where it asymptotes due to the instrumental effects. VN increases in magnitude across the heliosheath from -20 to -80 km/sec. The velocity components are strongly correlated, with magnitudes of all components moving together. The tendency of VR, VT, and VN to all increase and decrease simultaneously tends to keep the flow angles relatively constant.

The bottom panel of Figure 3 shows the flow angles in the RT and RN planes; the diamonds show the corrected RT values obtained by fitting Gaussian distributions to 6 months of data. The flow angles in the solar wind are near zero in the outer heliosphere. At the termination shock, the RT angle increased to about 20° and RT continues to increase across the heliosheath to the present value above 60 °. The RN angle increased at the termination shock to about 10°, fluctuated in the beginning of 2008 then decreased monotonically to about 30° in 2014. The RN angle is consistently about 1/2 of the RN angle, indicating most of the plasma flow is moving around the sides, not over the poles, of the heliosphere.

3. Discussion

As mentioned above, one puzzle has been the very different plasma flows observed at V1 and V2. The plasma instrument on V1 does not work, so the V1 flows are derived from LECP and CRS data using a Compton-Getting analysis on the low-energy particle data [5,6,19]. The LECP instrument scans in the RT plane and thus the flow speeds can be derived in those planes. At V1, the spacecraft is rolled occasionally to allow LECP to determine VN [20]; the CRS instrument also provides VN values during magnetometer rolls which occur every few months [8]. The Compton-Getting analysis assumes the ion composition is known and that the observed anisotropies arise from the convective flow.

One way to test the LECP speeds is to compare the LECP flows derived from V2 data with those observed by the V2 plasma instrument. Figure 5 shows a comparison of the flows observed by the PLS and LECP instruments in the RT, R, and T directions [9]. The LECP speeds are derived from 28-43 keV ion (generally proton) angular data using a linear Compton-Getting analysis. The Compton-Getting analysis assumes that the ion composition is known and that the observed anisotropies arise from the convective flow. The VR speeds are similar after the termination shock until 2009.3, then from 2009.3-2010.5 the LECP values are much larger (period A in the VR panel). The speeds match fairly well after 2010.5 except from 2012.7-2013.3 and near 2013.7 where the LECP values are higher then the PLS speeds. The LECP and PLS VT data both show increases through the heliosheath and similar speeds except from 2012.7-2013.3 (period B in the VT panel). We think the differences in the LECP and PLS VR values during 2009.3-2010.5 are caused by the LECP detector observing pickup oxygen ions. The ion detector measures the total energy, so the 28-43 keV proton channel will also measure 4-7 keV oxygen ions [21]. The Compton-Getting analysis assumes the ions were protons to derive the flow estimates in Fig. 5. If some of these ions were oxygen, the predicted convection speed would be too high because protons have a higher speed than pickup oxygen ions at the same energy. We do not understand why these heavy ions are prevalent at V2 during 2009.3-2010.5. However, pickup oxygen ions have been measured by the V1 and V2 LECP low-energy ion channels in high speed solar wind [22]. A small acceleration at the termination shock would enable pickup oxygen detection in the low speed heliosheath flow. From 2012.7 to 2013.3 ions from at least 28 keV to several MeV stream are observed to stream along the magnetic field. Thus convective flow does not cause all the observed anisotropy and the Compton-Getting analysis is unreliable. Except for regions A and B, where LECP speeds cannot be reliably determined, the PLS and LECP speeds are comparable. Both LECP and PLS observe much higher values of VR than LECP observed at V1. They both observe a very slow decrease in VR across the heliosheath, not the rapid decrease to zero observed at V1. Both PLS and LECP observe an increase in VT across the heliosheath, whereas at V1 a constant low-speed VT was observed. We note that where LECP and PLS disagree the LECP values are larger than the PLS values; similar problems with Compton-Getting at V1 would also give larger values of VT and VR, not lower values as observed. The data thus are consistent with the flow speeds in the V1 and V2 directions being very different; this difference does not appear to be an instrumental effect.

We also note that models generally do not match both the V1 and V2 speed profiles. Pogorelov et al. [23] use Ulysses data as input for their heliosphere model. Their model radial speed profile matches V2 VR observations well, but for V1 the model values are much higher than those observed. They find VT comparable to VN at V2, whereas observations find VT is twice VN, and they find less turning of the V2 flow than observed. Other models have similar

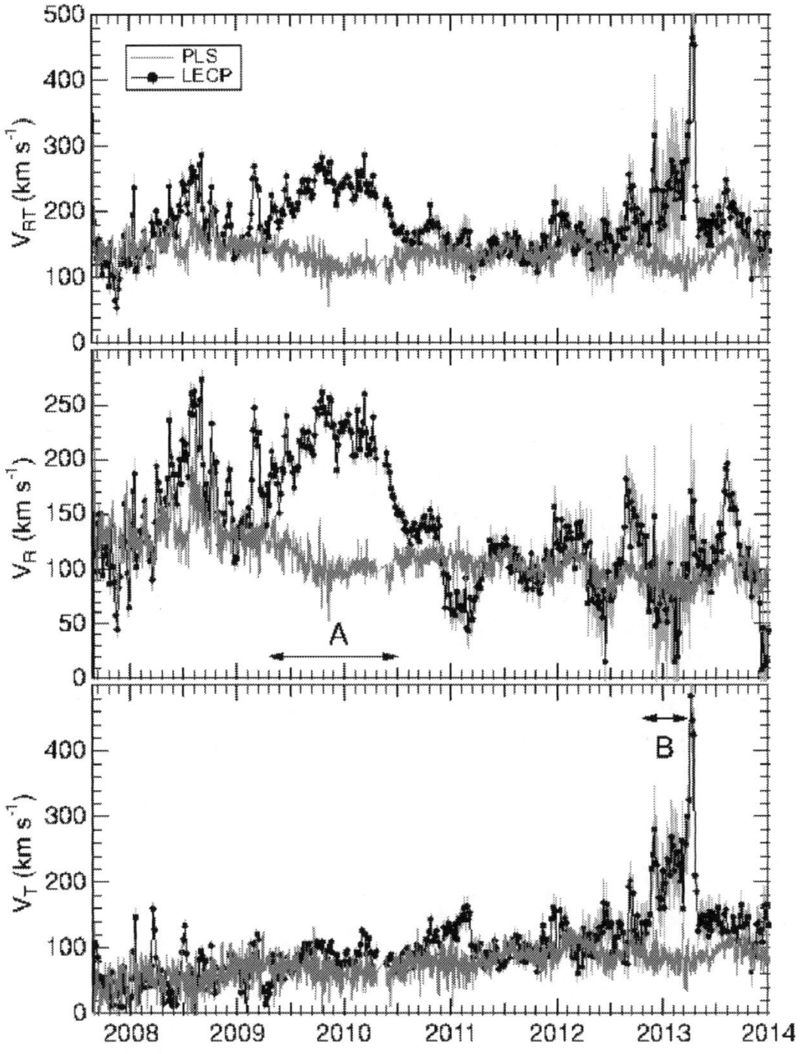

Figure 4. A comparison of V2 speeds observed by PLS and derived from LECP data using the Compton-Getting effect.

problems [24]

The timing of the V2 crossing of the heliopause is of great interest. Knowledge of the plasma parameters should provide conclusive evidence of whether the boundary observed by V1 is the heliopause. We will also learn if the stagnation region observed by V1 [7] is a transient or permanent feature of the outer heliosphere. The location of the heliopause itself will greatly enhance our understanding of heliospheric asymmetries. The plasma flow data provides some clues as to when the heliopause crossing may occur; at the heliopause we expect the flow to become tangential to the boundary, mostly in the T and N directions. The dashed line in the bottom of Figure 3 shows a linear fit to the corrected RT angles. The fit has a slope of 6.3 ±

1.0 °/year, so the flow angle would reach 90° (and V2 would reach the heliopause) in about 4 years.

3.1. Acknowledgements

JDR was supported under NASA contract 959203 from the Jet Propulsion Laboratory to the Massachusetts Institute of Technology RBD was supported at JHU/APL by NASA contract NNN06AA01C.

4. References

[1] Burlaga L F, Ness N F, Acuna M H, Lepping R P, Connerney J E P and Richardson J D 2008 *Nature* **454** 75
[2] Decker R B, Krimigis S M, Roelof E C, Hill M E, Armstrong T P, Gloeckler G, Hamilton D C and Lanzerotti L J 2008 *Nature* **454** 67
[3] Richardson J D, Kasper J C, Wang C, Belcher J W and Lazarus A J 2008 *Nature* **454** 63
[4] Stone E C, Cummings A C, McDonald F B, Heikkila B C, Lal N, Webber W R 2008 *Nature* **454** 71
[5] Decker R B, Krimigis S M, Roelof E C, Hill M E, Armstrong T P, Gloeckler G, Hamilton D C and Lanzerotti L J 2005 *Science* **309** 2020
[6] Decker R B, Krimigis S M, Roelof E C and Hill M E 2010 *AIP Conf. Proc.* **1302** 51
[7] Krimigis S M, Roelof E C, Decker R B, Hill M E 2011 *Nature* **474** 359
[8] Stone E C and Cummings A C 2012 *Proceedings of the 32ND International Cosmic Ray Conference, Beijing 2011* **12** 23
[9] Richardson J D and Decker R B 2014 *Astrophys. J.* **submitted**
[10] Burlaga L F, Ness N F and Stone E C 2013 *Science* **341** 147
[11] Krimigis S M, Decker R B, Roelof E C, Hill M E, Armstrong T P, Gloeckler G, Hamilton D C and Lanzerotti L J 2013 *Science* **341** 144
[12] Stone E C, Cummings A C, McDonald F B, Heikkila B C, Lal N and Webber W R 2013 *Science* **341** 150
[13] Webber W R and McDonald F B 2013 *Geophys. Res. Lett.* **40** 1665
[14] Gurnett D A, Kurth W S, Burlaga L F, and Ness N F 2013 *Science* **341** 1489
[15] Bridge H S, Belcher J W, Butler R J, Lazarus A J, Mavretic A M, Sullivan J D, Siscoe G L and Vasyliunas V M 1977 *Space Sci. Rev.* **21** 259
[16] Richardson J D, Stone E C, Kasper J C, Belcher J W, and Decker R B 2009 *Geophys. Res. Lett.* **36** L10102
[17] Richardson J D and Wang C 2012 *Astrophys. J. Lett.* **759** L19
[18] Richardson J D 2011 *Astrophys. J.* **740** 113
[19] Gleeson L J and Axford W F 1968 *Astrophys. Sp. Sci.* **2** 431
[20] Decker R B, Krimigis S M, Roelof E C and Hill M E 2012 *Nature* **489** 124
[21] Decker R B, Krimigis S M, Armstrong T P, Mosley C J, Hamilton D C and Gloeckler G 2003 *Adv. Sp. Res.* **32** 591
[22] Krimigis S M, Carbary J F, Keath E P, Bostrom C O, Axford W I, Gloeckler G, Lanzerotti L J, and Armstrong T P 1981 *J. Geophys. Res.* **86** 8227
[23] Pogorelov N V, Suess S T, Borovikov S N, Ebert R W, McComas D J and Zank G P 2013 *Astrophys. J.* **772** 2
[24] Provornikova E, Opher M, Izmodenov V V, Richardson J D, Decker R B and Toth G 2014, submitted to *Astrophys. J.*

Diffuse X-ray Background Constraints on Models of the Local Interstellar Medium

S L Snowden
NASA/GSFC, Code 662, Greenbelt, MD, 20771, USA

E-mail: steven.l.snowden@nasa.gov

Abstract. There is a flux of soft X-rays (0.07-0.284 keV) of diffuse origin observable over the entire sky. As X-rays of this energy are strongly absorbed by the interstellar medium (ISM), one optical depth is 10^{19} - 10^{20} H cm^{-2}, they provide a unique probe of neutral material in the solar vicinity. However, to be an effective probe requires that the distribution of emission be well understood, a requirement that is currently unfulfilled (although progress is being made), with unclear fractions originating in the Galactic halo, Local Hot Bubble, heliosphere, and Earth's magnetosheath. The various available data, their consistency (or lack thereof), and their implications for understanding the very local ISM are briefly discussed.

1. Introduction

The canonical view of the diffuse X-ray background has it comprised of a complicated combination of emission and absorption regions. Working inwards from cosmological distances, there is the extragalactic background dominated by emission from unresolved AGN, external galaxies, groups and clusters of galaxies, and the cosmic web. Closer to the Sun there is emission from the halo and central bulge of the Milky Way. The halo emission likely consists of a hotter extended (large scale height) component (kT~0.25 keV) and a cooler (kT~0.1 keV) clumpy component lying closer to the disk. Within the disk of the Galaxy, sometimes extending into the halo, are supernova remnants, superbubbles, and other regions of X-ray emitting plasma. In the Sun's backyard, the heliosphere is surrounded by the Local Cloud (LCl), which in turn is located within a bubble of hot (kT~0.1 keV) plasma called the Local Hot Bubble (LHB). Within the heliosphere there is extensive X-ray emission from solar wind charge exchange with interstellar neutrals passing through the solar system, as well as with exospheric hydrogen in Earth's magnetosheath. In addition, all of the emission originating beyond the LHB is moderated by absorption by the foreground neutral or molecular material in the ISM. Nearly all diffuse emission in the 0.07-1.5 keV band is thermal or charge exchange in origin, and therefore dominated by line emission, for the most part the same lines. However, observation technology is insufficient for the use of velocity information to constrain the distance to the emission, or to examine the difference between thermal and charge exchange emission. To further complicate the issue, the thermal emission regions have a variety of temperatures and absorption is strongly energy dependent.

Figure 1 shows images of the ¼ keV and ¾ keV sky. The structure of the X-ray sky appears quite different in the two energy bands due both to different absorption cross sections as well as the differences in the distributions of the emitting plasma at different temperatures. However, the ¼ keV band serves as better probe of the local ISM as absorption reaches one optical depth at ~10^{20} H I cm^{-2}, a better match with nearby column densities.

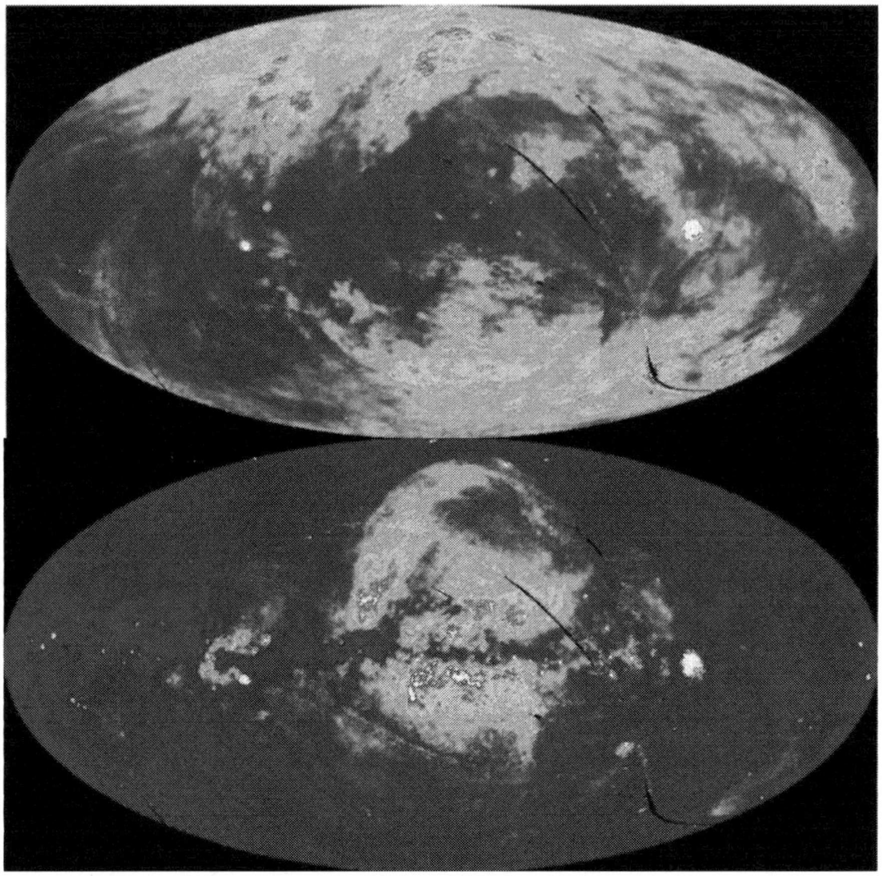

Figure 1. ROSAT All-Sky Survey maps of the ¼ keV (upper) and ¾ keV (lower) diffuse X-ray background. The images are in Aitoff-Hammer equal-area projections in Galactic coordinates, centred on the Galactic centre, with longitude increasing to the left and the north Galactic pole at the top. The false colour images show dim areas as purple and bright areas as red/white [1].

2. Setting the stage – the cool part of the local ISM

There is a deficiency of H I in the local ISM that is easily seen in plots of the average column density as a function of Galactic latitude (e.g., [2], see Figure 2). Studies using interstellar absorption lines [3][4][5] have shown that the deficiency is due to a cavity in the H I of the Galactic disk surrounding the Sun (called the Local Cavity, LCa). Recently, extensive absorption line and reddening measurements have been combined to provide a rather good three-dimensional map of the neutral material within the nearest ~500 pc from the Sun [6]. From [6] it is clear that the H I of the local Galactic disk is in no way distributed smoothly; there are both relatively dense regions as well as extensive voids.

3. Setting the stage – the warm part of the local ISM

Directly surrounding the heliosphere is a region of partially ionized gas which has been called the Local Cloud (LCl), the Very Local ISM (VLISM), or my personal favorite, the Local Fluff (LF, e.g., [8]). The LF was identified by interstellar absorption line measurements of nearby stars and has an extent of up to ~15 pc from the Sun (placing it well inside the LCa). It has a temperature of T~10^3 – 10^4 K and an H I density of ~0.1 cm^{-3}. (See other papers in these proceedings for much more extended discussions of the LF.) Because there are a variety of distinct velocity components observed in the absorption lines it has been thought that the LF is made up of ~15 cloudlets [8]. As there is some relative motion between the LF and the Sun, the neutral part of the LF penetrates through the heliospheric magnetic field and flows through the heliosphere with a speed of ~25 km s^{-1}. From [9] there is no indication of expansion or contraction implying pressure equilibrium between the LF and its surroundings. Note, however, that there is a new interpretation of the absorption line data by [10] which suggests that the LF may actually be a single monolithic cloud which is experiencing some compression along the flow direction and expansion in the cross direction.

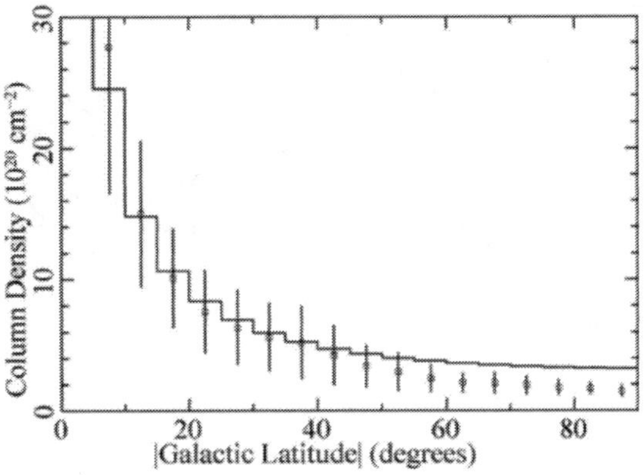

Figure 2. The average column density of Galactic H I in 5° latitude annuli as a function of Galactic latitude. The vertical error bars show the RMS range of values in each latitude bin while the model curve shows a csc|b| approximation to the data with the scaling roughly set by intermediate-latitude data.

4. Setting the stage – the hot part of the local ISM

The sky has been at least partially mapped at ¼ keV by four surveys (Figure 3) with different observation geometries and at different times during the solar cycle. The Wisconsin [11] and ROSAT [1] surveys are in good agreement while there is a minor zero-level offset between them and the SAS-3 [12] and HEAO-1 A2 [13] maps. The offset is likely due to an unremoved background contamination from solar wind charge exchange X-ray emission either in the heliosphere or Earth's magnetosheath. However, the angular structure of the surface brightness variations is consistent between the various maps, limited only by the statistical significance and angular resolution of the different surveys and instruments.

The ¼ keV and lower energy bands are those most relevant for studies of the local ISM. One optical depth for their band-averaged cross sections are N_H~10^{20} cm^{-2} for the C band (0.1-0.284 keV, historically named for the carbon filter used in early proportional counter observations of the SXRB), N_H~0.5x10^{20} cm^{-2} for the B band (0.1-0.188 keV, defined by the use of a boron filter), and N_H~0.1x10^{20} cm^{-2} for the Be band (0.07-0.111 keV, defined by the use of a beryllium filter). The B and C bands are primarily sensitive to emission in front of the nearest walls of the LCa (although the C band can "see" beyond the thinner walls and into the Galactic halo) and the Be band is sensitive to neutral material within the LCa.

The ¼ keV X-ray maps show a non-zero flux from the Galactic plane. With the short mean free path of these X-rays in the nominal ISM, the emission must originate locally, and as the LCa was a convenient place to confine a hot plasma, the Local Hot Bubble (LHB) [14] was born. As the plasma is confined within the LCa, it surrounds the Sun with an extent of ~50 to ~150 pc, and emits X-rays characteristic of a temperature kT~0.1 keV. The LHB, and the requirements indicating its existence, can be considered the first constraint on the local ISM.

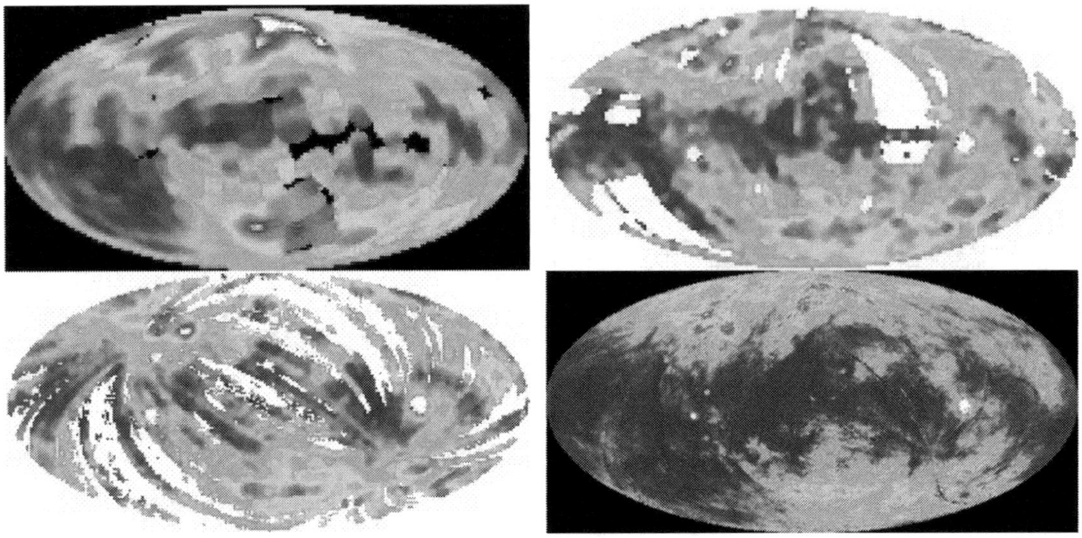

Figure 3. Maps of the ¼ keV diffuse X-ray background from the Wisconsin [11] (upper left, published in 1983), SAS-3 [12] (upper right, published in 1984), HEAO-1 A-2 [13] (lower left, published in 1992), and ROSAT [1] (lower right, published in 1995) surveys. Note that while the statistics and angular resolution improved over time, the basic structure of the ¼ keV SXRB remained unchanged.

While the LHB accounted for much of the observed emission at ¼ keV, there was clear evidence for a flux of a more distant origin (significantly above the relatively minor extrapolation of the extragalactic power law observed at higher energies) as soon as ROSAT began operations in 1990. The first clear "shadowing" feature (the absorption of a flux of distant origin by a distinct foreground object) was by the Draco Nebula [15]. Figure 4 shows the IRAS 100 μm and ROSAT All-Sky Survey (RASS) images of the Draco region where the negative correlation between the two images is rather spectacular.

5. Limits on neutral material within the LHB

The Be band data were used in the 1980's to place strong limits on the amount of diffuse neutral material similar to the LF present within the LHB [16]. Because the cross sections are a factor of 5 different between the Be and B bands, their hardness ratio is a strong function of an absorbing column density. However, the sample directions, which were scattered over the northern Galactic hemisphere (but away from the Loop I/Galactic center region), showed essentially no variation of the Be band to B band ratio although the intensities varied by more than a factor of three. This leads to constraint number 2 on the LISM: there is very little material (2σ upper limit of $< 10^{19}$ cm^{-2}) in the LHB similar to the LF (at least in the northern hemisphere).

6. Things get complicated by magnetosheath SWCX

Although the attribution to solar wind charge exchange (SWCX) came several years later, it was clear from the RASS raw data that we had a problem with contamination in our measurements of the SXRB. Long-term background enhancements (LTEs) lasting from hours to days left a strong signature in the data. The emission was clearly diffuse in origin, and with the duration of the episodes it was assumed to originate in the near-Earth environment.

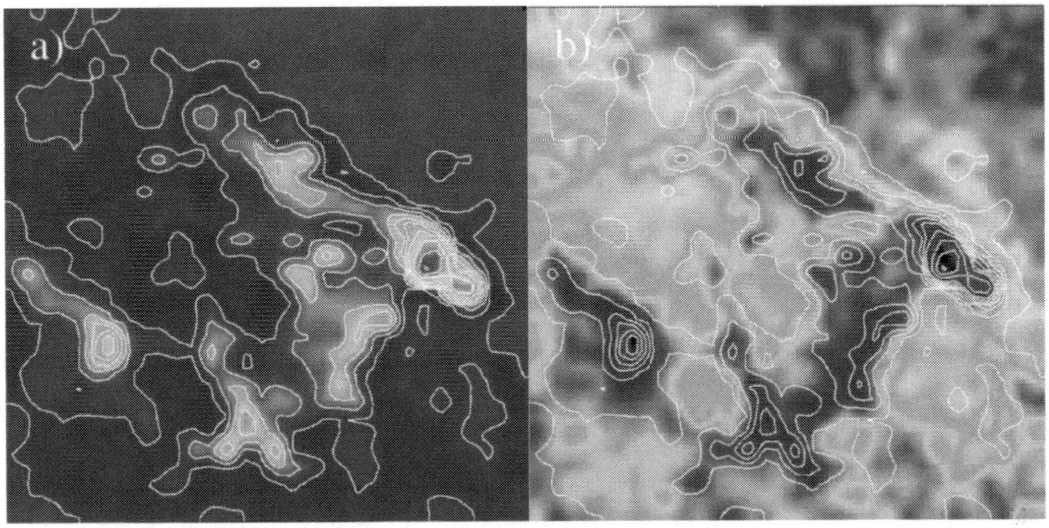

Figure 4. Maps of the IRAS 100 μm (left) and RASS ¼ keV (right) data of the Draco Nebula. In both cases purple indicates low intensity while red indicates high intensity. The contours in both images are those of the 100 μm data. The negative correlation between the two images is obvious indicating shadowing by the Draco Nebula of an X-ray flux of more distant origin. Considering the Galactic latitude of the nebula and its distance, the distant emission must originate in at least the lower halo.

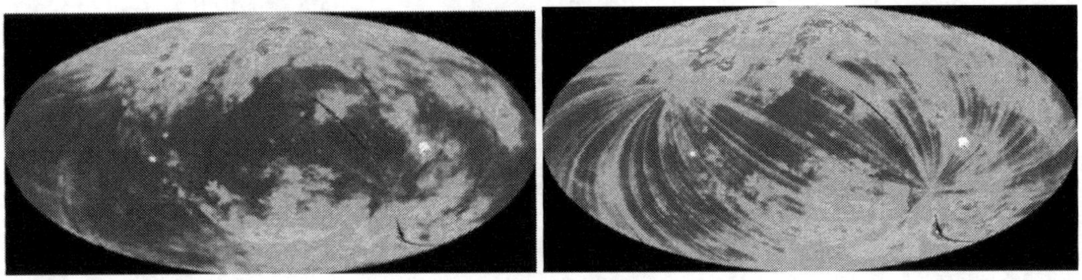

Figure 5. Maps of the cleaned (left) and uncleaned (of the LTE background, right) RASS ¼ keV data. The striping in the unceaned map is due to solar wind charge exchange with exospheric H in Earth's magnetosheath.

When the LTE count rate was compared to the solar wind flux during the ROSAT survey [17], the correlation was obvious, and while the correlation had considerable scatter it adhered to a linear relation. However, while modeling the light curve could address the temporally varying part (on time scales of ~1 day or less), any constant contribution would remain. Plotting the relationship as a scatter plot and extrapolating a linear fit to a zero solar wind flux provides an estimate for the residual SWCX contamination from the near-Earth environment of ~0.31 counts s^{-1} FOV^{-1}[18]. This provides constraint number 3, while there is considerable temporally variable SWCX emission from Earth's magnetosheath, after "cleaning" it probably accounts for only about 5-15% of the observed surface brightness of the ¼ keV band (for dim and bright regions, respectively).

7. Heliospheric SWCX – now life really gets complicated

Heliospheric SWCX emission is produced by solar wind ions charge exchanging with interstellar neutrals flowing through the solar system. As the emission is integrated out to the heliopause, most temporal variations are smoothed out but there are geometric variations due to purturbed flow of interstellar neutrals and the solar latitude variation of the solar wind.

Given that: 1) The flow of interstellar neutrals through the solar system is reasonably well understood, 2) the average solar wind flux throughout the solar system is reasonably well understood, and 3) the SWCX cross sections relevant to the ¼ keV band are understood, it is straight forward to model the heliospheric SWCX flux. While these assumptions get progressively less certain, [19] did put limits on the heliospheric SWCX contribution to the observed X-ray flux. If the model geometric distribution of SWCX emission is scaled so that it accounts for all of the observed flux in the Galactic plane there is "left over" emission above and below the plane but still within the Local Cavity. Given that a picture of the local ISM would look more than a slightly odd with hot bubbles floating above and below the plane in the LCa, heliospheric SWCX probably does not provide all of the ¼ keV flux observed in the Galactic plane. Very recently there has been a confirmation of this view [20]. Using a sounding rocket flight to measure the SWCX flux from the heliosphere He focusing cone it was determined that only 40±7% of the observed flux from a direction of minimal SXRB surface brightness in the RASS ¼ keV map in the Galactic plane was due to heliospheric and magnetosheath SWCX emission. This leads to the fourth and final contraint: The heliosphere contributes only a limited amount of flux to the surface brightness of the RASS ¼ keV background.

8. Summary

The four constraints that have been presented are: 1) Local emission leads to the idea of the Local Hot Bubble; 2) There is very little absorbing gas within the LHB (the LF may be relatively alone); 3) While there is considerable temporally variable SWCX emission from Earth's magnetosheath, after "cleaning" it probably accounts for only ~5% of the observed maximum SXRB flux and ~15% of the observed minimum SXRB flux; 4) The heliosphere contributes a limited flux to the observed SXRB, at least at ¼ keV, ~40% at low Galactic latitude and ~15% at high Galactic latitudes.

9. References

[1] Snowden, S. L., et al. 1997, ApJ, **485**, 125
[2] Knapp, G. R. 1975, AJ, **80**, 111
[3] Frisch, P. C., & York, D. G. 1983, ApJL, **271**, L59
[4] Paresce, F. 1984, AJ, **89**, 1022
[5] Lallement, R,; Welsh, B. Y., Vergely, J. L., Crifo, F., & Sfeir, D. 2003, A&A, **411**, 447, 3D Mapping of the Dense
[6] Lallement, R., Vergely, J.-L., Valette, B., Puspitarini, L., Eyer, L., & Casagrande, L. 2014, A&A, **561**, A91
[7] Snowden, S. L., Cox, D. P., McCammon, D., & Sanders, W. T. 1990, ApJ, **354**, 211
[8] Redfield, S., & Linsky, J. L. 2008, ApJ, **673**, 283
[9] Frisch, P. C., Redfield, S., & Slavin, J. D. 2011, ARA&A, **49**, 237
[10] Gry, C., & Jenkins, E. B. 2014, ArXiv e-prints, 2014arXiv1404.0326G
[11] McCammon, D., Burrows, D. N., Sanders, W. T., & Kraushaar, W. L. 1983, ApJ, **269**, 107
[12] Marshall, F. J., & Clark, G. W. 1984, ApJ, **287**, 633
[13] Garmire, G. P., Nousek, J. A., Apparao, K. M. V., Burrows, D. N., Fink, R. L., & Kraft, R. P. 1992 ApJ, **399**, 694
[14] Cox, D. P, & Snowden, S. L. 1986, AdSR, **6(2)**, 97
[15] Snowden, S. L., et al. 1991, Sci, 252, 1529
[16] Juda, M., et al. 1991, ApJ, **367**, 182
[17] Cravens, T., et al. 2001, JGR, **106**, 24883
[18] Kuntz, K. D., et al. 2014, in preparation
[19] Lallement, R., 2004, A&A, **418**, 143
[20] Galleazi, M., et al. 2014, Nature, in press

Journal of Physics: Conference Series **577** (2015) 012023 doi:10.1088/1742-6596/577/1/012023

Anisotropies in TeV Cosmic Rays Related to the Local Interstellar Magnetic Field from the IBEX Ribbon

N. A. Schwadron[1,2], F. C. Adams[3], E. Christian[4], P. Desiati[5], P. Frisch[6], H. O. Funsten[7], J. R. Jokipii[8], D. J. McComas[2,9], E. Moebius[1], and G. P. Zank[10]

[1] University of New Hampshire, Durham, AH, 03824
[2] Southwest Research Institute, San Antonio, TX 78228, USA
[3] University of Michigan, Ann Arbor, MI, 48109
[4] Goddard Space Flight Center, Greenbelt, MD 20771
[5] IceCube Research Center and Astronomy Department, University of Wisconsin, Madison, WI 53706
[6] University of Chicago, Department of Astronomy and Astrophysics, Chicago, IL 60637
[7] Los Alamos National Laboratory, Los Alamos, NM, 87545
[8] University of Arizona, Department of Planetary Sciences, Tucson, AZ 85721
[9] University of Texas, San Antonio, TX 78228
[10] University of Alabama, Huntsville, AL 35805

E-mail: n.schwadron@unh.edu

Abstract. The Interstellar Boundary Explorer (IBEX) observes enhanced Energetic Neutral Atoms (ENAs) emission in the keV energy range from a narrow ($\sim$20° wide) "ribbon" in the sky that appears to be centered on the direction of the local interstellar (LIS) magnetic field. The Milagro collaboration, the Asγ collaboration and the IceCube observatory have recently made global maps of cosmic ray fluxes in the TeV energy range, revealing anisotropic structures ordered in part by the local interstellar magnetic field and the interstellar flow. This paper following from a recent publication in *Science* makes the link between these disparate observations by developing a simple model of the magnetic structure surrounding the heliosphere in the Local Interstellar Medium (LISM) that is consistent with both IBEX ENA fluxes and TeV cosmic ray anisotropies. The model also employs the revised velocity direction of the LIC derived from neutral He observations by IBEX. By modeling the propagation of cosmic rays through this magnetic field structure, we specifically show that (1) the large-scale TeV anisotropy provides a roughly consistent orientation for the local interstellar magnetic field at the center of the IBEX Ribbon and corroborates the $\sim 3\ \mu$G magnitude of the local interstellar magnetic field derived from IBEX observations of the global heliosphere; (2) and small-scale structures in cosmic rays (over $< 30°$ angular scales) are influenced by the interstellar field interaction with the heliosphere at energies < 10 TeV. Thus, we provide a link between IBEX ENA observations, IBEX neutral observations of interstellar He, and TeV cosmic ray anisotropies, which are strongly influenced by the interactions between the local interstellar magnetic field, the flow of the local interstellar plasma, and the global heliosphere.

Journal of Physics: Conference Series **577** (2015) 012023 doi:10.1088/1742-6596/577/1/012023

1. Introduction

The Interstellar Boundary Explorer Mission (IBEX) was launched October 19, 2008, with the objective to discover the global interaction between the solar wind and the local interstellar medium [1]. IBEX measures Energetic Neutral Atoms (ENAs) created predominantly from charge-exchange between neutral interstellar hydrogen and plasma protons in the solar wind and interstellar medium [1; 2]. First results from IBEX include global maps of the heliosphere in ENAs [3–6], the first measurements of low energy interstellar oxygen and hydrogen atoms, and the first measurements of helium at 1 AU [7]. Higher energy maps from INCA were also revealed [8]. The IBEX and INCA sky maps showed striking differences between observations and predictions from models [6].

The IBEX maps revealed the existence of a narrow ribbon ($\sim$ 20° wide) of elevated ENA emissions [3–5] that forms a circular structure centered on (RA, DEC) $\sim$ (48.5°±1.5, -21.2°±1.6), which is likely near the local interstellar (LISM) magnetic field direction [6].

This paper discusses another observational signature of this LISM magnetic field in TeV cosmic ray anistropies [9]. Figure 1 shows the cosmic ray gyroradius for a typical field strength in the heliosheath (1 μG, blue) and LISM (3 μG, red). In the energy range from 0.5 to 10 TeV, cosmic rays with gyroradii of 100's of AU probe the scales in the LISM that are larger than the heliosphere. Recall that the deformation of the LIsMF also extends to 100's of AU. Therefore, it is precisely in the $\sim$ 0.5-10 TeV energy range where cosmic rays may probe the same large scale LIsMF structures apparently manifested in the IBEX Ribbon.

2. Cosmic Ray Anisotropy Observations

The arrival direction distribution of cosmic rays from tens of GeV to tens of TeV is found to have an asymmetry which is not compatible with the convection anisotropy expected by the motion of the solar system through the galactic medium [10; 11]. A broad relative excess is observed between equatorial right ascension $\alpha \approx 18$ hr and $\alpha \approx 8$ hr, with an amplitude of $10^{-4} - 10^{-3}$. The global structure of the anisotropy depends on the cosmic ray energy, but is found to change weakly in the energy range from about 1 TeV to a few tens of TeV. Fig. 2 (panel A) shows the map in equatorial coordinates of relative intensity in the cosmic ray arrival direction, as observed by Tibet ASγ in the northern equatorial hemisphere at a median energy of about 5 TeV [12], and by the IceCube Observatory in the southern equatorial hemisphere at a median energy of about 20 TeV [13]. In each declination band of about 5° the relative intensity represents the modulation of the counting rate with respect to the mean rate in that declination. The two observations reported in the figure correspond to different median energies, although the observations span a wide energy range to collect the large event sample needed for the determination of the small anisotropy amplitude. In particular IceCube reported that 68% of the data sample with median energy of about 20 TeV is between 4 TeV and 63 TeV [14]. Assuming an equivalent resolution for the Tibet ASγ, it turns out that the two data samples overlap significantly in cosmic ray energy, and therefore they can be directly compared. In Fig. 2 (panel A) it appears that the relative excess in arrival direction is oriented toward the region of the sky that comprises the downstream direction of the interstellar wind and the downfield direction for the LIsMF.

In order to accentuate the small scale features that appear on the global anisotropy sky map, the Milagro Collaboration subtracted a 30°-averaged sky map from that of the global anisotropy [see 15, for details]. The residual map highlights all the fine structures with characteristic angular scales that are smaller than 30°. A similar analysis was performed by the IceCube Observatory, where the averaging procedure was performed within 60°, meaning that only features smaller than such an angular scale are highlighted. Fig. 2 (panel B) shows the corresponding map of the statistical significance with respect to the averaged maps and with a 10° smoothing. The northern hemisphere shows the result from Milagro at a mean energy of

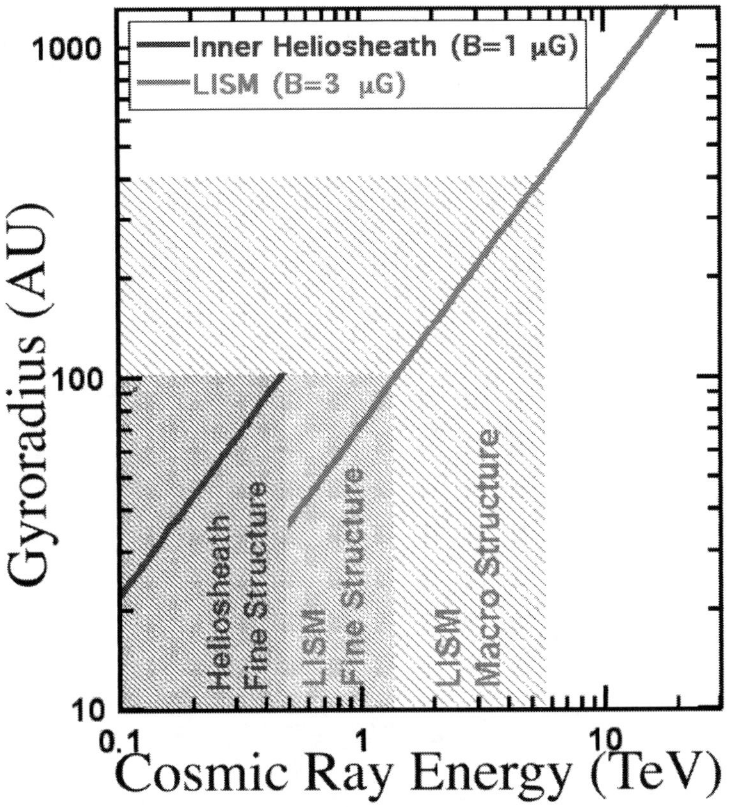

Figure 1. The gyroradius of cosmic rays in the heliosheath (1 μG, blue) and LISM (3 μG, red) probe increasingly large structures with increasing energy. Cosmic rays probe the heliosheath structure in the $\sim$ 0.1-0.5 TeV energy range, LISM fine structure < 100 AU in the $0.5 - 1.5$ TeV energy and LISM macrostructure including the distortion of the LISM field around the heliosphere of 100's of AU for energies between $\sim 1.5 - 10$ TeV. It is over this energy TeV energy range where cosmic rays probe the same magnetic structures from the LISM that likely produce the IBEX Ribbon. [From 9].

$\sim$1 TeV and the southern hemisphere shows the IceCube observations at a median energy of $\sim$ 20 TeV. Once again, although the characteristic energies of the two observations are different, the energy ranges of the observations significantly overlap. Some interesting visual correlation between the two halves of the sky arise.

In both sky maps, the localized excess of cosmic ray flux observed at equatorial coordinates $(\alpha, \delta) \approx$ (21 hr, +32°) is due to gamma rays from the Cygnus region to which both Tibet ASγ and Milagro are sensitive. In the southern hemisphere, the most significant localized excess and deficit regions at $(\alpha, \delta) \approx$ (7 hr, -45°) and $(\alpha, \delta) \approx$ (15 hr, -45°), respectively, correspond to the center of the global anisotropy regions, meaning that some of the large scale structure influences features in the residual sky map. The Milagro observation of a localized fractional excess in the direction of the heliotail does not appear to have a correspondance in the southern sky. On the other hand, the filamentary feature at right ascension $\alpha \approx$ 8-10 hr, appears to continue throughout the South in a quasi-circular pattern, even though here its statistical significance is not as high. Given the width of the energy response functions, it is possible that cosmic rays

Journal of Physics: Conference Series **577** (2015) 012023 doi:10.1088/1742-6596/577/1/012023

contributing to different localized regions have different characteristic energy. More observations are needed to refine these maps.

Fig. 2 (panel C) shows the IBEX Ribbon observed at 1.1 keV [3–6] in J2000 coordinates. We include labels for the upwind and downwind directions [16], the Voyager 1 (V1) and Voyager 2 (V2) directions and the upfield and downfield directions associated with the unperturbed LISM magnetic field. The black line shows the magnetic equator of the external LIsMF in the direction of the IBEX Ribbon center. The shift of the Ribbon to the inside of the LIsMF magnetic equator is likely due to flaring of the field on the noseward side of the heliosphere [6]. The global structure in all three panels is ordered about the LIsMF magnetic equator, which reinforces the concept that the LIsMF influences both TeV cosmic rays and the ENA fluxes observed by IBEX.

3. Local Interstellar Structure

The heliosphere interacts with the local interstellar flow that originated from the center of Loop I. Typical temperatures are $\sim 3000 - 13000$ K and proton densities $\sim 0.05 - 0.1$ cm^{-3} [18]. Deviations of the velocities of individual clouds from a rigid-body flow indicate that the flow is decelerating [19]. A decelerating flow sets the conditions for collisions between the local clouds. The Sun is inside of the LIC and no more than 0.1 pc away from the LIC edge in the direction of the galactic center hemisphere [20; 21]. In the direction of the anticenter hemisphere, the closest cloud to the LIC is the 'Blue Cloud' [BC, 22; 23]. Linsky *et al.* [24] show that the conditions for colliding clouds exist within ~ 10 pc, with relative velocities between overlapping clouds of up to 50 km s^{-1} (Figure 3). Among the closest of possible collisions would be a collision between the LIC and BC.

Cloud collisions in this region are supported by observations of short-timescale scintillation towards two compact radio sources, PKS 0405-385 and PSR J0437-47, which are located behind the BC [25]. Models of the timescales and frequency dependence of the scintillation show that the scattering screens, consisting of partially ionized gas such as the LIC, are within ~ 10 pc. Towards PSR J0437-47, the local turbulence occurs over spatial scales of ~ 0.02 AU [26]. PKS 0405-385 and PSR J0437-47 are located within 29° of the downfield direction.

The decelerating flow of local interstellar clouds past the Sun appears to be ordered by an evolved superbubble associated with Loop I [18]. An interstellar magnetic field interacting with the compressed, perhaps inhomogeneous, shell provides a natural explanation for inhomogeneities in the LIsMF. Figure 4 shows the flow of local clouds through the LSR for a meridianal cut through the galactic plane where the Z-direction is directed toward galactic north and the X-direction is toward galactic center. Measurements of interstellar He I inside of the heliosphere provide the LIC velocity [27; 28], while velocities for 15 local clouds have been determined with the triangulation of the radial velocity component towards different stars behind the clouds [25]. The Loop I superbubble shell is represented by the thick arc, which is based on the S1 shell of Wolleben [29] that is centered ≈ 78 pc away at galactic coordinates of L,B$\approx$346°,3° [29].

Pressure equilibrium between the gas and magnetic field in the LIC give a LIC magnetic field strength of ~ 2.7 μG, which is comparable to the ~ 3 μG field strength obtained from the globally distributed ENA fluxes [30].

4. Imprint of the heliosphere on the global anisotropy in TeV cosmic rays

Schwadron *et al.* [9] utilized the observations of IBEX to estimate the direction of the magnetic field in the LISM (Figure 5). Knowing in addition the flow velocity of the LISM based on neutral He measurements, we developed model estimates for the global anisotropy of TeV cosmic rays in the LISM associated with modulation by the local interstellar flow. The LISM flow vector in the heliocentric (HC), the Local Standard of Rest (LSR) frame, and the magnetic field vector are listed in Table 1. The flow configuration of the LISM has a flow velocity almost perpendicular

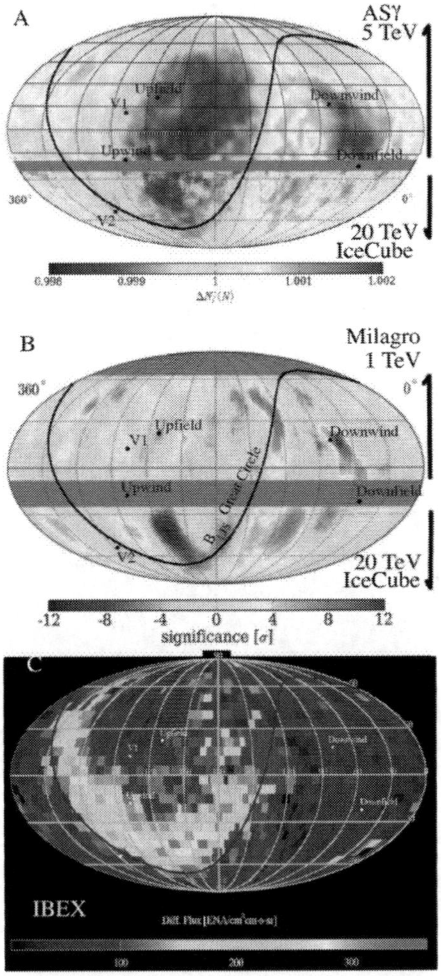

Figure 2. *Panel A*: relative intensity observed by Tibet Asγ collaboration at a median energy of ~ 5 TeV [17] in the north and by IceCube in the south at a median energy of ~ 20 TeV [13]. Both portions of the maps are smoothed with 3°-5°. *Panel B*: significance map that smoothes (over 30°-60°) and filters out the large scale anisotropy to reveal only the small scale anisotropy. The northern hemisphere is from the Milagro project, centered on ~ 1 TeV median energy [15]. The southern hemisphere is the map from the IceCube collaboration at $\sim$20 TeV median energy [13]. *Panel C*: The IBEX Ribbon observed at ~ 1.1 keV [3–6] in J2000 coordinates and the TeV Milagro/Ice Cube data. We include labels for the upwind and downwind directions [16], the Voyager 1 (V1) and Voyager 2 (V2) directions and the upfield and downfield directions associated with the unperturbed LIsMF. The TeV cosmic ray data appears to be ordered by the LISM magnetic equator. These plots are in equatorial coordinates with $\alpha = 0°$ at the right, and increasing towards the left.

to the local interstellar magnetic field and consistent with the outward expansion of plasma from the Loop I S1 superbubble (Fig. 5). The model also includes the propagation effects associated with deformation of the local interstellar magnetic field about the heliosphere. The resulting global anisotropy map in TeV cosmic rays is shown in Fig. 6. The full calculation is

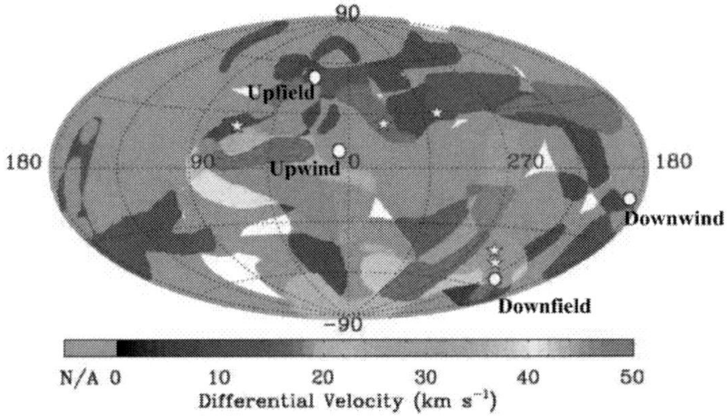

Figure 3. Zones of LISM cloud interaction with superposed upwind, downwind, upfield and downfield directions. Colored regions show the directions in which multiple LISM clouds are detected along the line of sight. The white regions indicate where no LISM cloud is detected, and gray regions where only a single cloud is detected. The color coding signifies the maximum magnitude of the differential velocity between clouds along the same line of sight. Five scintillation sight lines are shown here by star symbols. This plot is in galactic coordinates, and centered on the galactic center with longitude increasing towards the left. [Based on 24].

detailed by Schwadron et al. [9]. The main free parameter in these estimates of the global TeV anisotropy is the level of perpendicular diffusion for which we use typical parameters of the ISM that $\kappa_\perp/\kappa_\parallel \sim 0.003$ [32].

Table 1. The directions and magnitudes of the LISM velocity in the heliocentric (HC) rest frame, in the Local Standard of Rest (LSR), and the LISM magnetic field field inferred from the IBEX ribbon.

	Magnitude (km/s)	Galactic Long. (°)	Galactic Lat. (°)	Equatorial Rt. Asc. (°)	Equatorial Dec. (°)
LISM (HC frame[1])	23.2 ± 0.3	185.25 ± 0.24	-12.0 ± 0.5	78.5 ± 0.6	18.0 ± 0.5
Solar apex motion in LSR [2]	18.0 ± 0.9	47.9 ± 2.9	23.8 ± 2.0	267.0 ± 3.0	23.2 ± 3.1
LISM (LSR frame)	15.6 ± 0.9	138.7 ± 3.2	9.0 ± 2.2	55.3 ± 3.6	66.5 ± 1.1
LISM Mag. Field [3]		210.5 ± 2.6	-57.1 ± 1.0	48.5 ± 1.5	-21.2 ± 1.6

[1] Heliocentric (HC) rest frame [33]
[2] Solar apex motion in Local Standard of Rest, LSR [31]
[3] Field directions and uncertainty inferred from the IBEX highest energy steps (1.79 58 keV and 2.73 keV [34]) in which the ribbon maintains coherence and has the largest Line-of-Sight.

Supernova remants (SNR) also contribute to the anisotropy, at least in localized regions. While the direction of LISM magnetic field found by IBEX must play a significant role in ordering the cosmic ray anisotropy map, the sign of the dipolar anisotropy component could either align with or oppose the component that arises from local interstellar modulation. Further, the magnitude of the dipolar component of the cosmic ray anisotropy is poorly constrained by observations.

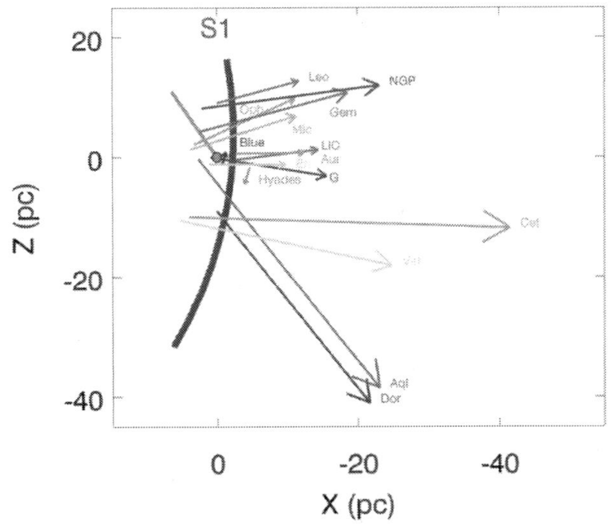

Figure 4. The arrows show the velocities of local interstellar clouds through the LSR in the X-Z meridian plane, where the X-axis is directed towards the galactic center and the Z-axis is directed towards the galactic north pole [based on heliocentric velocities in 25]. The LSR velocities are calculated using the solar apex motion [from 31]. The arrow origin is located at the cloud-center and the distance of the nearest star (that is not the Sun) in the cloud. Arrows are labeled with cloud names. The thick gray arc shows the location of the Loop I S1 superbubble, based on a thin shell with radius of 78 pc that is centered at galactic coordinates of L,B=346°,3° [29]. The purple line gives the direction of the LIsMF [4] and the orange dot shows the solar location.

5. Conclusions

The comparison shown in Figure 5 confirms the direction of the LISM magnetic field from the center of the IBEX ribbon. The result is somewhat surprising since the LISM magnetic field from the IBEX ribbon is determined in a very local interstellar environment, within 100's of AU from the heliosphere. Turbulence in the LISM might be expected to destroy signatures of the direction inferred from IBEX over 1-10 pc correlation scales. Such effects would obscure any relationship between global TeV cosmic ray anistropies and the IBEX ribbon. The fact that a relationship is observed suggests that the LISM magnetic field remains relatively coherent over scales of at least 1-10 pc. This in turn suggests that our heliosphere is currently moving through a magnetically ordered portion of the LISM. This reinforces the concept that the LISM magnetic field has sufficient pressure to play a significant role in LISM dynamics.

Thus, we have shown that TeV cosmic rays provide independent confirmation of the direction of the LISM magnetic field inferred from the center of the IBEX ribbon. The correlation is remarkable given that these independent observations are taken over vastly different energy ranges. IBEX measures energetic neutral atoms at thousands of eV and TeV anisotropy maps are formed from particles at higher energies by ten orders of magnitude. Our increasing understanding of the local interstellar environment is particularly important considering that the Voyager 1 spacecraft is now entering interstellar space.

Journal of Physics: Conference Series **577** (2015) 012023 doi:10.1088/1742-6596/577/1/012023

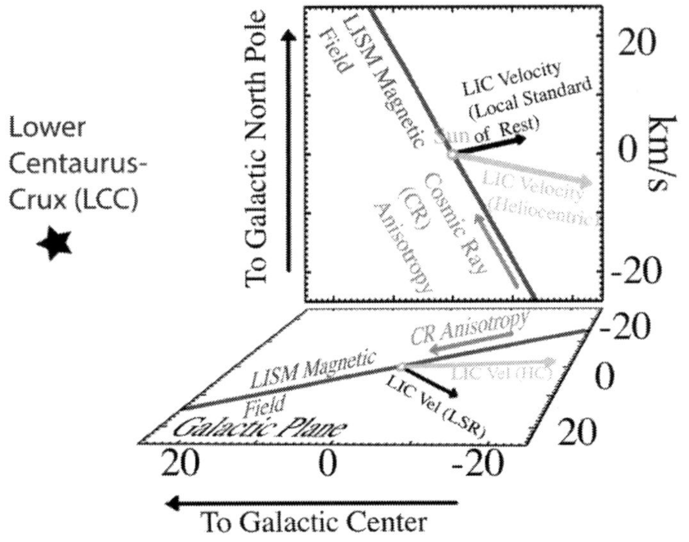

Figure 5. Directions and magnitudes of the LIC velocities inferred from IBEX in the heliocentric (HC; green vector) rest frame and the local standard of rest (LSR; black vector) in the galaxy. These vectors are shown in the galactic plane (lower plane) and the plane containing the galactic poles (upper plane) together with the magnetic field of the local interstellar medium (blue lines). The LIC velocity and magnetic field are a part of the Loop I superbubble roughly centered on the Lower Centaurus Crux (LCC). The cosmic ray anisotropy (grey vector) indicates average streaming of cosmic rays.

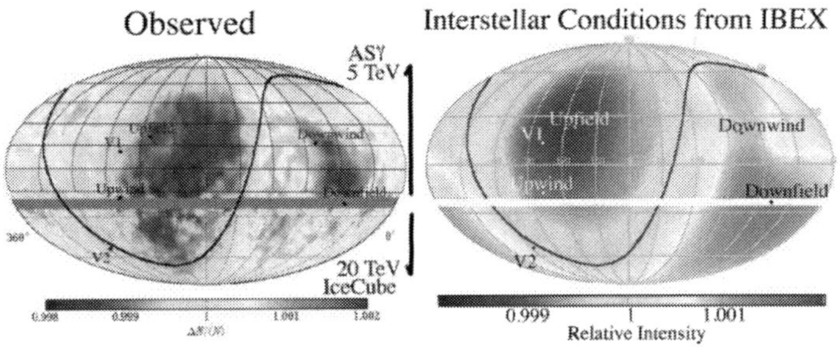

Figure 6. Comparison between observed (left) and modeled (right) cosmic ray intensities across the sky (J2000 coordinates). Black curves show the magnetic equator given a magnetic field direction derived from the center of the IBEX ribbon. (Left) The regions below 25° S latitude show the anisotropy map from IceCube with a median energy of 20 TeV and above 20° S latitude is the anisotropy map from AS-γ with 5 TeV median energy. (Right) The modeled map at 20 TeV is shown below 25° S latitude and at 5 TeV above 20° S latitude. [From 9].

5.1. Acknowledgments
We are deeply indebted to all of the outstanding people who have made the IBEX mission possible and have contributed to the Asγ, IceCube and Milagro projects. This work was carried out as a part of the IBEX project, with support from NASAs Explorer Program.

References

[1] McComas D J, Allegrini F, Bochsler P, Bzowski M, Collier M, Fahr H, Fichtner H, Frisch P, Funsten H O, Fuselier S A, Gloeckler G, Gruntman M, Izmodenov V, Knappenberger P, Lee M, Livi S, Mitchell D, Möbius E, Moore T, Pope S, Reisenfeld D, Roelof E, Scherrer J, Schwadron N, Tyler R, Wieser M, Witte M, Wurz P and Zank G 2009 *Space Sci. Rev.* **146** 11–33

[2] McComas D J, Funsten H O, Fuselier S A, Lewis W S, Möbius E and Schwadron N A 2011 *Geophys. Res. Lett.* **38** L18101

[3] McComas D J, Allegrini F, Bochsler P, Bzowski M, Christian E R, Crew G B, DeMajistre R, Fahr H, Fichtner H, Frisch P C, Funsten H O, Fuselier S A, Gloeckler G, Gruntman M, Heerikhuisen J, Izmodenov V, Janzen P, Knappenberger P, Krimigis S, Kucharek H, Lee M, Livadiotis G, Livi S, MacDowall R J, Mitchell D, Möbius E, Moore T, Pogorelov N V, Reisenfeld D, Roelof E, Saul L, Schwadron N A, Valek P W, Vanderspek R, Wurz P and Zank G P 2009 *Science* **326** 959–

[4] Funsten H O, Allegrini F, Crew G B, DeMajistre R, Frisch P C, Fuselier S A, Gruntman M, Janzen P, McComas D J, Möbius E, Randol B, Reisenfeld D B, Roelof E C and Schwadron N A 2009 *Science* **326** 964–

[5] Fuselier S A, Allegrini F, Funsten H O, Ghielmetti A G, Heirtzler D, Kucharek H, Lennartsson O W, McComas D J, Möbius E, Moore T E, Petrinec S M, Saul L A, Scheer J A, Schwadron N and Wurz P 2009 *Science* **326** 962

[6] Schwadron N A, Bzowski M, Crew G B, Gruntman M, Fahr H, Fichtner H, Frisch P C, Funsten H O, Fuselier S, Heerikhuisen J, Izmodenov V, Kucharek H, Lee M, Livadiotis G, McComas D J, Moebius E, Moore T, Mukherjee J, Pogorelov N V, Prested C, Reisenfeld D, Roelof E and Zank G P 2009 *Science* **326** 966–

[7] Möbius E, Bochsler P, Bzowski M, Crew G B, Funsten H O, Fuselier S A, Ghielmetti A, Heirtzler D, Izmodenov V V, Kubiak M, Kucharek H, Lee M A, Leonard T, McComas D J, Petersen L, Saul L, Scheer J A, Schwadron N, Witte M and Wurz P 2009 *Science* **326** 969–

[8] Krimigis S M, Mitchell D G, Roelof E C, Hsieh K C and McComas D J 2009 *Science* **326** 971–

[9] Schwadron N A, Adams F C, Christian E R, Desiati P, Frisch P, Funsten H O, Jokipii J R, McComas D J, Moebius E and Zank G P 2014 *Science* **343** 988

[10] Compton A H and Getting I A 1935 *Physical Review* **47** 817–821

[11] Gleeson L J and Axford W I 1968 *Astrophys. J.* **154** 1011

[12] The Tibet Asγ Collaboration, Amenomori M, Bi X J, Chen D, Cui S W, Danzengluobu, Ding L K, Ding X H, Fan C, Feng C F, Feng Z, Feng Z Y, Gao X Y, Geng Q X, Guo H W, He H H, He M, Hibino K, Hotta N, Hu H, Hu H B, Huang J, Huang Q, Jia H Y, Kajino F, Kasahara K, Katayose Y, Kato C, Kawata K, Labaciren, Le G M, Li A F, Li J Y, Lou Y Q, Lu H, Lu S L, Meng X R, Mizutani K, Mu J, Munakata K, Nagai A, Nanjo H, Nishizawa M, Ohnishi M, Ohta I, Onuma H, Ouchi T, Ozawa S, Ren J R, Saito T, Saito T Y, Sakata M, Sako T K, Shibata M, Shiomi A, Shirai T, Sugimoto H, Takita M, Tan Y H, Tateyama N, Torii S, Tsuchiya H, Udo S, Wang B, Wang H, Wang X, Wang Y, Wang Y G, Wu H R, Xue L, Yamamoto Y, Yan C T, Yang X C, Yasue S, Ye Z H, Yu G C, Yuan A F, Yuda T,

Zhang H M, Zhang J L, Zhang N J, Zhang X Y, Zhang Y, Yizhang, Zhaxisangzhu, Zhou X X and The Tibet ASγ Collaboration 2011 *Advances in Space Research* **47** 629–639

[13] Abbasi R, Abdou Y, Abu-Zayyad T, Adams J, Aguilar J A, Ahlers M, Andeen K, Auffenberg J, Bai X, Baker M and et al 2010 *Astrophys. J. Lett.* **718** L194–L198 (*Preprint* `1005.2960`)

[14] Abbasi R, Abdou Y, Abu-Zayyad T, Ackermann M, Adams J, Aguilar J A, Ahlers M, Allen M M, Altmann D, Andeen K and et al 2012 *Astrophys. J.* **746** 33 (*Preprint* `1109.1017`)

[15] Abdo A A, Allen B, Aune T, Berley D, Blaufuss E, Casanova S, Chen C, Dingus B L, Ellsworth R W, Fleysher L, Fleysher R, Gonzalez M M, Goodman J A, Hoffman C M, Hüntemeyer P H, Kolterman B E, Lansdell C P, Linnemann J T, McEnery J E, Mincer A I, Nemethy P, Noyes D, Pretz J, Ryan J M, Parkinson P M S, Shoup A, Sinnis G, Smith A J, Sullivan G W, Vasileiou V, Walker G P, Williams D A and Yodh G B 2008 *Physical Review Letters* **101** 221101 (*Preprint* `0801.3827`)

[16] Bzowski M, Kubiak M A, Moebius E, Bochsler P, Leonard T, Heirtzler D, Kucharek H, Sokol J M, Hlond M, Crew G B, Schwadron N A, Fuselier S and McComas D J 2012 *Astrophys. J. Suppl.* **In Press**

[17] Amenomori M, Bi X J, Chen D, Cui S W, Danzengluobu, Ding L K, Ding X H, Fan C, Feng C F, Feng Z, Feng Z Y, Gao X Y, Geng Q X, Guo H W, He H H, He M, Hibino K, Hotta N, Hu H, Hu H B, Huang J, Huang Q, Jia H Y, Kajino F, Kasahara K, Katayose Y, Kato C, Kawata K, Labaciren, Le G M, Li A F, Li J Y, Lou Y Q, Lu H, Lu S L, Meng X R, Mizutani K, Mu J, Munakata K, Nagai A, Nanjo H, Nishizawa M, Ohnishi M, Ohta I, Onuma H, Ouchi T, Ozawa S, Ren J R, Saito T, Saito T Y, Sakata M, Sako T K, Shibata M, Shiomi A, Shirai T, Sugimoto H, Takita M, Tan Y H, Tateyama N, Torii S, Tsuchiya H, Udo S, Wang B, Wang H, Wang X, Wang Y, Wang Y G, Wu H R, Xue L, Yamamoto Y, Yan C T, Yang X C, Yasue S, Ye Z H, Yu G C, Yuan A F, Yuda T, Zhang H M, Zhang J L, Zhang N J, Zhang X Y, Zhang Y, Yizhang, Zhaxisangzhu, Zhou X X and The Tibet ASγ Collaboration 2011 *Proc. of 32nd ICRC* (Beijing, China) p paper 0379

[18] Frisch P C, Redfield S and Slavin J D 2011 *Ann. Rev. Astron. Astrophys.* **49** 237–279

[19] Frisch P C, Grodnicki L and Welty D E 2002 *Astrophys. J.* **574** 834–846

[20] Wood B E, Linsky J L and Zank G P 2000 *Astrophys. J.* **537** 304–311

[21] Frisch P C and Mueller H R 2011 *Space Sci. Rev.* 130 (*Preprint* `1010.4507`)

[22] Lallement R, Bertin P, Ferlet R, Vidal-Madjar A and Bertaux J L 1994 *Astron. Astrophys.* **286** 898–908

[23] Hébrard G, Mallouris C, Ferlet R, Koester D, Lemoine M, Vidal-Madjar A and York D 1999 *Astron. Astrophys.* **350** 643–658 (*Preprint* `arXiv:astro-ph/9909061`)

[24] Linsky J L, Rickett B J and Redfield S 2008 *Astrophys. J.* **675** 413–419

[25] Redfield S and Linsky J L 2008 *Astrophys. J.* **673** 283–314 (*Preprint* `arXiv:0709.4480`)

[26] Smirnova T V, Gwinn C R and Shishov V I 2006 *Astron. Astrophys.* **453** 601–607 (*Preprint* `arXiv:astro-ph/0603490`)

[27] Bzowski M, Kubiak M A, Möbius E, Bochsler P, Leonard T, Heirtzler D, Kucharek H, Sokół J M, Hłond M, Crew G B, Schwadron N A, Fuselier S A and McComas D J 2012 *Astrophys. J. Suppl.* **198** 12 (*Preprint* `1202.0415`)

[28] Möbius E, Bochsler P, Bzowski M, Heirtzler D, Kubiak M A, Kucharek H, Lee M A, Leonard T, Schwadron N A, Wu X, Fuselier S A, Crew G, McComas D J, Petersen L, Saul L, Valovcin D, Vanderspek R and Wurz P 2012 *Astrophys. J. Suppl.* **198** 11

[29] Wolleben M 2007 *Astrophys. J.* **664** 349

[30] Schwadron N A, Allegrini F, Bzowski M, Christian E R, Crew G B, Dayeh M, DeMajistre R, Frisch P, Funsten H O, Fuselier S A, Goodrich K, Gruntman M, Janzen P, Kucharek H, Livadiotis G, McComas D J, Moebius E, Prested C, Reisenfeld D, Reno M, Roelof E, Siegel J and Vanderspek R 2011 *Astrophys. J.* **731** 56

[31] Schönrich R, Binney J and Dehnen W 2010 *MNRS* **403** 1829–1833 (*Preprint* 0912.3693)

[32] Shalchi A, Büsching I, Lazarian A and Schlickeiser R 2010 *Astrophys. J.* **725** 2117–2127

[33] McComas D J, Alexashov D, Bzowski M, Fahr H, Heerikhuisen J, Izmodenov V, Lee M A, Möbius E, Pogorelov N, Schwadron N A and Zank G P 2012 *Science* **336** 1291–

[34] Funsten H O, DeMajistre R, Frisch P C, Heerikhuisen J, Higdon D M, Janzen P, Larsen B A, Livadiotis G, McComas D J, Möbius E, Reese C S, Reisenfeld D B, Schwadron N A and Zirnstein E J 2013 *Astrophys. J.* **776** 30

Journal of Physics: Conference Series **577** (2015) 012024 doi:10.1088/1742-6596/577/1/012024

Using numerical models of bow shocks to investigate the circumstellar medium of massive stars

A.J. van Marle[1], **L. Decin**[1], **N.L.J. Cox**[1], **Z. Meliani**[2,3]

[1]Institute of Astronomy, KU Leuven, Celestijnenlaan 200D, B-3001, Leueven, Belgium
[2]Observatoire de Paris, 5 place Jules Janssen 92195 Meudon, France
[3] APC, Université Paris Diderot, 10 rue Alice Domon et Léonie Duquet, 75205 Paris Cedex 13, France

E-mail: `allardjan.vanmarle@ster.kuleuven.be`

Abstract. Many massive stars travel through the interstellar medium at supersonic speeds. As a result they form bow shocks at the interface between the stellar wind. We use numerical hydrodynamics to reproduce such bow shocks numerically, creating models that can be compared to observations. In this paper we discuss the influence of two physical phenomena, interstellar magnetic fields and the presence of interstellar dust grains on the observable shape of the bow shocks of massive stars.

We find that the interstellar magnetic field, though too weak to restrict the general shape of the bow shock, reduces the size of the instabilities that would otherwise be observed in the bow shock of a red supergiant. The interstellar dust grains, due to their inertia can penetrate deep into the bow shock structure of a main sequence O-supergiant, crossing over from the ISM into the stellar wind. Therefore, the dust distribution may not always reflect the morphology of the gas. This is an important consideration for infrared observations, which are dominated by dust emission.

Our models clearly show, that the bow shocks of massive stars are useful diagnostic tools that can used to investigate the properties of both the stellar wind as well as the interstellar medium.

1. Introduction

When a star moves through the interstellar medium (ISM), its wind collides with the interstellar gas. If the motion of the star is supersonic with respect to the sound speed in the ISM, this collision leads to the formation of a *bow shock* in the region ahead of the star. Because the size and shape of a bow shock is determined by the balance between the two ram pressure (of the stellar wind on the inside and of the motion of the ISM relative to the star on the outside) stellar wind bow shocks are powerful diagnostic tools that can help us determine the properties of the ISM, and the stellar wind. In addition, the instabilities that can occur in the bow shock can help us analyse the physical process that are taking place.

1.1. The general shape of stellar-wind bow shocks

Assuming that the interaction is supersonic with respect to both the stellar wind and the ISM (which is generally the case), the bow shock structure consists of four layers: the free-streaming stellar wind, the shocked stellar wind, the shocked ISM and the unshocked ISM. The free-streaming wind is separated from the shocked wind by the wind-termination shock. Similarly,

Journal of Physics: Conference Series **577** (2015) 012024 doi:10.1088/1742-6596/577/1/012024

the forward shock separates the shocked ISM from the unshocked ISM. Between the shocked wind and the shocked ISM lies a contact discontinuity.

The stand-off distance (R_D) between the star and the bow shock is determined by the ram pressure of the stellar wind and the ISM. These are in balance at a distance of

$$R_D = \sqrt{\frac{\dot{M}\, v_w}{4\pi \rho_{ISM}\, v_\star^2}}, \tag{1}$$

with $\dot{M}$ and v_w the mass loss rate and velocity of the stellar wind, ρ_{ISM} the density of the ISM and $v_\star$ the velocity of the star with respect to the local ISM [1]. Because R_D denotes the distance at which the stellar wind and the ISM are in equilibrium with one-another it actually gives us the location of the contact discontinuity, rather than either the wind termination shock, or the forward shock. In addition, we can use analytical approximations to determine the opening angle of the bow shock, which [1] described as

$$\frac{R(\theta)}{R_D} = \frac{1}{\sin\theta}\sqrt{3\left(1 - \frac{\theta}{\tan\theta}\right)} \tag{2}$$

with θ the angle between the direction of motion of the star and a line from the star to a particular point along the bow shock.

2. Numerical method

Although it is possible to predict the morphology of a bow shock analytically, including the general 2-D structure (E.g. [2, 3]), such analytical models are inherently limited in the amount and type of physical processes they can include as well as in their inability to quantitatively reproduce instabilities. Therefore, it becomes necessary to use numerical simulations to model the stellar wind bow shocks. We use the `MPI-AMRVAC` magneto-hydrodynamics code [4, 5], which solves the conservation equations for mass, momentum and energy on an adaptive mesh grid. For our calculations we include radiative cooling using the method described in [6], with a cooling curve for solar metallicity.

We model each bow shock in the co-moving frame of the star. We start our simulations by filling a 2-D cylindrical grid in the R,Z-plane with a constant density interstellar medium, which has a constant velocity parallel to the Z-axis and is kept constant by allowing material to flow in at the outer Z-boundary. At the lower Z-boundary, the material is allowed to flow out of the grid. The stellar wind is introduced by filling a small half sphere, centred on the origin, with gas according to a free-streaming stellar wind profile (constant velocity and density decreases with the radius squared).

Table 1. Physical parameters of the wind of α-Orionis and the local ISM, used as input for our simulations.

Mass loss rate	$\dot{M}$	=	3.0×10^{-6} $M_\odot$ yr^{-1}
Wind velocity	v_∞	=	15.0 km s^{-1}
Velocity w.r.t. ISM	$v_\star$	=	28.3 km s^{-1}
ISM density	ρ_{ISM}	=	$10^{-23.5}$ g cm^{-3}
ISM temperature	T_{ISM}	=	10 K
ISM magnetic field	B	=	3.0 μG

3. The influence of interstellar magnetic fields on the bow shock of α-Orionis

3.1. Background
Recent *Herschel* observations [7] show us that the bow shock of α-Orionis, a red supergiant (RSG) type evolved star is smooth, without large instabilities. However, both analytical predictions [2] and numerical models [8–10] indicate that a bow shock of this kind, where the stellar velocity through the ISM is larger than the wind velocity, should show large scale instabilities. Several explanations for this discrepancy have been offered, such as the possibility that the bow shock is too young [11, 12], or that the presence of ionizing photons reduces the instability [13]. An alternative explanation, as shown in [14], is that the interstellar magnetic field inhibits the growth of instabilities.

3.2. Interstellar magnetic fields
The ISM in the galaxy contains magnetic fields than can stretch out over large distances ($\simeq$ 100 pc) [15–18]. Estimates for the magnetic field in the Orion arm of the Galaxy at a distance of 8 000 kpc from the Galactic centre (corresponding to the approximate location of α-Orionis) range from $1.4 \pm 0.3\,\mu$G [19] through 2-3 μG in the region near our solar system [20] to 3.7-5.5 μG as obtained from *Voyager* measurements [21]. These values coincide with the values for the interstellar field in the galaxy at large obtained from *WMAP* data [22, 23].

3.3. Numerical setup
We use the basic numerical set-up described in Section 2, but include a magnetic field parallel to the Z-axis (and therefore parallel to the direction of motion of the star). This direction is chosen to preserve the 2-D symmetry of the problem and avoid the necessity to run the simulation in 3-D. For our computational grid we choose an initial size of of 160×160 grid cells, covering a physical domain of 2×2 pc. Using the adaptive mesh option of `MPI-AMRVAC`, we allow the code to refine this grid up to four times, depending on variations in the local gas velocity, which gives us $2\,560 \times 2\,560$ grid cells effectively. All physical input parameters for our model are give in Table 1 and are based on observational data from [24]. We run these simulations twice, once with and once without the magnetic field in order to compare the results.

3.4. Results
The result of our simulations, shown in Figs. 1 and 2, clearly demonstrate the influence of the magnetic field. Figure 1 shows the model without a magnetic field after 100 000 years physical time. The bow shock is clearly unstable at the contact discontinuity with both Kelvin-Helmholtz and Rayleigh-Taylor instabilities. These instabilities, which start out small in the region directly ahead of the star increase in size when they move down stream until they grow so large that they distort the general shape of the bow shock. For the model that includes an interstellar magnetic field (Fig. 2, the result is completely different. Instabilities do form initially at the front of the bow shock. However, rather than growing in size, they remain small as they move downstream. Also, the small scale structures that are clearly visible in the non-magnetic model are completely absent and the instabilities are limited to a single wavelength. This behaviour corresponds to the predictions of, for example, [25, 26], which show that a magnetic field inhibits the growth of instabilities with short wavelengths, with the critical wavelength determined by the magnetic field strength, the density contrast across the discontinuity and the angle between the wave-vector and the magnetic field. This was further demonstrated by [14], which showed models for a range of magnetic field strengths.

3.5. Discussion
It is clear from our models that the presence of an interstellar magnetic field can suppress the growth of instabilities in the bow shock of a RSG type star. Of course, this model is limited

by the fact that we have to align the magnetic field with the direction of motion of the star. Further study (in 3D) will be necessary to quantify this effect for magnetic fields at other angles.

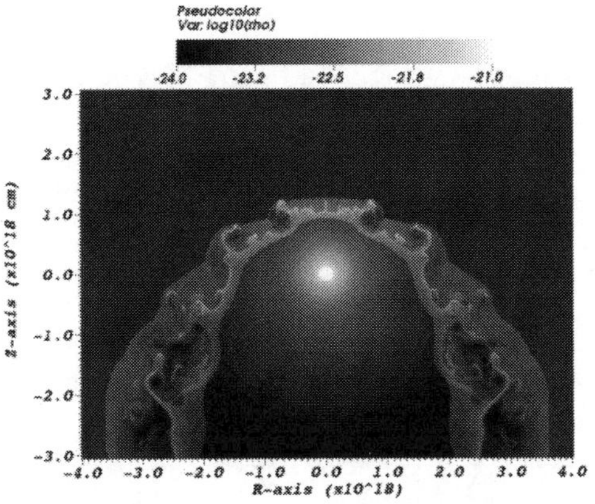

Figure 1. Density in $\mathrm{g\,cm^{-3}}$ for the non-magnetic model of the bow shock of α-Orionis. Note that the 2-D model has been reflected in the Z-axis to show a complete bow shock. The interface between the shocked wind and shocked ISM is clearly unstable.

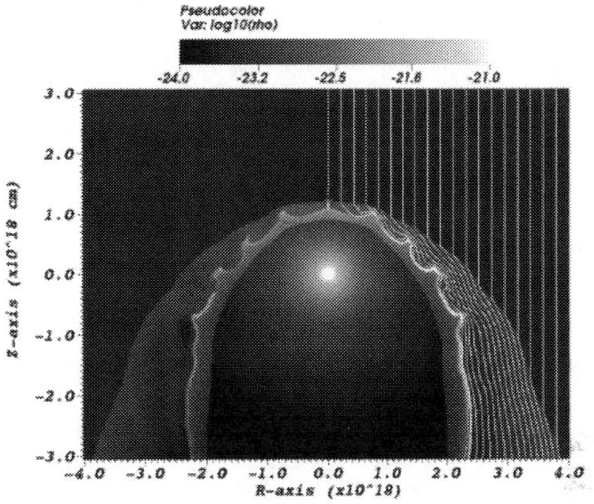

Figure 2. Similar to Fig. 1, but with an interstellar magnetic field of 3 μm. The field lines of the magnetic field are shown in the right side of the panel. The instabilities are much smaller than for the non-magnetic model.

Table 2. Physical parameters of the wind of an O-supergiant and the local ISM, used as input for our simulations.

Mass loss rate	$\dot{M}$	=	$1.0 \times 10^{-5}\ \mathrm{M_\odot\ yr^{-1}}$
Wind velocity	v_∞	=	$2300\ \mathrm{km\,s^{-1}}$
Velocity w.r.t. ISM	$v_\star$	=	$77\ \mathrm{km\,s^{-1}}$
ISM density	ρ_{ISM}	=	$10^{-23.5}\,\mathrm{g\,cm^{-3}}$
ISM temperature	T_{ISM}	=	$10\,000\,\mathrm{K}$
dust grain sizes	a_1, a_2, a_3	=	0.071 μm, 0.19 μm, 0.37 μm

4. Dust and gas in circumstellar bow shocks

4.1. Background

The introduction of satellites like *Spitzer* and *Herschel*, has allowed us to resolve circumstellar structures,such as bow shocks, in the infrared. However, rather than observing the morphology of the gas directly, these infared observations actually show us the distribution of dust grains, which are the primary source of infrared radiation. Therefore, it is absolutely necessary to investigate whether such dust grains, typically more than 0.5% of the total circumstellar mass, are representative for the gas. This was done for a red supergiant type star by [10] where the star itself was assumed to be the primary source of dust. These simulations showed that larger dust grains (> 0.045 μm) tend to decouple from the gas, once the gas is decelerated by the bow shock.

We now investigate the behaviour of dust grains in the bow shock of a hot star, where the situation is reversed. The wind of a hot massive star does not contain dust, therefore the main source for dust is the interstellar medium.

4.2. Numerical setup

We use the basic numerical set-up described in Section 2 and include the presence of the interstellar dust grain by filling the interstellar medium with dust grains of three different radii (radii of 0.071 μm, 0.19 μm, and 0.37 μmrespectively), each representing a 'bin' of grain sizes. These dust grains are treated as pressure-less gasses according to the same method described in [10], with the interaction between dust and gas included in the form of a drag force [27]. We assume that the total dust mass equals 0.5% of the ISM gas mass. The number densities of the three dust types are scaled in such a way that al three 'bins' contain an equal amount of mass and that the dust follows the size distribution of $n(a) \propto a^{-3.5}$ with n the particle density and a the grain radius described by [28]. The stellar wind and ISM parameters for our simulation are given in Table 2, based on the O4 supergiant BD+43 3654 [29], with stellar wind properties for such a star estimated according to [30]. We assume a warm ISM (10 000 K) because the stellar radiation can be expected to ionize the surrounding hydrogen, creating an HII region that extends well beyond the bow shock [31].

For this simulation we use a basic grid of 160×160 cells covering a physical domain of 10×10 parsec. The adaptive mesh is allowed four additional levels, giving us a maximum effective grid of $2\,560 \times 2\,560$ cells.

4.3. Results

The result of our simulation is shown in Figs. 3 and4. The gas (left side of Fig. 3) shows that the forward shock is highly radiative, causing the shocked ISM to be compressed into a thin shell. The smallest dust grains (right side of Fig. 3) show much less compression. These grains, which originate in the ISM, penetrate the shocked ISM shell and enter the shocked wind region behind it. They only come to a stop at the wind termination shock due to the increases drag force generated by the unshocked wind moving in the opposite direction. the intermediate and large dust grains (left and right side of Fig. 4) show that these grains, which have a larger momentum compared to their surface area are able to penetrate into the unshocked wind.

4.4. Discussion

Our simulations show that under the circumstances of our particular model, the dust distribution can deviate considerably from the gas morphology. This can have serious consequences for observations of such bow shocks. In this particular case, observations at visual wavelengths, dominated by gas emission would show a thin, highly compressed bow shock (the high density shell of shocked ISM), whereas infrared observations, dominated by the dust, would show a thick shell.

5. Conclusions

We have shown that numerical models can be used to reproduce the bow shocks of massive stars. These models can also be used to investigate the effect of physical phenomena, such as magnetic fields and the presence of dust grains, which can influence the structure of the bow shock and/or the manner in which it will appear in observations. In the future we hope to continue our research in this field by including additional physical effects, such as thermal conduction and the interaction between dust grains and the magnetic field. We also intend to extend our models to 3-D in order to investigate what occurs/ when the 2-D symmetry is broken. E.g. when magnetic field and direction of motion are not aligned.

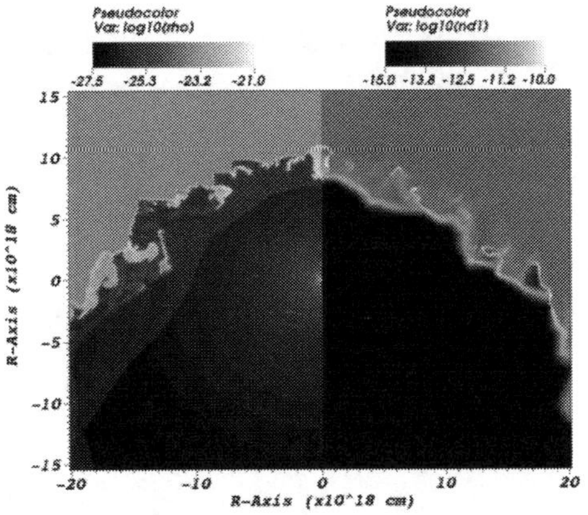

Figure 3. Gas density in $\mathrm{g\,cm^{-3}}$ of the gas (left) and number density for small dust grains (right) for the bow shock of a massive O4 supergiant. The gas shows a thin, unstable shell of shocked ISM, whereas the dust grains are spread out over the entire shocked gas region with the highest concentration at the wind termination shock.

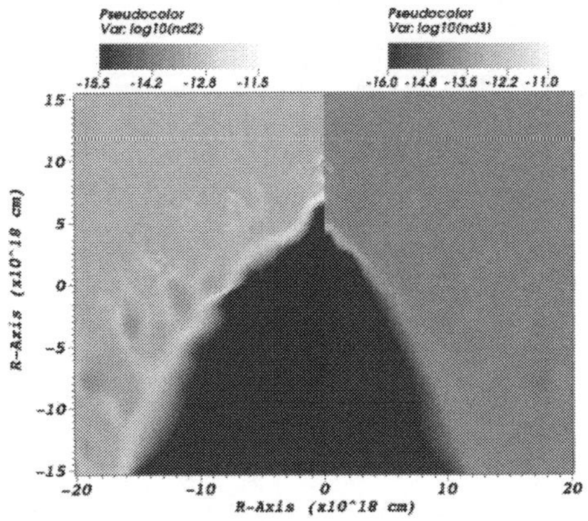

Figure 4. Similar to Fig. 3, but for the number density of intermediate (left) and large (right) dust grains. Owing to their larger inertia, these grains cross both the shocked ISM and the shocked wind and enter the free-streaming wind region before the drag-force stops them, with the largest grains showing the deepest penetration.

Acknowledgments
A.J.v.M. acknowledges support from FWO, grant G.0277.08, K.U.Leuven GOA/2008/04 and GOA/2009/09.

References
[1] Wilkin F P 1996 *ApJL* **459** L31
[2] Dgani R, van Buren D and Noriega-Crespo A 1996 *ApJ* **461** 927
[3] Schulreich M M and Breitschwerdt D 2011 *A&A* **531** A13
[4] van der Holst B, Keppens R and Meliani Z 2008 *CoPhC* **179** 617
[5] Keppens R, Meliani Z, van Marle A J, Delmont P, Vlasis A and van der Holst B 2012 *JCoPh* **231** 718
[6] van Marle A J and Keppens R 2011 *C&F* **42** 44
[7] Decin, L et al 2012 *A&A* **548** A113
[8] Brighenti F and D'Ercole A 1995 *MNRAS* **277** 53
[9] Comeron F and Kaper L 1998 *A&A* **338** 273
[10] van Marle A J, Meliani Z, Keppens R and Decin L 2011 *ApJL* **734** L26
[11] Mohamed S, Mackey J and Langer N 2012 *A&A* **541** A1
[12] Mackey J, Mohamed S, Neilson H R, Langer N and Meyer D M A 2012 *ApJL* **751** L10
[13] Meyer D M A, Gvaramadze V V, Langer N, Mackey J, Boumis P and Mohamed S 2014 *MNRAS* **439** L41
[14] van Marle A J, Decin L and Meliani Z 2014 *A&A* **561** A152
[15] Rand R J and Kulkarni S R 1989 *ApJ* **343** 760
[16] Ohno H and Shibata S 1993 *MNRAS* **262** 953
[17] Beck R 2009 *ASTRA* **5** 43
[18] Shabala S S, Mead J M G and Alexander P 2010 *MNRAS* **405** 1960
[19] Frick P, Stepanov R, Shukurov A and Sokoloff D 2001 *MNRAS* **325** 649
[20] Heerikhuisen J and Pogorelov N V 2011 *ApJ* **738** 29
[21] Opher M, Bibi F A, Toth G, Richardson J D, Izmodenov V V and Gombosi T I 2009 *Natur* **462** 1036
[22] Jansson R and Farrar G R 2012 *ApJ* **757** 14

[23] Jansson R and Farrar G R 2012 *ApJL* **761** L11
[24] Ueta T, Izumiura H, Yamamura I, Nakada Y, Matsuura M, Ita Y, Tanabé T, Fukushi H, Matsunaga N and Mito H 2008 *PASJ* **60** 407
[25] Jun B I, Norman M L and Stone J M 1995 *ApJ* **453** 332
[26] Breitschwerdt D, Freyberg M J and Egger R 2000 *A&A* **361** 303
[27] Kwok S 1975 *ApJ* **198** 583
[28] Draine B T and Lee H M 1984 *ApJ* **285** 89
[29] Comerón F and Pasquali A 2007 *A&A* **467** L23
[30] Muijres L E, Vink J S, de Koter A, Müller P E and Langer N 2012 *A&A* **537** A37
[31] McKee C F, van Buren D and Lazareff B 1984 *ApJL* **278** L115

13th Annual International Astrophysics Conference: Voyager, IBEX, and the Interstellar Medium IOP Publishing
Journal of Physics: Conference Series **577** (2015) 012025 doi:10.1088/1742-6596/577/1/012025

Plasma Physics of the Very Local Interstellar Medium

G P Zank, P Hunana, and P Mostafavi

Center for Space Plasma and Aeronomic Research (CSPAR) and Department of Space Science, University of Alabama in Huntsville, Huntsville, AL 35805, USA

E-mail: `garyp.zank@gmail.com`

Abstract. Models of the large-scale heliosphere show that the very local interstellar medium (VLISM), bounded by the heliopause and either a bow wave or bow shock, experiences significant heating by the deposition of neutral hydrogen (H) originating from the hot inner heliosheath. As hot neutrals stream into the interstellar medium, they experience charge exchange with the background cooler interstellar protons, creating a population of energetic ($\sim$ 1 keV) pickup ions. Similarly, fast neutrals created in the supersonic solar wind stream into the local interstellar medium and create a pickup ion population in the VLISM. The importance of these pickup ions is thought to manifest itself in the creation of the IBEX ribbon. Like the outer regions of the supersonic solar wind and the inner heliosheath, the VLISM is a pickup ion mediated plasma. Here we derive a model of a pickup ion mediated plasma using an approach analogous to a Chapman-Enskog expansion. We derive the anomalous heat flux and obtain a three-fluid model comprising electrons, thermal protons, and pickup ions. We investigate waves in a pickup ion mediated VLISM plasma, comparing the basic properties to those of the better known two-fluid model.

1. Introduction

The interstellar plasma upwind of the heliopause is mediated by energetic PUIs. It was noted already by Zank et al., 1996 [1] that energetic neutral H created via charge exchange in the inner heliosheath (IHS) and fast solar wind could "splash" back into the VLISM where they would experience a secondary charge exchange. The secondary charge exchange of hot and/or fast neutral H with cold ($\sim$ 6300 K [2]) VLISM protons leads to the creation of a hot or suprathermal PUI population in the VLISM. The heating of the VLISM has been discussed in detail by Zank et al., 2013 [3] who pointed out that the heating of the VLISM plasma would result in an increase of the sound speed with a concomitant weakening or even elimination of the bow shock (yielding instead a bow wave). Zirnstein et al., 2014 [4] have extended the Zank et al., 2010 [5] model by including the multiple PUI populations that contribute to the heating of the VLISM [3]. Here, we restrict out attention to neutrals created in both the IHS and supersonic solar wind, i.e., with typical speeds of $\sim$ 100 km/s or $\sim$ 400 km/s, that experience secondary charge exchange in the VLISM. The PUIs form a tenuous ($n_p \simeq 5 \times 10^{-5}$ cm^{-3} [4]) suprathermal component in the VLISM.

Coulomb collisions can equilibrate the background VLISM thermal plasma and the PUI protons. Assume the background VLISM is a Maxwellian plasma comprised of thermal protons and electrons. PUIs then satisfy the ordering

$$v_{ts} \ll v_p < v_{te},$$

where $v_{ts/e}$ denotes the background proton/electron thermal speed respectively and v_p the PUI speed. For PUIs scattering collisionally off a Maxwellian distribution of background protons, the collision frequency becomes

$$\nu_s^{ps} = \frac{n_s e^4 \ln\Lambda}{2\pi\varepsilon_0^2 m_p^2 v^3} s^{-1}, \tag{1}$$

illustrating the well-known v^{-3} dependence with PUI speed. By contrast, PUI collisional scattering off a Maxwellian electron background yields a larger collision frequency which is given by

$$\nu_s^{pe} = \frac{n_e e^4 \ln\Lambda m_e^{1/2}}{2(2\pi)^{3/2}\varepsilon_0^2 (kT_e)^{3/2} m_p}. \tag{2}$$

If the collisional time scale exceeds the characteristic flow time of the plasma region of interest, $\tau_f \simeq L/U$, where L is the size of the region and U the characteristic velocity, then the PUI distribution will not equilibrate with the background plasma. Expressions (1) and (2) are necessary to determine whether one needs to introduce a model that distinguishes the PUIs from the background plasma protons. In the case of the supersonic solar wind, Isenberg, 1986 [6] used related arguments to justify the introduction of a form of multi-fluid model to describe a coupled solar wind - PUI plasma. Isenberg [6] assumed instantaneous isotropization of the PUI distribution at the time of creation. While a reasonable approximation when considering the effects of PUIs on a large-scale flow such as the supersonic solar wind, to understand the impact of PUIs on the basic plasma physical system requires that the PUI-scattering time scale not be neglected. As we discuss below, PUI scattering yields the dominant heat flux and dissipative (collisionless viscosity) processes in a collisionless coupled thermal plasma - PUI system.

For the VLISM plasma, we take the background proton and electron number density to be approximately $n_{s/e} \sim 0.08$ cm^{-1} and the PUI speed to be $\sim$ 100 km/s (i.e., H created in the IHS). This yields a lower limit on the PUI-proton collision time of $\tau_s^{ps} = (\nu_s^{ps})^{-1} \sim 1.3 \times 10^9$ s. For a length scale of 75 AU, and typical flow speed of 15 km/s, the characteristic flow time is $\tau_f \sim 7.5 \times 10^8$ s. For PUI-electron collisions, we assume an electron temperature, $T_e \sim 10,000$ K. Consequently, $\tau_s^{pe} \sim 3.2 \times 10^8$, which is comparable to the characteristic flow time τ_f on a scale of 75 AU. Thus, neither proton nor electron collisions can equilibrate a PUI-thermal plasma in the VLISM on scales smaller than at least 75 AU.

2. Governing model

In deriving a multi-fluid model that includes PUIs, we shall assume that the distribution functions for the background protons and electrons are each Maxwellian, which ensures the absence of heat flux or stress tensor terms for the background plasma. The exact continuity, momentum, and energy equations governing the thermal electrons and protons are therefore given by

$$\frac{\partial n_e}{\partial t} + \nabla \cdot (n_e \mathbf{u}_e) = 0; \tag{3}$$

$$m_e n_e \left(\frac{\partial \mathbf{u}_e}{\partial t} + \mathbf{u}_e \cdot \nabla \mathbf{u}_e \right) = -\nabla P_e - e n_e (\mathbf{E} + \mathbf{u}_e \times \mathbf{B}); \tag{4}$$

$$\frac{\partial P_e}{\partial t} + \mathbf{u}_e \cdot \nabla P_e + \gamma_e P_e \nabla \cdot \mathbf{u}_e = 0, \tag{5}$$

for the electrons, and

$$\frac{\partial n_s}{\partial t} + \nabla \cdot (n_s \mathbf{u}_s) = 0; \tag{6}$$

$$m_p n_s \left(\frac{\partial \mathbf{u}_s}{\partial t} + \mathbf{u}_s \cdot \nabla \mathbf{u}_s \right) = -\nabla P_s + e n_s \left(\mathbf{E} + \mathbf{u}_s \times \mathbf{B} \right); \tag{7}$$

$$\frac{\partial P_s}{\partial t} + \mathbf{u}_s \cdot \nabla P_s + \gamma_s P_s \nabla \cdot \mathbf{u}_s = 0, \tag{8}$$

for the protons. Here $n_{e/s}$, $\mathbf{u}_{e/s}$, and $P_{e/s}$ are the macroscopic fluid variables for the electron/proton number density, velocity, and pressure respectively, $\gamma_{e/s}$ the electron/proton adiabatic index, $\mathbf{E}$ the electric field, $\mathbf{B}$ the magnetic field, and e the charge of an electron.

Pickup ions initially form an unstable distribution that excites Alfvénic fluctuations. The self-generated fluctuations and *in situ* turbulence serve to scatter PUIs in pitch-angle. The Alfvén waves and magnetic field fluctuations both propagate and convect with the bulk velocity of the system $\mathbf{U} = \mathbf{U}(\mathbf{u}_e, \mathbf{u}_s, \mathbf{u}_p, n_e, n_s, n_p, m_e, m_p)$, where n_p and $\mathbf{u}_p$ refer to PUI variables. The PUIs are governed by the Fokker-Planck transport equation with a collisional term $\delta f / \delta t|_c$,

$$\frac{\partial f}{\partial t} + \mathbf{v} \cdot \nabla f + \frac{e}{m_p} \left(\mathbf{E} + \mathbf{v} \times \mathbf{B} \right) \cdot \nabla_v f = \left. \frac{\delta f}{\delta t} \right|_c, \tag{9}$$

for average electric and magnetic fields $\mathbf{E}$ and $\mathbf{B}$. We assume that the velocity $\mathbf{v}$ of PUIs is always non-relativistic. On transforming the transport equation (9) into a frame that ensures there is no change in PUI momentum and energy due to scattering, assuming that the cross-helicity is zero, and introducing the random velocity $\mathbf{c} = \mathbf{v} - \mathbf{U}$, we obtain

$$\frac{\partial f}{\partial t} + (U_i + c_i) \frac{\partial f}{\partial x_i} + \left[\frac{e}{m_p} \left(\mathbf{E} + \mathbf{U} \times \mathbf{B} \right)_i + \frac{e}{m_p} \left(\mathbf{c} \times \mathbf{B} \right)_i - \frac{\partial U_i}{\partial t} - (U_j + c_j) \frac{\partial U_i}{\partial x_j} \right] \frac{\partial f}{\partial c_i} = \left. \frac{\delta f}{\delta t} \right|_c. \tag{10}$$

The velocity $\mathbf{U}$ is still unspecified so we choose $\mathbf{U}$ such that $\mathbf{E}' \equiv \mathbf{E} + \mathbf{U} \times \mathbf{B} = 0$. This assumption corresponds to choosing

$$\mathbf{U}_\perp = \mathbf{U} - \mathbf{U}_\parallel = \frac{\mathbf{E} \times \mathbf{B}}{B^2} \equiv \mathbf{U}, \tag{11}$$

since we choose $\mathbf{U}_\parallel = 0$ ($\mathbf{U}_\parallel$ is parallel to $\mathbf{B}$ and therefore arbitrary). This corresponds to expressing (9) in the guiding center frame. The use of the velocity $\mathbf{U}$ then yields

$$\frac{\partial f}{\partial t} + (U_i + c_i) \frac{\partial f}{\partial x_i} + \left[\frac{e}{m_p} \left(\mathbf{c} \times \mathbf{B} \right)_i - \frac{\partial U_i}{\partial t} - (U_j + c_j) \frac{\partial U_i}{\partial x_j} \right] \frac{\partial f}{\partial c_i} = \left. \frac{\delta f}{\delta t} \right|_c. \tag{12}$$

By taking moments of (12), we can derive the evolution equations for the macroscopic PUI variables, such as the number density $n_p = \int f d^3c$, velocity $n_p u_{p_i} = \int c_i f d^3c$, and so on. Although unspecified for now, we shall assume that moments of the collisional term $\delta f / \delta t|_c$ are zero. This can be checked against the particular scattering model that we use below. The zeroth moment of (12) yields the continuity equation for PUIs,

$$\frac{\partial n_p}{\partial t} + \frac{\partial}{\partial x_i} \left(n_p \left(U_i + u_{p_i} \right) \right) = 0, \tag{13}$$

where $\mathbf{u}_p$ is the PUI bulk velocity in the guiding center frame. For the first moment, we multiply (12) by c_j and integrate over velocity space. This yields, after a little algebra,

$$\frac{\partial}{\partial t} \left(n_p \left(U_j + u_{p_j} \right) \right) + \nabla \cdot \left[n_p \mathbf{U} \left(U_j + u_{p_j} \right) + n_p \mathbf{u}_p U_j \right] + \frac{\partial}{\partial x_i} \int c_i c_j f d^3c = \frac{e}{m_p} n_p \varepsilon_{jkl} u_{p_k} B_l, \tag{14}$$

where ε_{ijk} is the Levi-Civeta tensor.

To close equation (14), we need to evaluate the PUI distribution function f, which requires that we solve (12). In solving (12), we assume 1) that the PUI distribution is gyrotropic, and 2) that scattering of PUIs is sufficiently rapid to ensure that the PUI distribution is nearly isotropic. We can therefore average (12) over gyrophase, obtaining the "focused transport equation" for non-relativistic particles [8]. Details of the derivation can be found in Ch. 5 of [7]. To solve the gyrophase-averaged transport equation requires that we specify the scattering or collisional operator. We make the simplest possible choice, which is the isotropic pitch-angle diffusion operator,

$$\frac{\partial}{\partial \mu}\left(\nu_s(1-\mu^2)\frac{\partial f}{\partial \mu}\right), \tag{15}$$

where $\mu = \cos\theta$ is the cosine of the particle pitch-angle θ, and $\nu_s = \tau_s^{-1}$ is the scattering frequency. The form of the scattering operator (15) allows us to solve the focused transport equation using a Legendre polynomial expansion of the distribution function f. The first-order correct solution to the gyrophase-averaged form of equation (12) is

$$f \simeq f_0 + \mu f_1; \tag{16}$$

$$f_0 = f_0(\mathbf{x}, c, t); \tag{17}$$

$$f_1 = -\frac{c\tau_s}{3} b_i \frac{\partial f_0}{\partial x_i} + \frac{DU_i}{Dt}\frac{\tau_s}{3} b_i \frac{\partial f_0}{\partial c}, \tag{18}$$

where $c = |\mathbf{c}|$ is the particle random speed, $\mathbf{b} \equiv \mathbf{B}/B$ is a directional unit vector defined by the magnetic field, and $D/Dt \equiv \partial/\partial t + U_i \partial/\partial x_i$ is the convective derivative. Both f_0 and f_1 are functions of position, time, and particle random speed c i.e., independent of pitch-angle μ (and of course gyrophase ϕ). Of particular importance is the retention of the large-scale velocity, acceleration, and shear terms. These terms are often neglected in the derivation of the transport equation describing f_0 (for relativistic particles, the transport equation is the familiar cosmic ray transport equation). Thus, the second term in (18) is typically neglected. However, the velocity derivative term in the moment equation (18) is well known in relativistic transport theory for cosmic rays and is known as the relativistic heat inertia term [e.g. 11,12,13]. In the context of deriving a multi-fluid model, retaining these terms is essential to derive the correct multi-fluid formulation for PUIs. By introducing

$$\begin{aligned}\int c_i c_j f d^3c &= \int (c_i - u_{p_i})(c_j - u_{p_j}) f d^3c + n_p u_{p_i} u_{p_j} \equiv \int c_i' c_j' f d^3c + n_p u_{p_i} u_{p_j} \\ &\simeq \int c_i' c_j' (f_0 + \mu f_1) d^3c + n_p u_{p_i} u_{p_j},\end{aligned}$$

we show that

$$\frac{\partial}{\partial x_i}\int c_i' c_j' f_0 d^3c = \frac{1}{m_p}\frac{\partial}{\partial x_i}\left(\delta_{ij} P_p\right), \qquad \text{and} \qquad \int c_i' c_j' \mu f_1 d^3c = 0, \tag{19}$$

where

$$P_p \equiv m_p \frac{4\pi}{3}\int c'^2 f_0 c'^2 dc. \tag{20}$$

Consequently, the PUI stress tensor is identically zero at the first-order and the pressure tensor is isotropic, $\delta_{ij} P_p$. Including the second-order terms in the Legendre polynomial expansion of the gyrophase-averaged transport equation does in fact yield a non-zero collisionless stress tensor. The PUI momentum equation to first-order can therefore be expressed as

$$\frac{\partial}{\partial t}\left(n_p\left(U_j + u_{p_j}\right)\right) + \frac{\partial}{\partial x_i}\left[n_p\left(\mathbf{U} + \mathbf{u}_p\right)\left(U_j + u_{p_j}\right) + \frac{1}{m_p}\delta_{ij} P_p\right] = \frac{e}{m_p} n_p \varepsilon_{jkl} u_{p_k} B_l. \tag{21}$$

To derive the transport equation for P_p, we multiply (12) by $\frac{1}{2}c^2$ and integrate over d^3c. We then use (16) - (18) to evaluate the various integrals. Introducing $\mathbf{c}' \equiv \mathbf{c} - \mathbf{u}_p$ as before, we find

$$\int \frac{1}{2}c^2 f_0 d^3c = \frac{3}{2}\frac{1}{m_p}P_p + \frac{1}{2}n_p u_p^2,$$

for example. Similarly, we find that the heat flux $\mathbf{q}(\mathbf{x}, t)$ can be expressed as

$$q_i(\mathbf{x}, t) \equiv \int \frac{1}{2}c'^2 c'_i f d^3c' = \frac{1}{2}\int c^2 c_i f d^3c - \frac{5}{2}\frac{1}{m_p}u_{p_i}P_p - \frac{1}{2}n_p u_p^2 u_{p_i}. \quad (22)$$

It then follows that

$$\int \frac{1}{2}c'^2 c'_i f_0 d^3c = \pi \int c'^3 \mu b_i f_0 c'^2 dc' = 0,$$

and

$$\int \frac{1}{2}c'^2 c'_i \mu f_1 d^3c' = -\frac{2\pi}{3}\int c'^2 \kappa_{ij}\frac{\partial f_0}{\partial x_j}c'^2 dc' = -\frac{1}{2}\bar{\kappa}_{ij}\frac{\partial P_p}{\partial x_j} = q_i(\mathbf{x}, t). \quad (23)$$

In (23), we introduced the spatial diffusion coefficient

$$\kappa_{ij} \equiv b_i \frac{c^2 \tau_s}{3} b_j, \quad (24)$$

together with PUI speed-averaged form $\bar{\kappa}_{ij} \equiv K_{ij}$. The collisionless heat flux for PUIs is therefore described in terms of the PUI pressure gradient and consequently the averaged spatial diffusion introduces a PUI diffusion time and length scale into the multi-fluid system.

We obtain, after some algebra, the transport equation for the PUI pressure

$$\frac{\partial P_p}{\partial t} + (\mathbf{u}_p + \mathbf{U}) \cdot \nabla P_p + \frac{5}{3}P_p \nabla \cdot (\mathbf{u}_p + \mathbf{U}) = \frac{1}{3}\nabla \cdot (\mathbf{K} \cdot \nabla P_p), \quad (25)$$

illustrating that the PUI heat flux yields a spatial diffusion term in the PUI equation of state. The PUI system of equations is properly closed and correct to the first-order. The PUI total energy equation may be expressed as

$$\frac{\partial}{\partial t}\left(\frac{3}{2}P_p + \frac{1}{2}n_p (u_p + U)^2\right) + \frac{\partial}{\partial x_i}\left[\frac{1}{2}n_p (u_p + U)^2 (u_{p_i} + U_i) + \frac{5}{2}P_p (u_{p_i} + U_i) - \frac{1}{2}K_{ij}\frac{\partial P_p}{\partial x_j}\right]$$
$$= \frac{e}{m_p}\varepsilon_{ijk} n_p u_{p_j} B_k (u_{p_i} + U_i) . (26)$$

The full system of PUI equations is given by (13), (21), and (25) or (26). It is not particularly illuminating to work in the guiding center frame, and we simplify (13), (21), and (25), (26), by letting

$$\mathbf{U}_p = \mathbf{u}_p + \mathbf{U}.$$

The RHS of equations (21) and (26) is proportional to $\mathbf{u}_p \times \mathbf{B}$, which becomes

$$(\mathbf{U}_p - \mathbf{U}) \times \mathbf{B} = \mathbf{E} + \mathbf{U}_p \times \mathbf{B},$$

since $\mathbf{E}$ was perpendicular to $\mathbf{B}$ by construction. Hence the PUI fluid equations can be written in the more familiar form

$$\frac{\partial n_p}{\partial t} + \nabla \cdot (n_p \mathbf{U}_p) = 0; \quad (27)$$

$$\frac{\partial}{\partial t}(n_p \mathbf{U}_p) + \nabla \cdot [n_p \mathbf{U}_p \mathbf{U}_p + \mathbf{I}P_p] = \frac{e}{m_p}n_p (\mathbf{E} + \mathbf{U}_p \times \mathbf{B}); \quad (28)$$

$$\frac{\partial}{\partial t}\left(\frac{3}{2}P_p + \frac{1}{2}n_p U_p^2\right) + \nabla \cdot \left[\frac{1}{2}n_p U_p^2 \mathbf{U}_p + \frac{5}{2}P_p \mathbf{U}_p - \frac{1}{2}\mathbf{K} \cdot \nabla P_p\right] = \frac{e}{m_p}n_p \mathbf{U}_p \cdot (\mathbf{E} + \mathbf{U}_p \times \mathbf{B}), \quad (29)$$

which is the form we use below. Similarly, we have

$$\frac{\partial P_p}{\partial t} + \mathbf{U}_p \cdot \nabla P_p + \frac{5}{3} P_p \nabla \cdot \mathbf{U}_p = \frac{1}{3} \left(\nabla \cdot \mathbf{K} \cdot \nabla P_p \right). \tag{30}$$

The full thermal electron-thermal proton-PUI multi-fluid system is therefore given by equations (3) - (8) and (27) - (29) or (30), together with Maxwell's equations

$$\frac{\partial \mathbf{B}}{\partial t} = -\nabla \times \mathbf{E}; \tag{31}$$
$$\nabla \times \mathbf{B} = \mu_0 \mathbf{J}; \tag{32}$$
$$\nabla \cdot \mathbf{B} = 0; \tag{33}$$
$$\mathbf{J} = e \left(n_s \mathbf{u}_s + n_p \mathbf{U}_p - n_e \mathbf{u}_e \right), \tag{34}$$

where $\mathbf{J}$ is the current and μ_0 the permeability of free space. The diffusion tensor is assumed to be of a simple diagonal form (i.e., we do not include the off-diagonal terms associated with drift and curvature - see the discussion in [7]) and we specify

$$\mathbf{K} = \begin{pmatrix} \kappa_\perp & 0 & 0 \\ 0 & \kappa_\perp & 0 \\ 0 & 0 & \kappa_\parallel \end{pmatrix}; \tag{35}$$
$$\mathbf{k} \cdot \mathbf{K} \cdot \mathbf{k} = k^2 (\kappa_\perp \sin^2 \theta + \kappa_\parallel \cos^2 \theta); \tag{36}$$
$$\kappa_\perp = \eta \frac{1}{3\Omega_p} C_0^2, \qquad \kappa_\parallel = \frac{1}{3\Omega_p} C_0^2, \tag{37}$$

and $\eta = 0.01$. In estimating the diffusion coefficients (37) from (24), we choose a typical PUI speed for the VLISM and assume that the scattering time can be approximated by the PUI gyrofrequency.

3. Linear wave modes in the VLISM

Here we consider wave modes admitted in the PUI mediated VLISM as described by the multi-fluid equations (3) - (8) and (27) - (34). We linearize the PUI multi-fluid equations using

$$n_{e/s/p} = n_{e0/s0/p0} + \delta n_{e/s/p}; \quad \mathbf{u}_{e/s} = \delta \mathbf{u}_{e/s}; \quad \mathbf{U}_p = \delta \mathbf{U}_p; \quad P_{e/s/p} = P_{e0/s0/p0} + \delta P_{e/s/p};$$
$$\mathbf{B} = \mathbf{B}_0 + \delta \mathbf{B} = B_0 \hat{\mathbf{z}} + \delta \mathbf{B}; \quad \mathbf{E} = \delta \mathbf{E},$$

and seek plane wave solutions $\propto \exp[i(\omega t - \mathbf{k} \cdot \mathbf{x})]$, where ω is the frequency, and $\mathbf{k} = (k_x, 0, k_z)$ the wave number. The neglect of electron inertia yields an 11th-order polynomial dispersion relation,

$$\sum_{n=0}^{11} A_n \left(\frac{\omega}{\Omega_p} \right)^n = 0, \tag{38}$$

where the coefficients are cumbersome and given elsewhere. We introduce the proton gyrofrequency $\Omega_p = eB_0/m_p$, the electron number density n_0, the density ratio $\alpha \equiv n_{s0}/n_0$ and thus $n_{p0} = n_0(1 - \alpha)$. The square of the Alfvén speed is defined with the electron number density n_0 as

$$V_A^2 = \frac{B_0^2}{\mu_0 n_0 m_p},$$

and the thermal proton, electron, and PUI sound speeds are defined as

$$C_s^2 = \frac{\gamma_s P_{s0}}{\rho_{s0}} = \frac{\gamma_s P_{s0}}{n_{s0} m_p},$$
$$C_e^2 = \frac{m_e}{m_p}\frac{\gamma_e P_{e0}}{\rho_{e0}} = \frac{\gamma_e P_{e0}}{n_0 m_p},$$
$$C_p^2 = \frac{\gamma_p P_{p0}}{\rho_{p0}} = \frac{\gamma_p P_{p0}}{n_{p0} m_p},$$

and $\gamma_{e/s/p}$ refers to the appropriate adiabatic index.

By way of example, the dispersion relation for parallel propagating waves factors into three polynomials. The first corresponds to dispersive Alfvén waves, i.e., ion-cyclotron or whistler modes, satisfying

$$\omega^4 - (2k^2V_A^2 + \frac{k^4V_A^4}{\Omega_p^2})\omega^2 + k^4V_A^4 = \left(\omega^2 + \frac{k^2V_A^2}{\Omega_p}\omega - k^2V_A^2\right)\left(\omega^2 - \frac{k^2V_A^2}{\Omega_p}\omega - k^2V_A^2\right) = 0, \quad (39)$$

with solutions

$$\omega = \pm\frac{k^2V_A^2}{2\Omega_p} + kV_A\sqrt{1 + \left(\frac{kV_A}{2\Omega_p}\right)^2} \quad \text{(yields } \omega > 0); \quad (40)$$
$$\omega = \pm\frac{k^2V_A^2}{2\Omega_p} - kV_A\sqrt{1 + \left(\frac{kV_A}{2\Omega_p}\right)^2} \quad \text{(yields } \omega < 0). \quad (41)$$

The ion-cyclotron mode satisfies $\omega \to \pm\Omega_p$ in the limit that $k \to \infty$, whereas the whistler mode satisfies $\omega \to \pm\infty$ in the same limit (since electron inertia is neglected).

The second factorization corresponds to a hybrid proton-PUI mode

$$\omega = \pm\Omega_p. \quad (42)$$

The remaining polynomial dispersion relation couples the proton and PUI sound modes, and is given by

$$\omega^5 - \frac{i}{3}\mathbf{k}\cdot\mathbf{K}\cdot\mathbf{k}\omega^4 - (C_e^2 + C_s^2 + C_p^2)k^2\omega^3 + \frac{i}{3}\mathbf{k}\cdot\mathbf{K}\cdot\mathbf{k}(C_e^2 + C_s^2)k^2\omega^2$$
$$+\Big((1-\alpha)C_e^2C_s^2 + C_p^2(C_s^2 + \alpha C_e^2)\Big)k^4\omega - \frac{i}{3}\mathbf{k}\cdot\mathbf{K}\cdot\mathbf{k}(1-\alpha)k^4C_e^2C_s^2 = 0. \quad (43)$$

For the VLISM, we have approximately $\alpha \simeq 1$. The dispersion relation is therefore usefully rewritten as

$$\left(\omega^4 - \frac{i}{3}\mathbf{k}\cdot\mathbf{K}\cdot\mathbf{k}\omega^3 - (C_e^2 + C_s^2 + C_p^2)k^2\omega^2 + \frac{i}{3}\mathbf{k}\cdot\mathbf{K}\cdot\mathbf{k}(C_e^2 + C_s^2)k^2\omega + C_p^2(C_e^2 + C_s^2)k^4\right)\omega$$
$$= (1-\alpha)k^4C_e^2C_s^2\left(\frac{i}{3}\mathbf{k}\cdot\mathbf{K}\cdot\mathbf{k} - \omega\right). \quad (44)$$

The diffusion term $\mathbf{k}\cdot\mathbf{K}\cdot\mathbf{k}$ can be expressed as $\kappa_\| k^2$ for parallel wave propagation.

For a characteristic speed V_0, we may exploit the length scale introduced by the diffusion coefficient $\kappa_\|$ to consider long wavelength and short wavelength solutions of (44). For the long wavelength limit, we obtain five solutions, one of which is a purely damped mode. Another solution is is the PUI acoustic mode that is coupled weakly to the thermal plasma and is diffusively damped,

$$\omega \simeq \pm C_p k\left(1+(1-\alpha)\frac{C_e^2(C_p^2-C_s^2)}{2C_p^2(C_p^2-C_s^2-C_e^2)}\right)+\frac{i}{6}\kappa_\| k^2\left(1-(1-\alpha)\frac{C_e^2(C_p^2-C_s^2)^2+C_s^2C_e^4}{C_p^2(C_p^2-C_s^2-C_e^2)^2}\right). \tag{45}$$

The final long wavelength limit mode admitted by (43) is a two-fluid acoustic mode that is coupled weakly to the PUIs and is weakly diffusively damped, with

$$\omega \simeq \pm(C_s^2+C_e^2)^{1/2}k\left(1-\frac{(1-\alpha)C_e^2(C_p^2-C_s^2)}{2(C_s^2+C_e^2)(C_p^2-C_s^2-C_e^2)}\right)+\frac{i}{6}\kappa_\| k^2\frac{(1-\alpha)C_p^2C_e^4}{(C_s^2+C_e^2)(C_p^2-C_s^2-C_e^2)^2}. \tag{46}$$

The converse short wavelength limit shows that there is no PUI mode. Only the two-fluid acoustic mode persists and it is very weakly coupled to the PUIs via the concentration $(1-\alpha)$, and this mode is also damped weakly by PUIs. Thus, we have

$$\omega \simeq \pm(C_s^2+C_e^2)^{1/2}k\left(1-(1-\alpha)\frac{C_s^2C_e^2}{2(C_s^2+C_e^2)^2}\right)+\frac{3i}{2\kappa_\|}(1-\alpha)\frac{C_p^2C_e^4}{(C_s^2+C_e^2)^2}. \tag{47}$$

A second slowly propagating mixed thermal acoustic mode exists, which is also damped by PUIs, together with a purely damped non-propagating mode.

Related investigations of the dispersion relation for perpendicularly propagating modes have been undertaken, but we do not show these results here. Instead, we present in Fig. 1 numerical solutions to the full dispersion relation in the VLISM. These solutions are completely general in that they include the electron inertial terms.

Considering only positive frequencies, we introduce the following colors to correspond to the following (approximate) designation of wave modes. 1) The slow (red) and fast (blue) two-fluid magnetosonic modes:

$$\omega_{s,f}=k\sqrt{\frac{1}{2}(V_S^2+V_A^2)\pm\frac{1}{2}\sqrt{(V_S^2+V_A^2)^2-4V_S^2V_A^2\cos^2\theta}};$$

2) the Alfvén mode (green):

$$\omega_A = kV_A\cos\theta;$$

3) the slow pickup ion mode (cyan):

$$\omega_{sp} = kC_p\cos\theta;$$

and the fast pickup ion mode (black):

$$\omega_{fp}=\Omega_p.$$

For the VLISM, the PUI sound speed C_p is larger than both the Alfvén speed V_A and the combined thermal plasma sound speed $V_S=\sqrt{C_s^2+C_e^2}$. Thus, for small propagation angles, the order of the modes will always be $\omega_s<\omega_A<\omega_f<\omega_{sp}<\omega_{fp}$, i.e., red, green, blue, cyan, black. However, as θ increases, the frequency of the fast proton (blue) mode increases too with $\omega_f\to k\sqrt{V_S^2+V_A^2}$ and the frequency of the slow pickup ion (cyan) mode decreases with $\omega_{sp}\to 0$. Therefore, beyond some critical angle θ_c, the fast proton mode is faster than the slow pickup ion mode (i.e., the blue mode will find itself above the cyan mode). This happens at the critical angle

$$\cos^2\theta_c=\frac{V_A^2}{C_p^2}\left(1+\frac{V_S^2}{V_A^2}-\frac{V_S^2}{C_p^2}\right). \tag{48}$$

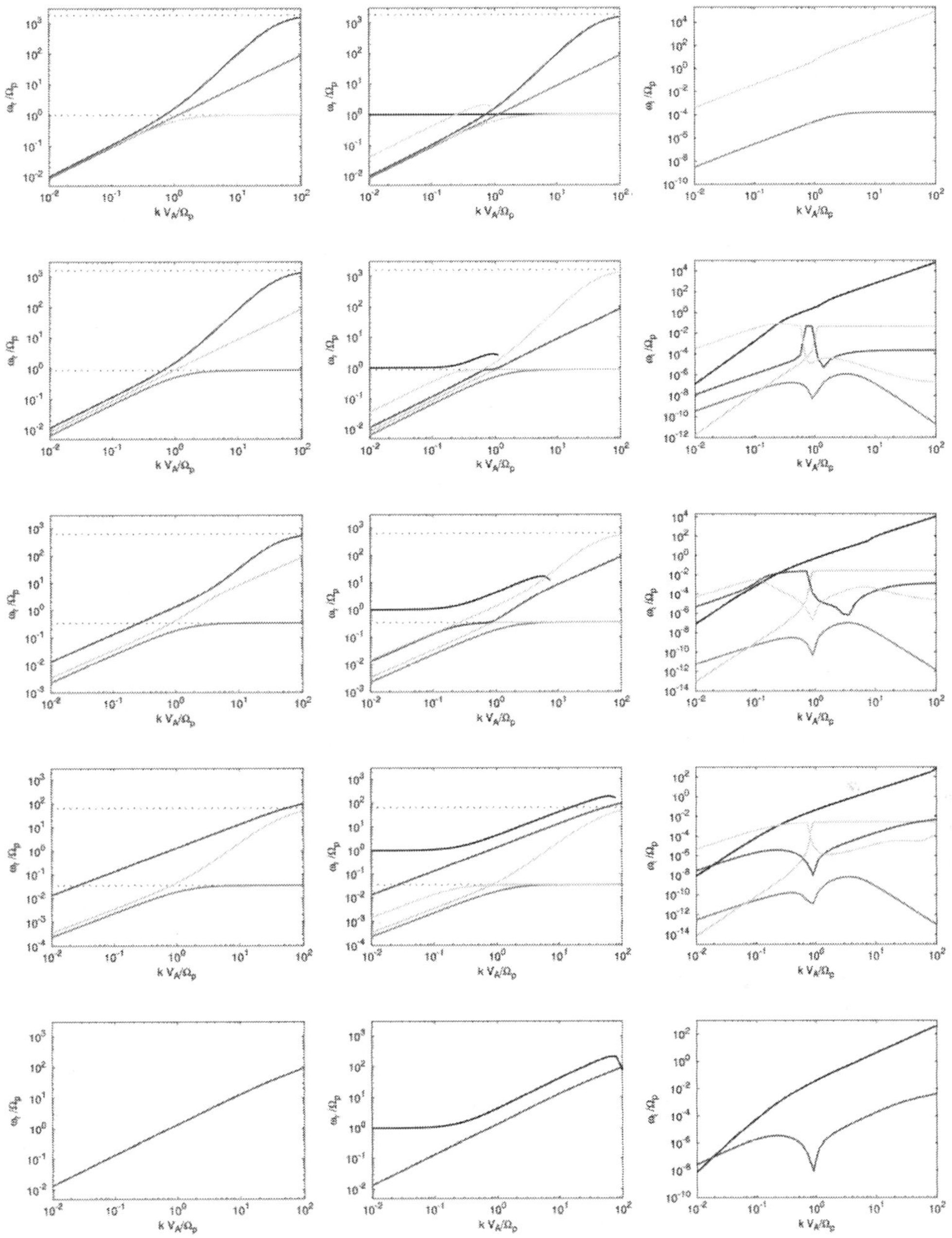

Figure 1. In the left and right panels, the horizontal dotted lines correspond to the asymptotes $\omega = \Omega_p \cos\theta$ and $\omega = \Omega_e \cos\theta$. **Top panel:** $\theta = 0°$, 2-fluid wave modes (left); 3-fluid with diffusion wave modes showing the real frequency (middle panel), and the damping rate (right panel). **Second panel:** $\theta = 30°$, 2-fluid (left); 3-fluid real frequency (middle), and damping rate (right). Note that the fast PUI mode (black) is evanescent at wave numbers greater than about $k \geq \Omega_p/V_A$. **Third panel:** $\theta = 70°$, 2-fluid (left); 3-fluid real frequency (middle), and damping rate (right). **Fourth panel:** $\theta = 88°$, 2-fluid (left); 3-fluid real frequency (middle), and damping rate (right). **Bottom panel:** $\theta = 90°$, 2-fluid (left); 3-fluid real frequency (middle), and damping rate (right).

Furthermore, because $\omega_A < \omega_{sp}$ for all angles θ beyond the critical angle θ_c the order of the modes will always be $\omega_s < \omega_A < \omega_{sp} < \omega_f < \omega_{fp}$, i.e., red, green, cyan, blue, black. Of course, this is only approximate since we are discussing the strict limits $\alpha \to 1$ and $\mathbf{k} \cdot \mathbf{K} \cdot \mathbf{k} \to 0$.

For numerical solutions of the dispersion relation in the VLISM, we assume a PUI density $n_p = 5 \times 10^{-5}\text{cm}^{-3}$ and an electron density $n_0 = 0.08\text{cm}^{-3}$, so that $n_p/n_0 = 6.25 \times 10^{-4}$. Thus, the parameter $\alpha = 1 - n_p/n_0 = 0.999375$. We assume a VLISM magnetic field $B_0 = 3\mu G = 0.3$ nT, which yields the Alfvén speed as $V_A = B_0/\sqrt{\mu_0 n_0 m_p} = 23.14$ km/s. The proton temperature is taken to be $T_s = 15{,}000$ K, the electron temperature $T_e = T_s$ and the polytropic indices $\gamma_s = \gamma_e = \gamma_p = 5/3$. The proton and electron sound speeds are $C_s^2 = \gamma_s k_B T_s/m_p$ and $C_e^2 = \gamma_e k_B T_e/m_p$, which yields $C_s = C_e = 14.37$ km/s. Since we are assuming that fast neutrals from the supersonic solar wind and hot neutrals from the IHS are the origin of the PUIs in the VLISM, we use $U_0 \sim 200$ km/s to compute the pickup ion sound speed $C_p = U_0\sqrt{\gamma_p/7} = 97.59$ km/s. For the diffusion tensor, we use a characteristic PUI speed of $C_0 \sim 200$ km/s to evaluate the diffusion coefficient. The perpendicular diffusion is suppressed relative to the parallel diffusion coefficient, and is given by $\kappa_\perp = \eta\kappa_\parallel$ with parameter $\eta = 0.01$.

Three columns of panels are shown in Fig. 1. The leftmost column shows numerical solutions of the standard two-fluid plasma model [9, 10] i.e., thermal electrons and protons only. The middle column shows plots of the real frequency as a function of wave number for the PUI-mediated 3-fluid plasma. The rightmost column shows the corresponding imaginary part of the frequency as a function of wave number of the PUI-mediated VLISM. The rows correspond to specific wave propagation directions, from parallel (top row) to perpendicular (bottom), as listed in the figure caption.

The 2-fluid dispersion relation is included for reference, so that the differences in the PUI-mediated model are apparent. The 2-fluid model for wave propagation angles $\theta < 90°$ shows that three waves exist in the system: the fast (blue) and slow (red) magnetosonic modes and the Alfvén mode (green). These are clearly distinguishable for $0 < \theta < 90°$. For large k, the parallel Alfvén mode tends to the asymptote $\omega = \Omega_p \cos\theta$, and the fast mode to $\omega = \Omega_e \cos\theta$. For oblique angles, the Alfvén mode (green) eventually resonates at the electron frequency $\Omega_e \cos\theta$. The 2-fluid perpendicular case is degenerate with only the fast mode remaining.

The PUI-mediated model possesses a richer set of wave modes than the 2-fluid case. Shown in the middle panels, the PUI-mediated dispersion relation now exhibits two further modes, a fast PUI (black) and slow PUI (cyan) mode. The PUI sound speed is larger than the thermal plasma sound speeds and so these dispersion curves lie above their thermal counterparts. For parallel propagation, the dispersive Alfvén waves (39) are unchanged from the 2-fluid case. The hybrid PUI mode ($\omega = \pm\Omega_p$) is now present (black curve). Three acoustic modes, the slow PUI mode (cyan, equation (45)), the fast thermal mode (blue, equation (46), and the slow thermal mode (red) exist at long wavelengths (small k), and are more clearly distinguished with increasing propagation angle. As illustrated in the top panel of the right column, and supported analytically by expressions (45) and (46), the PUI sound mode is damped and the fast thermal acoustic mode is not. The parallel propagating PUI acoustic mode exists only for $k \leq \Omega_p/V_A$, after which it is completely damped away and no longer propagates. By contrast, at short wave lengths, the slow thermal acoustic mode (equation (47) is coupled only weakly to the PUIs, the propagation speed being $\sim \sqrt{C_s^2 + C_e^2}$, but it is damped by PUIs as illustrated by the red curve in the top panel of the right column (see also equation (47)). For parallel propagation ($\theta = 0°$), we note that various modes intersect.

With increasing propagation angles θ, the five wave modes remain conceptually similar in terms of the classification but now, as illustrated in the panels of the leftmost column, all oblique modes are damped by PUIs, including the dispersive Alfvén mode. Also, despite not being damped when propagating parallel to $\mathbf{B}_0$, the fast thermal acoustic mode is much more strongly damped than the slow thermal acoustic mode, despite the latter experiencing damping at $\theta = 0°$. The fast and slow PUI modes are clearly distinguishable as θ increases. The fast PUI mode (black) deviates from the $\omega = \Omega_p \cos\theta$ asymptote with increasing k, but becomes evanescent shortly after peaking in frequency. The imaginary frequency of course becomes very large for this mode. The slow PUI acoustic mode frequency increases with increasing k, tending to the electron gyrofrequency asymptote $\omega = \Omega_e \cos\theta$. The fast and slow PUI acoustic modes can also intersect at certain propagation angles. As noted earlier, above a critical angle θ_c the fast thermal acoustic mode is faster than the slow PUI mode. This is illustrated in the fourth panel down of Fig. 1. Finally, for perpendicular propagation (last panel down), only the fast PUI and fast thermal acoustic modes remain, both experiencing damping. Above a certain value of k, the short wavelength fast PUI mode becomes evanescent.

4. Conclusions

A multi-fluid description of a PUI-mediated plasma has been developed. We have assumed fast scattering of PUIs by turbulence maintains a nearly isotropic PUI distribution. In deriving PUI closures, we followed a Chapman-Enskog-like procedure applied to a collisionless plasma. As a result, we find that PUIs introduce a non-negligible heat flux, which is expressed as spatial diffusion term in the PUI pressure with a diffusion tensor whose elements are proportional to the scattering time and reflect the magnetic field geometry of the system. Although not presented here, we have derived the collisionless viscosity transport coefficients for the PUI fluid.

By way of example, we have applied the new three-fluid model to investigate basic plasma physical processes in the VLISM. With Voyager 1 making in situ observations and IBEX making remote observations of the VLISM, this investigation of basic plasma properties of the VLISM is clearly important. Charge exchange of fast and hot neutral H created in the supersonic solar wind or IHS with VLISM protons leads to the creation of an energetic PUI population that is not equilibrated with the background thermal VLISM plasma within about 75 - 100 AU of the heliopause. Thus, in studying the VLISM, we need utilize a PUI-mediated plasma description.

We have investigated linear wave modes in the VLISM using the new multi-fluid model derived here. Instead of the usual three modes found in a 2-fluid plasma, a richer set of waves is found, which includes fast and slow PUI acoustic modes. Our numerical and analytical results show that PUIs damp all obliquely propagating wave modes, including some wave modes that are damped completely at short wavelengths. Our results suggest that PUIs are likely effective at damping waves in the VLISM. We do not take into account the possibility of Landau damping [14] of VLISM wave modes via the background thermal plasma, but this is likely to further dissipate interstellar medium fluctuations. However, we have not yet addressed the question of whether this efficient PUI damping manifests itself as increased dissipative heating of the VLISM.

Acknowledgments

This work was supported by IBEX NASA/SwRI award NNG05EC85C, subcontract A99132BT. GPZ appreciates discussions at the team meeting "Heliosheath Processes and Structure of the Heliopause: Modeling Energetic Particles, Cosmic Rays, and Magnetic Fields" supported by the International Space Science Institute in Bern, Switzerland.

References

[1] Zank, G. P., Pauls, H. L., Williams, L. L., & Hall, D. T. 1996, *J. Geophys. Res.*, 101, 21639
[2] McComas, D. J., et al. 2012, Science, 336, 1291
[3] Zank, G. P., Heerikhuisen, J., Pogorelov, N. V., Burrows, R., & McComas, D. 2010, ApJ, 708, 1092
[4] Zirnstein, E. J., Heerikhuisen, J., Zank, G. P., Pogorelov, N. V., McComas, D. J., & Desai, M. I. 2014, ApJ, 783, 129
[5] Zank, G. P., Heerikhuisen, J., Wood, B. E., Pogorelov, N. V., Zirnstein, E., & McComas, D. J. 2013, ApJ, 763, 20

[6] Isenberg, P. A. 1986, J. Geophys. Res., 91, 9965
[7] Zank, G. P. 2013, Transport Processes in Space Physics and Astrophysics (Springer New York Heidelberg Dordrecht London)
[8] Isenberg, P. A. 1997, J. Geophys. Res., 102, 4719
[9] Stringer, T. E. 1963, Journal of Nuclear Energy, 5, 89
[10] Sahraoui, F., Belmont, G., & Goldstein, M. L. 2012, ApJ, 748, 100
[11] Webb, G. M. 1985, ApJ., 296, 319
[12] Webb, G.M. 1987, ApJ, 321, 606
[13] Webb, G. M. 1989, ApJ, 340, 1112
[14] Hunana, P., Goldstein, M. L., Passot, T. Sulem, P. L., Laveder, D., & Zank, G. P. 2013, ApJ, 766, 93

Journal of Physics: Conference Series **577** (2015) 012026 doi:10.1088/1742-6596/577/1/012026

The Group Abundance Fraction: A statistically robust measure of particle composition and of spatial structure in images

Edmond C. Roelof
Johns Hopkins University/Applied Physics Laboratory, Laurel MD 20723-6099

E-mail: edmond.roelof@jhuapl.edu

Abstract. When two processes (A and B) are measured by counting instruments, we wish to present the results as an "abundance" along with an associated uncertainty. When the number of individual counts of either component is low, the usual ratio (A/B) is statistically ill-defined because there is a finite probability of zero counts in the denominator. However, the "Group Abundance Fraction" (F), *e.g.*, A/(A+B), is a statistically robust parameter (under the condition that there must be at least one total count between A and B in order to constitute a measurement). We rigorously derive expectation values (over a conditional Poisson statistical ensemble) for <F> and its standard deviation σF that are valid for all values of total counts greater than one. We then derive useful estimators for these ensemble expectation values in terms of simple algebraic functions of the actual observed accumulated counts. We also incorporate the practical effects of instrument response functions and background rates. The results are not restricted to particle species composition (or spectra), but are also applicable when A and B represent counts in adjacent "pixels" of spatial images formed by the particles.

1. Introduction

The new physics of the boundary regions of our solar system is being revealed by *in situ* measurements of energetic charged particles from Voyagers 1 and 2 (from ~50 keV/nucleon up to galactic cosmic rays), as well as by remote sensing of energetic neutral atoms (ENAs) by IBEX (~0.5-6 keV) and Cassini/INCA (5-55 keV). After crossing the heliopause, the outer boundary of the heliosheath at 122 AU in August 2012, Voyager 1 now appears to be measuring the composition of galactic cosmic rays down to energies of a few MeV/nucleon, but at rather low counting rates. Consequently, there is a need for a statistically robust measure of composition "ratios" between two chemical species, e.g., A and B at specified energies. There is also a related need in the analysis of spatial distributions of ENA intensity in all-sky maps of the heliosheath. In this second case, A and B could represent counts in adjacent "pixels" of a spatial map, or counts in the same pixels over different time intervals. In the least intense regions of the images, the count rates are so low that individual pixels may not even accumulate 100 useable counts over one year. Nonetheless, it is scientifically important to be able to decide whether there are statistically significant spatial gradients in these regions within a given accumulated skymap, or whether there are statistically significant time variations in the same pixel in the skymaps from one year to another.

Consequently we are going to be discussing the Poisson counting statistics of a rather extreme (but nonetheless important) class of "one time" measurements of rare (or sparse) events. To frame this discussion in quite general terms, we imagine instruments that accumulate counts from two independent Poisson processes (A and B). We don't require that these measurements be made simultaneously, or even with the same instrument. The sources could be cosmic ray particles, photons, or energetic neutral atoms. What counts are the counts. To emphasize this, we designate the counts obtained from process A by (M) and from process B by (N). One way of comparing the two processes is to form the "abundance ratio" R= M/N, but this is obviously a poorly behaved statistic, because when the counts are sparse, the denominator (N) could easily be zero because of fluctuations in the counts. On the other hand, the "group abundance fraction" F=M/S where S=M+N (the total counts in the "group") is much better behaved, as long as we can find a way to deal with the possibility that the denominator S=0. We still have to allow for the finite probability that there might be no counts at all (S=0, so M=N=0), and we will do that by introducing a properly defined *conditional* probability.

Here are a couple of comments before we get into the details of our approach. Firstly, physicists usually think in terms of expected *rates*, not counts, and they usually have some feeling for what the "true" rate should be, so a measurement with S=0 could possibly be used to estimate upper limits on the combined processes (A and B). This approach, if properly formulated, would lead us into Bayesian statistics and therefore beyond the scope of this paper. We will therefore not attempt the interpretation of the special case (S=0), but rather we will exclude it from our analysis by rigorously defining a *conditional* probability that requires $S \geq 1$. This conditional probability will provide the statistical description we need to extract quantitative information about the processes (A and B) from the one-time observation of the counts (M and N). Secondly, does it matter which species we choose for the numerator of the fraction F=M/(M+N)? It really doesn't. If our observations indicated that F<1/2, then we would be analyzing the *minor* component (M<N), and we will be most interested in developing formulas in the limit as $F \to 0$. On the other hand, if F>1/2, we would be analyzing the *major* component (M>N), so we should demand that our formulas also be valid as $F \to 1$. However, in the latter case, what we are really interested in quantifying is how close (F) is to unity. The difference from unity is 1-F=1-M/(M+N)=N/(M+N)<1/2, so in this latter case we should just revert to the former analysis by simply switching our choices for the labels of M and N and then utilize our original formulas for $F \to 0$. If the counts (M,N) are comparable, then it doesn't matter which component we analyze.

In order to proceed, we must be very clear in our nomenclature. Although it may be a bit redundant at times, we will indicate (when it is necessary to make the distinction) the *actual one-time measured counts* by M* and N*. Unlike experiments usually analyzed in statistics textbooks (*e.g.*, [1] which is particularly clear on nomenclature and illuminating examples), where one has (K) *multiple samples* of the same measureable quantity, like X_k ($1 \leq k \leq K$), we have only a *single sample* of the pair of numbers (M*,N*) from our two different processes (A and B). Nonetheless, we can actually obtain information on A and B (even with this paucity of observational information), if we know beforehand that A and B are independent *Poisson processes*, because each Poisson process is completely specified by a *single* parameter, *i.e.*, the *expected* number of accumulated counts (which we shall call C_A or C_B, respectively).

1.1. The simple Poisson process

Let us regress to a brief discussion of a simple (uni-variate) Poisson process in order to clarify some issues relevant to our more general study and (in doing so) to illustrate some important terminology. Explicitly, this Poisson distribution is the well-known (but not always well-understood) probability that one will obtain n* counts from a single actual measurement

$$P_n(c) = c^n/(n)! \exp(-c) \tag{1}$$

We will use n* to indicate a particular measurement of the counts, while we will use (n) as the summation variable when calculating *ensemble* averages over a theoretical ensemble of *all possible measurements* weighted by the probability distribution $P_n(c)$. Although by its definition as a real measurement, the *observed* number of counts (n*) is an integer, the *true* number of counts (c) that specifies the Poisson distribution need not be an integer. This is because in its most familiar formulations, the "true" number of counts c=rT is defined in terms of an average *true rate* (r) of counts accumulated over a time T. Neither the true rate (r) nor the true number of counts (c) can ever be precisely measured; they can only be estimated from the data. Indeed, that estimation process is the challenge that we face when interpreting data.

One calculates the *expectation* value of any function x(n) of the number of counts (n) by averaging x(n) over the Poisson distribution function $P_n(c)$. We will call x(n) a *count* function, because it can be immediately evaluated using the observed number of counts as x(n*). For general distribution functions, such a function of observables x(n*) is sometimes called a "statistic", because it may help us to estimate (in some way) an ensemble parameter like (c). However, we'll stick with the jargon-free terminology of "count function". For the Poisson distribution, the *expectation value* $y_x(c)$ is the ensemble average {x(n)} of count function x(n) given by the sum

$$\{x(n)\} = \sum_{n=0,\infty} x(n)P_n(c) = y_x(c) \tag{2}$$

Clearly the expectation value {x(n)} for the count function x(n) will be some corresponding function $y_x(c)$ of the (sole) ensemble parameter (c), and the latter will depend functionally on the form of the mathematical form of count function x(n). However, more generally one needs to speak of a (possibly non-linear) *ensemble* function z(c) formed from a combination of the expectation values $y_x(c)$, $y_{x'}(c)$, *etc.*, that are, in turn, the set of expectation values for the functions x(n), x'(n), *etc.* It is very important to note that if the relation defining the ensemble function z(c) is *non-linear* in the expectation values $y_x(c)$, $y_{x'}(c)$, *etc.*, then (except under the extraordinary circumstance to be described immediately below), z(c) may not itself be calculable as a simple expectation value (average) over the distribution function.

Let us illustrate these potentially confusing relationships with some well-known examples from the simplest Poisson distribution $P_n(c)$ of Eq. (1). For the count functions, let us take the number (n*) itself, so that $x_1(n)=n$, and its square (n^{*2}), so that $x_2(n)=n^2$. The familiar corresponding ensemble (unconditional) expectation values are $\{n\}=y_{x1}(c)=c$ and $\{n^2\}=y_{x2}(c)=c^2+c$. A widely used measure of the dispersion about the mean value of any count function is its *variance*, which we will write as $\sigma^2 n=(\sigma n)^2$ in order to avoid confusion with $\sigma(n^2)\neq(\sigma n)^2$, so that $\sigma n=(\sigma^2 n)^{1/2}$ is the familiar *standard deviation* (SD). In general, the variance is a *non-linear* combination of these expectation values: $\sigma^2 n=\{(n-\{n\})^2\}=\{n^2\}-\{n\}^2=y_{x2}(c)-y_{x1}(c)^2$. Thus for a general probability distribution function, we see that the variance is *not* generally expressible itself as a simple expectation value (obtained by a simple weighted sum over the probability distribution), because it is a non-linear function of expectation values. However, in the *special case* of the Poisson distribution, the variance reduces to the surprisingly simple expression $\sigma^2 n=c$, where we already know that c={n}. In other words, the value of non-linear variance is just the expectation value of (n*) itself. This peculiar (and defining) property of the Poisson distribution permits the deceptively simple estimation of the expectation value and SD of the mean from a *single* observation: all we have to *assume* is that the *observed* number of counts (n*) is likely to be "close" to the expectation value {n}=c averaged over the Poisson ensemble specified by the ensemble parameter (c), the "true" counts. It is the unique property of the Poisson distribution ($\sigma^2 n=c$) that allows us to make the familiar estimate with its uncertainty (from just a single observed value n*) of the "true" counts (c) as $n^*\pm(n^*)^{1/2}$, assuming the closeness of n* to c.

We can quantify the "closeness" just mentioned above by the concepts of "*unbiased*" and "*biased*" *estimators* of an ensemble function z(c), using a single measurement (n*) and assuming a Poisson distribution. For example, we call $x_1(n^*)=n^*$ an *unbiased* estimator of the true counts (c) because for its expectation value, we have $\{x_1(n)\}=y_{x1}(c)=c$. By this definition, the observed value (n*) is always an unbiased estimator of (c), the average of (n) over the ensemble. However $x_2(n^*)=n^{*2}$ is a *biased* estimator of the square of the true counts squared (c^2), because $\{x_2(n)\}=y_{x2}(c)=c^2+c$. The bias can either be expressed additively, as written, or multiplicatively, as $c^2(1-1/c)$. Either way, the biasing term becomes negligible for c>>1, so the biased estimator can still be useful in the limit of high counts. However, there is an *unbiased* estimator for (c^2); it is a new count function $x_{21}(n^*)=x_2(n^*)-x_1(n^*)=n^{*2}-n^*=n^*(n^*-1)$, because $\{n(n-1)\}=y_{x2}(c)-y_{x1}(c)=(c^2+c)-(c)=c^2$. This works, by the way, because the expectation value of the estimator {n(n-1)} is a *linear* combination of expectation values, $y_{x2}(c)-y_{x1}(c)$. In general, the trick is to find an unbiased (or weakly biased) estimator like $x_{21}(n^*)$ for the ensemble function of interest, like $z(c)=(c^2)$. That will be our approach in the following Sections where we will seek the mean and variance of the group abundance fraction for the full bivariate probablility distribution. Please bear in mind that the estimator for the variance, being a function of the observed counts, will itself still be a random number that may differ significantly from its expectation value, but that's often the best we can do. All such uncertainties are exacerbated when we are dealing with small numbers of counts.

2. Anticipation of results

All the foregoing discussion may seem to be formal overkill when applied to the simple case of a univariate unconditional Poisson distribution. However, the distinctions we have highlighted will prove to be crucial in deriving the properties of the group abundance fraction from a bivariate (conditional) Poisson distribution. The derivation involves a considerable amount of mathematical manipulation of infinite series, but surprisingly it leads to some results that are so simple and useful that we would like to anticipate them at the outset. If we observe M* counts from process A and N* counts from process B for a total of S*=M*+N* counts from both processes, then an *unbiased* estimator of the "true" fraction (f) of A-counts to the total (A and B)-counts is

$$f \rightarrow F^* = M^*/S^*$$

Table 1. Measurements, expectations values, and estimators

Unconditional Joint Distribution W(M,N)	**Ensemble Expectation value**	**Estimator (unbiased)**	**Estimator (biased)**
$\{M\}$	$C_A=fC$	M*	---
$\{N\}$	$C_B=(1-f)C$	N*	---
$\{S\}=\{M+N\}$	C	S*=M*+N*	---
$\sigma^2M=\{M^2\}-\{M\}^2$	C_A	M*	---
$\sigma^2N=\{N^2\}-\{N\}^2$	C_B	N*	---
Binomial Distribution D(f|S) (S≥1)			
$\langle M^0\rangle_S$	$[f+(1-f)]^S=1$	---	---
$\langle M^1\rangle_S$	fS	M*	---
$\langle M^2\rangle_S$	$fS+f^2S(S-1)$	---	---
$\langle M(S-M)\rangle_S$	f(1-f)S(S-1)	---	---
$\sigma^2F_S=\langle F^2\rangle_S-\langle F\rangle_S^2$	f(1-f)/S	F*(1-F*) /(S*-1)	
Conditional Joint Distribution U(M,S) (S≥1)			
$\langle F\rangle=\langle M/S\rangle$	f	M*/S*	---
$\langle M\rangle$	$fC/Q_0(C)$	---	M*
$\langle N\rangle$	$(1-f)C/Q_0(C)$	---	N*
$\langle S\rangle$	$C/Q_0(C)$	---	S*
$\sigma^2F=\langle F^2\rangle-\langle F\rangle^2$	$f(1-f)g_1(C)$	(?)	(?)
$\langle\sigma^2F'^*\rangle=\langle F(1-F)/S\rangle$	$\sigma^2F[1-g_2(C)/g_1(C)]$	---	F*(1-F*)/S*
$\langle\sigma^2F^*\rangle=\langle F(1-F)/(S-1)\rangle$	$\sigma^2F+[1-f(1-f)]C/[\exp(C)-1]$	---	F*(1-F*)/(S*-1)

A *biased* estimator for the variance of the true group abundance fraction σ^2F is

$$\sigma^2F \rightarrow \sigma^2F^* = F^*(1-F^*)/(S^*-1)$$

i.e., the appropriately defined ensemble expectation value of $\langle F(1-F)/(S-1)\rangle$ is not *exactly* equal to σ^2F. Although biased (particularly for low total counts), σ^2F^* on average will be *greater* than its ensemble expectation value σ^2F and hence will on average give a conservative estimate of the true variance. These formulas (to be developed in detail in the rest of this paper) are valid in the context of a *conditional* joint probability distribution for (M,N) wherein the occurrence of S=0 total counts is excluded.

Table 1 has been introduced at this point to aid the reader in following the inter-relationships among the formulas as they are introduced. The notation {x(M,N)} in the table is reserved for an average of a count function x(M,N) over the *unconditional* bivariate Poisson distribution W(M,N), while $\langle x(M,S)\rangle_S$ is the average of x(M,S), where S=M+N, over the *binomial* distribution $D_M(f|S)$ for the group abundance fraction in which the total counts (S) are assigned *a priori*. Finally $\langle x(M,S)\rangle$ will indicate the average of x(M,S), over the *conditional* Poisson distribution U(M,S) with the restriction $S\geq1$ (to be developed later). And so we begin.

3. The unconditional bivariate Poisson distribution

We turn to the main calculations of this paper, valid for sparse data where M* and/or N* are only a few counts. If the Poisson processes are separate, they will be statistically independent. Thus, from the definition in Eq. (1), the joint probability distribution of the counts will be given by

$$W(M,N) = P_M(C_A)\,P_N(C_B) = C_A^{\,M}/M!\,\exp(-C_A)\,C_B^{\,N}/N!\,\exp(-C_B) \qquad (3)$$

where C_A and C_B are the expected number of counts from processes A and B, respectively. We are not using the (M*,N*) notation in Eq. (3), because (M,N) will be the running indices when we calculate *ensemble averages* over the bivariate (joint) Poisson distribution. As is well known, the expectation values averaged over the *unconditional* joint probability distribution W(M,N) of the counts (M,N) and their variances are both equal to the "true" counts (C_A and C_B, respectively). See Table 1.

At this point, we take leave of these familiar formulas, because our *goal* is to calculate *exactly* the expectation values of the group abundance ratio F=M/(M+N) over a *conditional* joint probability distribution. We introduce the ensemble parameter for the "group abundance fraction" (f) into the notation by writing C_A=fC and C_B=(1-f)C for the expected counts from the A and B processes, respectively, and $C=C_A+C_B$ for the total expected counts from the two processes. We also write S=M+N for the sum of the observed counts, and therefore set N=S-M. Since we must have $N\geq0$, this requires that $0\leq M\leq S$. We thereby obtain a well-known transformation of the unconditional joint probability

$$W(M,S\text{-}M) = f^M/M!\,(1\text{-}f)^N/N!\,C^S\exp(-C)$$

$$= \exp(-C)\,C^S/M!N!\,f^M(1\text{-}f)^N = \exp(-C)\,C^S/S!\,B(M|S)\,f^M(1\text{-}f)^{S-M} \qquad (4)$$

where we have regrouped the remaining factors and introduced the notation B(M|S) for the binomial coefficient (that emphasizes the roles played by M and S).

$$B(M|S)=S!/M!(S\text{-}M)! \qquad (5)$$

One then recognizes that W(M,S-M) is the product of a Poisson distribution (in S) and a binomial distribution (in M):

$$W(M,S\text{-}M) = P_S(C)\,D_M(f|S) \qquad 0\leq M\leq S \qquad (6)$$

We have written $P_S(C)$, as in Eq. (1), for the *Poisson* distribution of the total counts (S) specified by the expected total counts (C) for the ensemble, and $D_M(f|S)$ for the binomial distribution of the A-counts (M) specified by the ensemble fraction (f) of the total counts (S), *i.e.*, $f=C_A/C$.

$$P_S(C) = \exp(-C)\,C^S/S! \qquad D_M(f|S)=B(M|S)\,f^M(1\text{-}f)^{S-M} \qquad 0\leq M\leq S \qquad (7)$$

Although we just derived the transformed unconditional bivariate distribution through a mathematical manipulation, we could have written it down directly, based on two well-known results from probability theory. Firstly, the additive combination (sum) of two independent Poisson processes is itself a Poisson process, *i.e.*, the simple probability distribution (in terms of the total number of counts S=M+N) must be the Poisson distribution $P_S(C)$, since it represents the combination of the two Poisson distributions $P_M(C_A)$ and $P_N(C_B)$, where M+N=S and $C_A+C_B=C$. Secondly, *if* we know *beforehand* that S total counts occurred from the combined processes (A+B) and that the A counts are a fraction (f) of the total number of counts, then the A-process counts (M) selected at random must be distributed according to the *binomial distribution* $D_M(f|S)$ generated by the individual coefficients of the expansion of $[f+(1-f)]^S$. Thus the bivariate probability distribution of W(M,S-M) must be the product of the conditional probability of obtaining (M) counts from the A process – *if* there were (S) counts from the combined A+B process (*binomial distribution*) – multiplied by the simple probability that (S) counts were actually obtained from the combined A+B process (*Poisson distribution*). That's why Eq. (4) has the form of Eq. (6) .

3.1. High-count limit

We can easily see how the joint distribution behaves in the high count limit (S>>1). As is well known, the simple Poisson distribution $P_S(C)$ approaches a sharply peaked Gaussian with mean value (C) and standard deviation ($C^{1/2}$<<C). In the limit $C\to\infty$, the Gaussian becomes an expression for the delta function, δ(S-C), and, moreover, the *observed* total counts (S*) very likely will give a good estimate for the expected total counts (C). Thus, in somewhat loose notation, we can consider that the high-count limit of the joint distribution must be

$$W(M,S-M)\to \delta(S-S^*)D_M(f|S) \tag{8}$$

This approximation was the starting point of a statistical analysis of the group abundance ratio by Gehrels [2] in which he developed *confidence limits* on the estimate of (f) based on the observed counts (M*,N*). His basic two assumptions were that he could estimate f by M*/S* and C by S*=M*+N*. The confidence limits were then developed from the properties of the remaining *binomial* distribution in Eq. (8), and the results were presented in numerical tables. His analysis is summarized in Appendix A. Further work along this line [3] introduced Bayesian statistics.

3.2. The properties of the binomial distribution

However, using the same Eq. (8), we could take an approach mathematically independent from the confidence limit formalism. Rather than computing confidence limits on F=M/S for the binomial distribution $D_M(f|S)$, we can calculate the *expectation values* $\langle F\rangle_S$ and the *variance* $\sigma^2F_S=\langle F^2\rangle_S-\langle F\rangle_S{}^2$ from the same *binomial* distribution $D_M(f|S)$. We have introduced the notation $\langle X\rangle_S$ for the expectation value for a count function X(S) over the binomial distribution for a pre-assigned value of S (such as the observed total counts S*). Then we will be able to express the uncertainties in our result as a "one-sigma" expression $\langle F\rangle_S\pm\sigma F_S$. Consider the expectation values for the binomial moments of the form (M^k).

$$\langle M^k\rangle_S = \sum_{M=0,S} M^k\, D_M(f|S) = \sum_{M=0,S} M^k\, B(M|S)\, f^M(1-f)^{S-M}$$

For the reader's convenience, these familiar results are re-derived in Appendix B and summarized in Table 1. The immediately relevant ones are $\langle M^1\rangle_S = fS$ and $\langle M^2\rangle_S = (fS)^2 + Sf(1-f)$.

$$\langle F\rangle_S = \langle M/S\rangle_S = \langle M\rangle_S/S = f \tag{9}$$

$$\sigma^2F_S = \langle (M/S)^2\rangle_S - \langle M/S\rangle_S{}^2 = \langle M^2\rangle_S/S^2 - \langle M\rangle_S{}^2/S^2= [f^2+f(1-f)/S] - f^2 = f(1-f)/S \tag{10}$$

Journal of Physics: Conference Series **577** (2015) 012026 doi:10.1088/1742-6596/577/1/012026

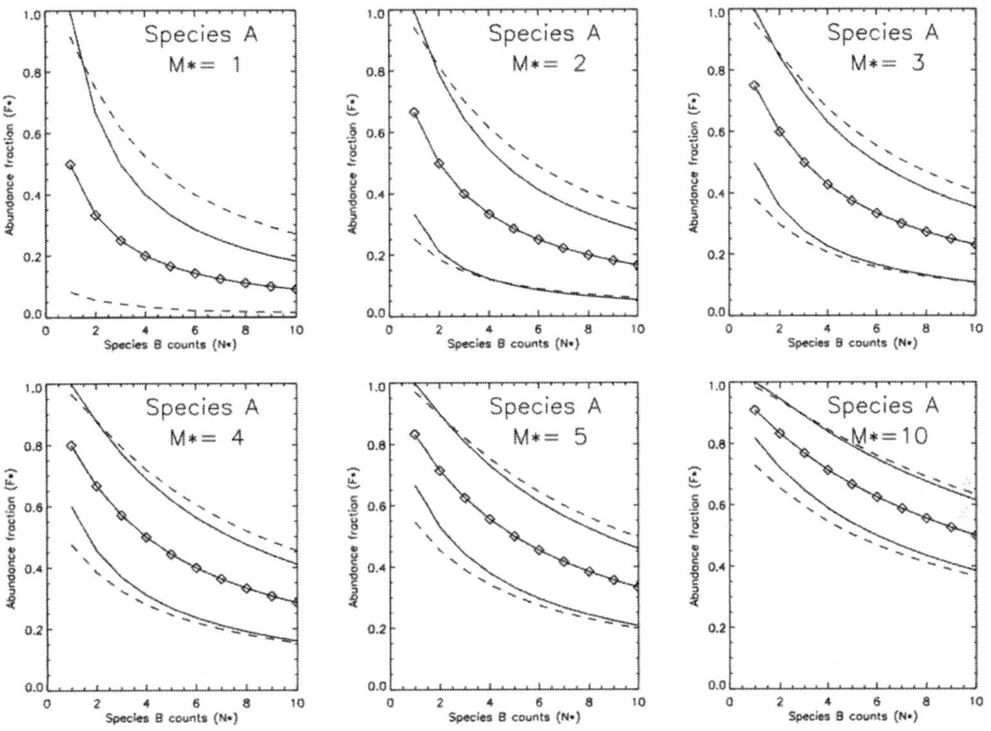

Figure 1. Comparison of expectation values for the group abundance fraction $F^*=M^*/(M^*+N^*)$ and its uncertainty for very low counts (M^*,N^*). Solid lines: $F^*_S \pm \sigma F^*_S$ from the "one-sigma" binomial estimators, Eq. (12) of this paper. Dashed lines: $F\pm CL$ from binomial distribution confidence limits, from [2]. The agreement between these two quite different estimation techniques is quite good.

It turns out that there is an *unbiased* estimator ($\sigma^2F^*_S$) for σ^2F_S in terms of the observed counts. Curiously, it is *not* $F^*(1-F^*)/S^*$, but rather it is $\sigma^2F^*_S= F^*(1-F^*)/(S^*-1)$, because (as we can see from the binomial moments in Table 1)

$$\langle\sigma^2F^*\rangle_S = \langle F(1-F)/(S-1)\rangle_S = \langle M\rangle_S/S(S-1) - \langle M^2\rangle_S/S^2(S-1)$$

$$= (fS)/S(S-1) - [(fS)^2 + Sf(1-f)]/S^2(S-1) = [fS^2-f^2S^2-Sf+Sf^2]/S^2(S-1) = f(1-f)/S \quad (11)$$

We can then represent the estimated results for the *binomial distribution* in terms of a two-sided "one-sigma" uncertainty

$$F^*_S \pm \sigma F^*_S = (M^*/S^*)\pm[M^*(S^*-M^*)/S^{*2}(S^*-1)]^{1/2} \quad (12)$$

In the six panels of Figure 1, we compare this simple algebraic formula for the estimator of $F^*_S\pm\sigma F^*_S$ with the confidence limits from the tables of Gehrels [2] for single-sided confidence levels of 0.8413 (corresponding to one-sigma for a Gaussian). Each of the six panels of Fig. 1 plots the $M^*/(M^*+N^*)$, the estimator for (f, in diamonds) for a fixed values of the A-counts ($M^*=1,2,3,4,5$, and 10). Each plot (whose abscissa covers the B-count range $1\leq N^*\leq 10$), also compares the confidence limits (CL from Table 6 of Gehrels [2], dashed curves) with the algebraic estimator for ($f\pm\sigma F$) from Eq. (12) (solid curves). Figure 1 reveals that the one-sigma estimates of $F^*\pm\sigma F^*$ using our biased algebraic estimator $\sigma^2F^*=F^*(1-F^*)/(S^*-1)$ agree rather well for even very small values of (M^*,N^*) with the confidence limits tabulated by Gehrels [2] in his Table 6. Our one-sigma upper limit almost always falls below his upper confidence level estimate (for $N^*>1$) and our one-sigma lower limit almost always lies above his lower confidence level estimate (except for $M^*=1$ where our lower limit is always identically equal to zero).

We would like to obtain similar results that are valid for the entire range of expected total counts ($C>0$), unencumbered by the high-count approximation $S^*>>1$ that approximates the joint probability distribution $W(M,S-M)$ by the delta function and the binomial distribution $D_M(f|S)$, as in Eq. (8). Gehrels [2] anticipated that his confidence levels should "give somewhat conservative limits" when S^* was a small integer. We shall see that this is indeed borne out by the following analysis in which we will obtain a one-sigma estimator valid for all values of total observed counts $S^*\geq 1$. We will also demonstrate that this more general estimator (σ^2F^*) will turn out to be identically the same as that just obtained for the *unbiased* binomial estimator, namely $\sigma^2F_S^*=F^*(1-F^*)/(S^*-1)$ in Eq. (11). However, for small values ($1\leq S^*\leq 5$), our algebraic estimator σ^2F^* will give a *biased overestimate* of σ^2F. Thus, the "true" expectation values of "one-sigma" uncertainties will be even smaller than indicated in Fig. 1, *i.e.*, they would lie between the solid curves and further inside the dashed curves (from Table 6 of [2]). Consequently, σ^2F^* will be a *conservative overestimate* of the true expectation value of the variance (σ^2F) for all allowed values ($S^*\geq 1$) of the total counts. It will require the following two Sections to establish quantitatively this qualitative result.

4. The conditional bivariate Poisson distribution ($S\geq 1$)

We would like to find good estimators (in terms of the observed counts M^* and N^*) not only the expectation value (f) for F, but also for its *variance* (σ^2F) about f. However, we immediately can see a difficulty in calculating the required moments of $F=M/S$. Any moment $\{F^k\}$ with $k>0$ over the unconditional bivariate Poisson distribution $W(M,S)$ will be undefined when $S=0$ because it will include the term $M=N=S=0$ for which F is indeterminate. The way out of the difficulty lies in acknowledging (see the related comment in Section 1), that we gain very little physically useful information on the abundance fraction if no counts at all were accumulated from either process (*i.e.*, 0/0 is indeterminate). Fortunately, within the basic laws of probability, there is a mathematical method for rigorously *excluding* the special case ($S=0$) from our analysis by re-casting it in terms of a *conditional probability* $U(M,S)$ which properly describes all the remaining cases ($S\geq 1, 0\leq M\leq S$). We don't ignore the possibility that $S=0$; rather we just do proper "book-keeping" in terms of probabilities. Such "book-keeping" of zeros is essential to the treatment of sparse data, *e.g.*, [4].

Since these two cases ($S=0$ and $S\geq 1$) are mutually exclusive, their probabilities must be additive. The *unconditional* joint probability $W(M,N)$ – with the variable change $N=S-M$ so that $0\leq M\leq S$ -- must be the sum of the simple probability that $S=0$ plus the *conditional* probability $U(M,S)$ multiplied by the simple probability that $S\geq 1$.

$$W(M,S-M) = \delta_{S,0}\,\mathrm{prob}(S=0) + (1-\delta_{S,0})\,\mathrm{prob}(S\geq 1)U(M,S)$$

However, in our notation given in Eq. (6),

$$W(M,S-M) = \delta_{S,0}\,P_0(C)D_0(f|0) + (1-\delta_{S,0})\,[1-P_0(C)D_0(f,0)]\,U(M,S) \tag{13}$$

because $prob(S=0)=P_0(C)=exp(-C)$ and $prob(S\geq1)=1-prob(S=0)=1-exp(-C)$. We have used the Kronecker logical symbol $\delta_{S,0}$ to clarify the relationship of the terms. By definition, $D_0(f|0)=1$. Also, because the combination will occur so often, we will designate

$$Q_0(C) = prob(S\geq1) = 1-P_0(C) = 1-exp(-C)$$

Eq. (13) then *defines* the *conditional* joint probability distribution function U(M,S) for the condition $S\geq1$.

$$U(M,S) = (1-\delta_{S,0})W(M,S-M)/Q_0(C)= (1-\delta_{S,0})/Q_0(C)\, P_S(C)\, D_M(f|S) \qquad S\geq1 \text{ and } 0\leq M\leq S \qquad (14)$$

Henceforward, we will suppress the factor $(1-\delta_{S,0})$, since is understood to be an intrinsic part of the definition of U(M,S), because it exists if and only if $S\geq1$.

4.1. *Conditional expectation value of (F) the group abundance fraction (S≥1)*

We are now prepared to calculate the *conditional* expectation value of the group abundance fraction $\langle F\rangle=\langle M/S\rangle$ under the condition $(S\geq1)$.

$$\langle F\rangle = \langle M/S\rangle = 1/Q_0 \sum_{S=1,\infty}\sum_{M=0,S} (M/S)\, U(M,S) = 1/Q_0 \sum_{S=1,\infty} P_S(C)\sum_{M=0,S} (M/S)\, D_M(f|S)$$

Because the separability of the conditional probability U(M,S), the inner summation will be over moments of the form $\langle M^k\rangle_S$ from the *binomial distribution* $D_M(f|S)$. These moments have been discussed above in Eqs. (9) and (10), calculated in Appendix B, and summarized in Table 1. For example the first conditional bivariate moment is

$$\langle F\rangle = \langle M/S\rangle = P_0/Q_0 \sum_{S=1,\infty} C^S/S!\,(1/S)\,\langle M\rangle_S = (P_0/Q_0)\, f \sum_{S=1,\infty} C^S/S!$$

This elementary summation will occur so often that we will assign it the symbol

$$I_0(C) = \sum_{S=1,\infty} C^S/S! = exp(C)-1$$

In fact, the other commonly occurring factor is

$$P_0(C)/Q_0(C)=exp(-C)/[1-exp(-C)]=1/I_0(C)$$

Substituting these functions, we have the important result that F* itself is an *unbiased* estimator of the ensemble parameter (f) over the bivariate conditional probability distribution U(M,S) because

$$\langle F\rangle = f \qquad (15)$$

4.2. *Discussion of <F>*

Eq. (15) contains our first important result: the Poisson expectation value $\langle M/S\rangle$ (under the condition $S\geq1$) *exactly equals* the ensemble fraction parameter (f). In other words, the abundance fraction of the *observed* counts ($F^*=M^*/S^*$) is an *unbiased* estimator of the ensemble fraction parameter (f). This lack of bias is a non-trivial result, as can be seen from the expectation values of the counts (M, N, and S). These are quickly obtained from Table 1.

$$\langle M\rangle = fC/Q_0 \qquad \langle N\rangle = \langle S-M\rangle = (1-f)C/Q_0 \qquad \langle S\rangle = C/Q_0$$

In other words, when we exclude the possibility that M=N=S=0, the conditional expectation value of <M> is no longer simply the abundance fraction (f) multiplied by the number of counts (C). It, as well as the expectation values for <N> and <S> are all *biased* by the factor $1/Q_0(C)$. This highlights the important fact that the lack of bias in <F>=f is actually a consequence of our use of the *conditional* probability. Interestingly enough, <F>=<M>/<S>, *i.e.*, the *unbiased* expectation value <F> equals the ratio of the *biased* expectation values for M and S.

The formulas reveal why there must be biasing for very few counts. Consider their limiting values as the expected counts $C\rightarrow 0$. Without the biasing denominator, we would have <M>, <N>, and <S> all three going to zero, in contradiction to our condition $S\geq 1$. However, because $C/Q_0\rightarrow 1$ when $C\rightarrow 0$, *i.e.*, for very rare events, we have instead.

$$lim_{(C\rightarrow 0)} \langle M\rangle = f \qquad lim_{(C\rightarrow 0)} \langle N\rangle = 1-f \qquad lim_{(C\rightarrow 0)} \langle S\rangle = 1$$

These limits make perfect sense. When the events are very rare, we must then have $S\approx 1$, because we are imposing the condition $S\geq 1$, *i.e.*, we must still measure at least a single count. Under this circumstance ($S\approx 1$), the conditional expectation value for <M> should be the ensemble abundance fraction (f) itself, and that of <N> should be (1-f), because the probability for a single *total* count coming from species A must be (f), and for a single *total* count coming from species B must be (1-f).

4.3. Conditional expectation value: variance (σ^2F) of the group abundance fraction ($S\geq 1$)

The variance of the abundance fraction is $\sigma^2F = \langle F^2\rangle - \langle F\rangle^2$, where

$$\langle F^2\rangle = \langle M^2/S^2\rangle = (1/Q_0)\sum_{S=1,\infty} P_S(C) \sum_{M=0,S} (1/S^2)\langle M^2\rangle_S$$

$$= (1/I_0)\sum_{S=1,\infty} C^S/S!\,(1/S^2)[\,fS+f^2S(S-1)] = (1/I_0)\sum_{S=1,\infty} C^S/S!\,[f(1-f)/S+f^2]$$

$$= f(1-f)/I_0 \sum_{S=1,\infty} C^S/S!S + f^2 (I_0/I_0) = f(1-f)g_1(C) + f^2$$

$$\sigma^2F = \langle F^2\rangle - \langle F\rangle^2 = f(1-f)g_1(C) \tag{16}$$

An important new function $g_1(C)$ has appeared in our formula for the variance of (F). It is actually only one of a more general set defined by the normalized sums $I_k(C)$

$$g_k(C) = I_k(C)/I_0(C) \qquad I_k(C) = \sum_{S=1,\infty} C^S/S!S^k \qquad k\geq 0 \tag{17}$$

where it so happens that for k=0, the summation reduces to $\exp(C)-1= I_0(C)$; in fact, that is why we chose that symbol for that function. Please note that, despite the similar notation, the infinite sums $I_k(x)$ are *not* modified Bessel functions. We need to know how $g_1(C)$ varies with (C), but it will be just as easy to examine the properties of the general set of functions $g_k(x)$ for $k\geq 1$ as it is for just k=1 alone. Some useful properties of these functions are derived, and they are evaluated in the two extreme ranges (C<<1 and C>>1) in Appendix C. The functions $g_k(C)$ actually behave quite simply, decreasing monotonically away from $g_k(0)=1$ for C<<1 and asymptotically towards zero as $1/C^k$ for C>>1. The behaviors of $g_1(C)$ and $g_2(C)$ are graphed in Figure 2, along with $1-g_2(C)/g_1(C)$, over the range $0\leq C\leq 20$.

4.4. Statistical significance: "n-sigma" ratios for the group abundance fraction

We can now introduce "n-sigma" statistical significance properties of the group abundance fraction (F) in terms of the ensemble expectation mean <F>=f and its expected variance (σ^2F). Let us designate the "n-sigma" ratio as (η) and define its square by

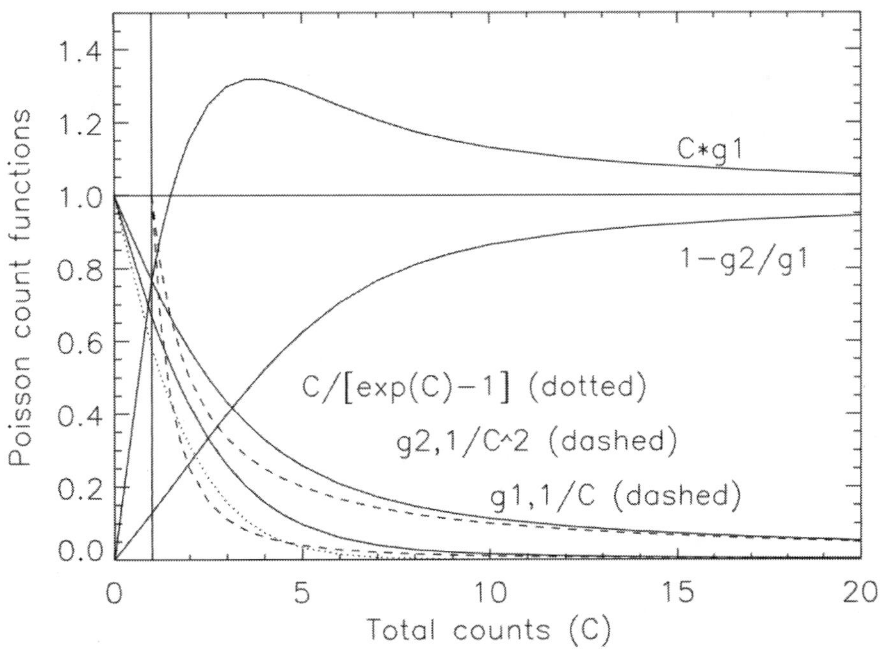

Figure 2. Various probability functions that appear in the expectation values of the variance of the Group Abundance Fraction (σ^2F) and its estimators. See Table 1 for a summary and specific discussions in the text.

$$\eta^2 = \langle F\rangle^2/\sigma^2F = f^2/[f(1\text{-}f)g_1(C)] \approx C/(1/f\text{-}1) \tag{18}$$

We could use function $g_1(C)$ in the subsequent discussion, but it is much easier to use the approximation $g_1(C)\approx 1/C$ which is good within ±20% for $1<C<3.5$ and then becomes $<+10\%$ for $C>10$, i.e. over our practical range of interest. See Fig. 2. As examples of its utility, we can quickly answer two important questions (relevant to making a low-count observation) in terms of the ensemble parameters. First, how many true counts (C) are required to measure a true abundance fraction (say f=0.1) to a significance ratio of η=1, 2, or 3 sigma? From Eq. (18), $C\approx\eta^2(1/f\text{-}1)= \eta^2(9)$, so the answer is C>9, 36, or 81 total counts, respectively. Second, what is the smallest true abundance fraction (f) that can be measured (statistically significantly) with C=9 counts for different significance ratios (η=1, 2, or 3)? Also from Eq. (18), $f\approx 1/(1+C/\eta^2)= 1/(1+9/\eta^2)$, so the answers are f=0.10(0.90), 0.31(0.69), or 0.50(0.50) respectively. There are double values for f, because inspection of Eq. (18) reveals that if (f) is a solution, so is its complement (1-f). See the related comment in Section 1. The degree to which these answers concerning (η) are practically meaningful obviously depends on how close our data-based estimators (F* and σ^2F^*, to be derived below) are, on average, to the ensemble expectation values for f and σ^2F, calculated in Eqs. (15) and (16), respectively. For now, a crude estimate on the uncertainty range of (η_{est}) can be obtained from Eq. (18) by ignoring any uncertainty in σ^2F and writing $\eta_-<\eta_{est}<\eta_+$ where $\eta_\pm=(f\pm\sigma f)/\sigma f$ so that

$$(f\text{-}\sigma f)/\sigma f < \eta_{est} < (f+\sigma f)/\sigma f \qquad \text{or} \qquad \eta\text{-}1 < \eta_{est} < \eta+1$$

5. Estimator (σ^2F^*) for the variance (σ^2F)

In order to make best use of our observed values (M*,N*), we need to obtain an estimator (σ^2F^*) from the data for the expectation value of the variance (σ^2F) given in Eq. (16). We already know from Eq. (15) that we can use F*=M*/S* as an *unbiased* estimator of (f). What we therefore need for practical applications is a "statistic" for σ^2F, *i.e.*, a count function (call it σ^2F^*) of the observed counts (M^*,S^*) whose expectation value (call it $<\sigma^2F^*>$), will perforce be "close" to the function $\sigma^2F=f(1-f)g_1(C)$, *i.e.*, we seek an *acceptably biased* estimator (σ^2F^*) for σ^2F in terms of the observed counts (M^* and N^*). If found, we could present our measurement of $F^*=M^*/S^*$ with a "one-sigma" uncertainty as $F^*\pm\sigma F^*$, where $\sigma F^*=(\sigma^2F^*)^{1/2}$.

We were encouraged that it might be possible to find a simple *algebraic* expression (in terms of M^* and N^*) for the estimator σ^2F^* by the fact that, for the *binomial* distribution (in which S is set equal to S^*), we found in Eq. (11) an *unbiased* estimator for $\sigma^2F_S=f(1-f)/S$ in terms of the observed counts, namely $\sigma^2F^*_{S^*}= F^*(1-F^*)/(S^*-1)$, because $<\sigma^2F^*>_S = \sigma^2F_S$. See Table 1. At first, the form of $\sigma^2F=f(1-f)g_1(C)$ suggested that we try a statistic of the form

$$\sigma^2F'^*= F^*(1-F^*)/S^* = M^*(S^*-M^*)/S^{*3}$$

because we have shown in Appendix C (and Fig. 2) that $g_1(C)\approx 1/C$ as long as $C>1$, *i.e.*, over a useful range of "small" numbers of total counts. However, there is an immediate difficulty for our *conditional* probability distribution (requiring $S\geq 1$) at the smallest limit $S^*=1$, because this estimator ($\sigma^2F'^*$) would yield *exactly* zero for either $M^*=0$ or $M^*=1$ (where $N^*=S^*-M^*=0$). This is an unacceptable limit for an estimator, because it says that there is no uncertainty at all in the estimate of σ^2F for very small counts, and that is nonsense. In fact, we can quickly compute its expectation value for all values of the ensemble parameters (f and C) by substituting the binomial moments $<M^k>_S$ in Table 1 into our summations over ($S\geq 1$) used above.

$$<\sigma^2F'^*>= <F(1-F)/S> = <M(S-M)/S^3>$$

$$= (1/I_0) \textstyle\sum_{s=1,\infty} C^S/S!S^3 <M(S-M)>_S = (1/I_0) \sum_{s=1,\infty} C^S/S!S^3 f(1-f)S(S-1)$$

$$= f(1-f)/I_0 \textstyle\sum_{s=1,\infty} C^S/S! (1/S-1/S^2) = f(1-f)[g_1(C)-g_2(C)]= \sigma^2F [1- g_2(C)/g_1(C)] \quad (19)$$

The resulting multiplicative aliasing factor $(1-g_2/g_1)$ is plotted in Fig. 2 and its limiting behavior is discussed in Appendix C. It asymptotically approaches unity as $(1-1/C)$ for $C>>1$ and zero as $(C/8)$ for $C<<1$. Thus it is always ≤ 1 for finite values of (C), so $<\sigma^2F'^*>$ always *underestimates* the true variance. Even though (as just mentioned above) its zero-trending behavior in the low-count limit is unacceptable, we shall see below that it still yields a useful *lower* bound for the true value σ^2F for $C>1$.

Because it must enter into the averaging summation over *all* $S\geq 1$ (including $S=1$), we need an estimator that makes "sense" and predicts a finite variance, even for the extreme value $S^*=1$ (one total count) when either $M^*=0$ or $M^*=1$. Remarkably, an estimator that does work has precisely the form of the *binomial* estimator σ^2F_S from Eq. (11) and Table 1.

$$\sigma^2F^*= \sigma^2F_{S^*} = F^*(1-F^*)/(S^*-1) = M^*(S^*-M^*)/S^{*2}(S^*-1) \quad (20)$$

At first glance, the estimator would appear to be singular at $S^*=1$ due to the factor (S^*-1) in the denominator. However, when $S^*=1$, M^* can only take on the values $M^*=0$ or $M^*=1$. In either case, the numerator is also zero. Actually, we can write $M^*=S^*-1$ in the first case, and $S^*-M^*=S^*-1$ in the second. Thus in both cases the zeros in the numerator and denominator form the ratio $(S^*-1)/(S^*-1)=1$. One could make this argument more rigorously in terms of "limits" $S^*\to 1$ and $M^*\to 0$ or $M^*\to 1$, but it doesn't really matter because, being an estimator, we are free to *define* the value of the fraction in these limiting cases, as long as we treat it consistently in our summations over the distribution function.

$$\sigma^2F^*|_{S^*=1,M^*=0} = (1-M^*)/S^{*2} [(S^*-1)/(S^*-1)] = 1$$

$$\sigma^2 F^*|_{S^*=1,M^*=1} = M^*/S^{*2}\,[(S^*-1)/(S^*-1)] = 1 \qquad (21)$$

Then, according to our convention, the variance $\sigma^2F^*=1$ for $S^*=1$ (a single count) for either $M^*=0$ and $N^*=1$ or $M^*=1$ and $N^*=0$. We argue that this makes sense, because for these values, either $F^*=0$ or $F^*=1$ and we have no way of knowing how correct either value is. The uncertainty $\sigma F^*=(\sigma^2F^*)^{1/2}$ is therefore well represented by unity.

5.1. *Calculation of the expectation value for the estimator* σ^2F^*

We can now apply this convention to an exact calculation of the expectation value for the estimator over our conditional probability distribution. In this case, though, it is necessary to split off the first term (with S=1) from the outer summation.

$$\langle\sigma^2F^*\rangle = 1/I_0 \sum_{s=1,\infty} C^S/S!S^2 \sum_{M=0,S} M(S-M)/(S-1)\, D_M(f|S)$$

$$= C/I_0 \sum_{M=0,1} M(1-M)/(S-1)|_{S=1}\, D_M(f|1) + 1/I_0 \sum_{s=2,\infty} C^S/S!S^2 \sum_{M=0,S} M(S-M/(S-1)\, D_M(f|S)$$

By the arguments given above, the ratio $M(S-M)/(S-1)$ should be set to (1) for either M=0 or M=1, whereas $D_0(f|1)=f^0(1-f)^1=1-f$ and $D_1(f|1)=f^1(1-f)^0=f$. Thus our convention leads to

$$\sum_{M=0,1} M(1-M)/(S-1)\, D_M(f|S) = (1-f) + (f) = 1$$

The remaining summation ($S\geq 2$) is easily evaluated from the formula in Table 1 and Appendix B.

$$1/I_0 \sum_{s=2,\infty} C^S/[S!\,S^2(S-1)]\,\langle M(S-M)\rangle_S/ = f(1-f)/I_0 \sum_{s=2,\infty} C^S/S!S = f(1-f)\,[g_1(C)-f(1-f)C/I_0]$$

which, when re-united with the S=1 term, yields our desired result.

$$\langle\sigma^2F^*\rangle = f(1-f)g_1(C) + [1-f(1-f)C/I_0(C)] \qquad (22)$$

Since the term $f(1-f)\leq 1/2$ and $C/[\exp(C)-1]\leq 1$, the bracketed quantity in Eq. (22), which constitutes an additive bias for the estimator of σ^2F, is always between 0 and 1/2. In other words, the expectation value of the estimator $\langle\sigma^2F^*\rangle=\langle F(1-F)/(S-1)\rangle$ always *overestimates* the exact value $\sigma^2F=\langle F^2\rangle-\langle F\rangle^2=f(1-f)g_1(C)$ by no more than an additive bias of 1/2 (for $C\rightarrow 0$), decreasing to a zero bias (as $C\rightarrow\infty$). The bias reflects the fact (demonstrated above) that the estimator for the abundance fraction $\sigma^2F^*=M^*N^*/S^{*2}(S^*-1)$ is observationally quite uncertain as $C\rightarrow 0$ under the condition $S^*\geq 1$, because then S^* must take the value $S^*=1$ in this limit of very low counts, where $F^*=M^*/S^*$ is then going to have to take either the value $F^*=0$ or $F^*=1$.

The behavior of the multiplicative bias in the two estimators $\langle\sigma^2F^*\rangle/\sigma^2F$ and $\langle\sigma^2F'^*\rangle/\sigma^2F$ is graphed in Figure 3 as the square root of the ratio of the biased estimator to the true value. Using the somewhat sloppy notation $\langle\sigma F^*\rangle=\langle\sigma^2F^*\rangle^{1/2}$ and $\langle\sigma F'^*\rangle=\langle\sigma^2F'^*\rangle^{1/2}$

$$\langle\sigma F^*\rangle/\sigma F = [\langle\sigma^2F^*\rangle/\sigma^2F]^{1/2} \quad \text{and} \quad \langle\sigma F'^*\rangle/\sigma F = [\langle\sigma^2F'^*\rangle/\sigma^2F]^{1/2}$$

The curves demonstrate that the former ratio always exceeds unity (for any value of f), while the latter ratio is always less than unity (independent of the value of f). Another interpretation of the plot is that the true expectation value (σ^2F) must lie in between the two *biased* expectation values $\langle\sigma^2F^*\rangle$ and $\langle\sigma^2F'^*\rangle$.

Journal of Physics: Conference Series **577** (2015) 012026 doi:10.1088/1742-6596/577/1/012026

Therefore we recommend the estimator

$$\sigma F^*=[F^*(1-F^*)/(S^*-1)]^{1/2} \tag{23}$$

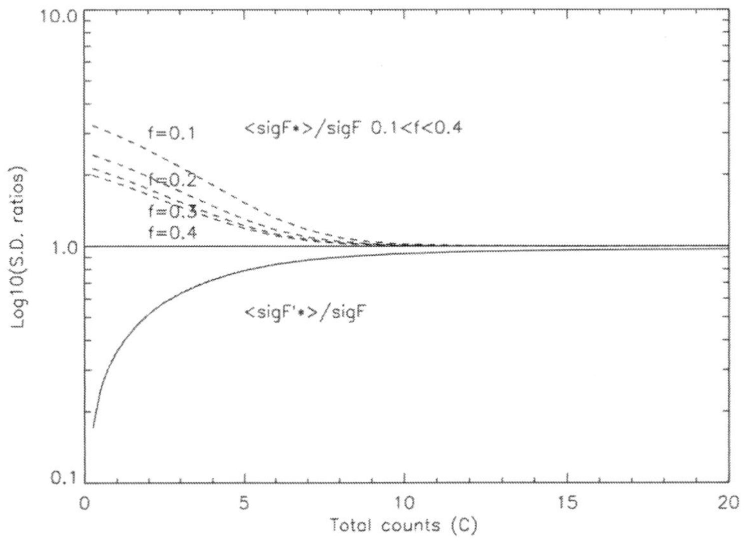

Figure 3. Ratios $<\sigma F^*>/\sigma F=[<\sigma^2 F^*>/\sigma^2 F]^{1/2}$ and $<\sigma F'^*>/\sigma F=[<\sigma^2 F'^*>/\sigma^2 F]^{1/2}$ as a function of the number of counts (C) and selected values for the group abundance fraction (f). The quantities $<\sigma^2 F^*>$ $<\sigma^2 F'^*>$ are *biased* estimators of the "true" variance ($\sigma^2 F$). See Table 1 and the discussion in the text.

as a conservatively biased estimator for the uncertainty (σF) in the estimated value $F^*=M^*/S^*$ for the group abundance fraction (f):

$$f\pm\sigma F \rightarrow M^*/S^* \pm [F^*(1-F^*)/(S^*-1)]^{1/2} \tag{24}$$

Having obtained in Eq. (24) our desired result, we next take up two issues that are essential in the analysis of actual observations. These go beyond the mathematics of Poisson probability, because they involve the inescapable physical characteristics of the instruments that obtain the counts.

6. Relationship between counts and intensities

Our statistical analysis concerned itself with the "true" counts (C_A and C_B) or equivalently (f and C), because those are the sole parameters that specify the Poisson distributions for species A and B. However, the physically significant quantities are the incident uni-directional differential *intensities* being measured (j_A and j_B). If the counts were both accumulated by the same detector over the same time interval, then the abundance fraction of the true counts (f) is also that of the intensities (h)

$$f=C_A/(C_A+C_B) \qquad h=j_A/(j_A+j_B)$$

The general relation between the counts and intensities is $C_A=\rho_A j_A$, where the proportionality constant $\rho_A=(TG\Delta E)_A$. Here T is the accumulation time, ΔE is the energy pass-band, and G is the efficiency-weighted geometry factor (*e.g.*, in units of $cm^2 sr$). Then there is generally a somewhat more complicated relation between f and h since $C_A=\rho_A j_A$ and $C_B=\rho_B j_B$

$$f = 1/(1+ \rho_B j_B/\rho_A j_A)$$

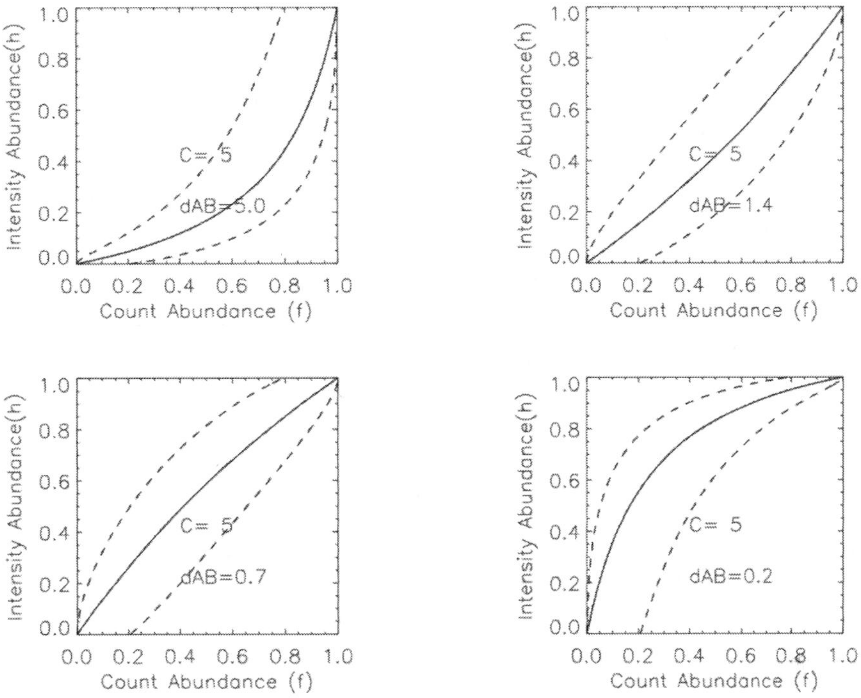

Figure 4. Solid curves: The group intensity abundance fraction $h=j_A/(j_A+j_B)$ *vs.* the group count abundance fraction $f= C_A/(C_A+C_B)$. Dashed curves: approximate one-sigma uncertainties in the fractions owing to the variance in the count fraction (σ^2F). The four panels are all for C=5 total counts, but the detector response difference (d_{AB}) takes different values.

Since $h=1/(1+j_B/j_A)$ we have $j_B/j_A=1/h-1$ and

$$1/f = 1+(\rho_B/\rho_A)(1/h-1)$$

This is easily solved for the *intensity* group abundance ratio $h=j_A/(j_A+j_B)$.

$$h = f/[f+(\rho_A/\rho_B)(1-f)]$$

The solution is well behaved, in that that f=0 implies h=0 and f=1 implies h=1. For intermediate values of f, we always have h<1 because the term in the denominator with the factor (1-f) is always positive.

But, what is the uncertainty in intensity fraction (h) due to the uncertainty in the group abundance fraction f? The most straightforward (although crude) two-sided "one-sigma" estimate is $h_-<h<h_+$ where

$$h_\pm = f_\pm/[f_\pm+(\rho_A/\rho_B)(1-f_\pm)] \qquad f_\pm=f\pm\sigma F(C) \qquad \sigma^2F=f(1-f)g_1(C) \tag{26}$$

These formulas are all expressed in terms of the ensemble parameters (f and C). They are plotted (h *vs.* f) in Figure 4 for choices of the total number of counts (C) and the channel detector response ratio ($d_{AB}=\rho_A/\rho_B$). To estimate from the actual data the group abundance ratio of the incident intensities $h=j_A/(j_A+j_B)$, we would use our estimators (based on the observed counts M* and N*) for the ensemble parameters

$$f \rightarrow F^*=M^*/(M^*+N^*)=M^*/S^* \qquad \text{and} \qquad \sigma^2F \rightarrow F^*(1-F^*)/(S^*-1)$$

7. Inclusion of counting backgrounds

Experimentally, there is still an additional complication, that of the backgrounds in actual detectors. We can easily include Poisson counting backgrounds (individually) into each of the processes (A and B), since the sum of two Poisson processes is a Poisson process. We simply add *background* counts (k_A and k_B) to the *foreground* counts (c_A and c_B) for two the two species (A and B). This will again yield a Poisson process. We will assume that the average rates for the backgrounds are known with high precision previous calibration measurements. Then, by the foregoing arguments, the ensemble group abundance fraction is

$$f = (c_A+k_A)/(c+k) \qquad \text{so that} \qquad 1-f=(c_B+k_B)/(c+k) \tag{27}$$

where $c=c_A+c_B$ is the total of the *foreground* counts and $k=k_A+k_B$ is the total of the background counts. We will assume that the expectation values for the background counts (k_A and k_B) are well-estimated by previous calibration measurements.

By the theorem for combining independent Poisson processes, the *observed* counts from process A (*before* imposing the condition $S=M+N\geq1$) will have the expectation value $C_A=c_A+k_A$, and those from process B will have $C_B=c_B+k_B$, with $C=c+k$ expected for the total observed counts. We therefore can directly write

$$\langle F\rangle = f = (c_A+k_A)/(c+k) \tag{28}$$

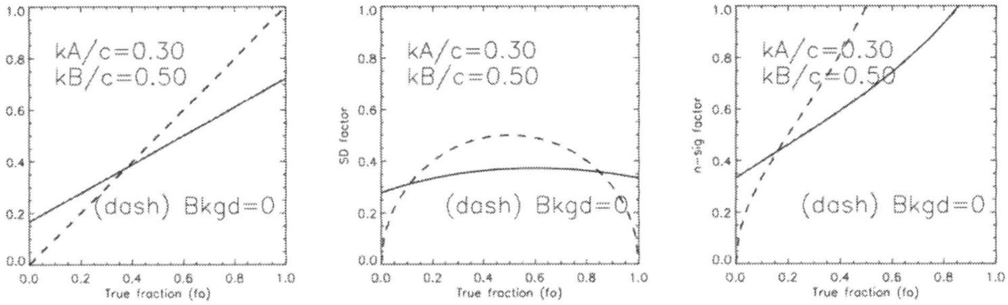

Figure 5. Background effects. *Left panel:* Ensemble expectation value for the group abundance fraction (f) in the presence of background counts (k_A and k_B) in the two channels (A and B) with total foreground counts (c). *Center panel:* Ensemble expectation value for the standard deviation of the fraction (σF) – multiplied by $c^{1/2}$ - in the presence of background counts (k_A and k_B). *Right panel:* The "n-sigma" ratio ($f/\sigma F$) – divided by $c^{1/2}$ - in the presence of background counts (k_A and k_B). Rather severe background contamination has been presented: $k_A/c=0.3$ and $k_B/c=0.5$.

Let us rewrite the expectation value of the background-affected abundance fraction (f) in terms of the (unobservable) ensemble *foreground* fraction $f_0=c_A/c$. Then

$$f=(f_0+k_A/c)/(1+k/c) \text{ so that}$$

$$f_0=(1+k/c)f-k_A/c=f+(kf-k_A)/c \tag{29}$$

The linear relation (f *vs.* f_0) reveals the effect of backgrounds on the expectation value of the group abundance fraction (as a function of the "true" foreground fraction f_0). The relation depends only on the ratios of the A and B backgrounds (k_A and k_B) to the total foreground counts (c). It is plotted in Figure 5 (left panel) for the particular background/foreground ratios $k_A/c=0.3$ and $k_B/c=0.5$. One sees immediately that the expectation value (f) cannot be smaller than $k_A/(k+c)$, nor can (f) exceed the ratio $(k_A+c)/(k+c)<1$. What about the expectation value of the variance $\sigma^2F=\langle F^2\rangle-\langle F\rangle$? By the just-cited properties of the sums of independent Poisson processes, the formula in the presence of backgrounds remains

$$\sigma^2F=f(1-f)g_1(C)=(c_A+k_A)(c_B+k_B)/(c+k)^2\,g_1(C)=[Cg_1(C)](f_0c+k_A)[(1-f_0)c+k_B]/(c+k)^3$$

$$c\sigma^2F/[Cg_1(C)]=(f_0+x_A)(1-f_0+x_B)/(1+x)^3 \tag{30}$$

We see that after multiplying the equation through by the number of *foreground* counts (c), the RHS involves only background/foreground ratios: $x_A=k_A/c$, $x_B=k_B/c$, $x=x_A+x_B=k/c$, and $\Delta x=k_B-k_A$. The divisor on the LHS $[Cg_1(C)]$ is nearly unity (within ±20%, see Fig. 1)), so if we set to unity we get the convenient approximation

$$c\sigma^2F\approx[-f_0^2+(1+\Delta x)f_0+x_A(1+x_B)]/(1+x)^3 \tag{31}$$

We will call the RHS of this approximation the "variance factor", and its square root the "SD factor". The LHS contains the only explicit dependence on the foreground counts (c). The SD factor is plotted in Figure 5 (center panel) for the specific parameter values $x_A=k_A/c=0.3$ and $x_B=k_B/c=0.5$. Again, as for f *vs.* f_0, the RHS depends only upon the ratios of the backgrounds (k_A and k_B) to the total foreground counts (c). The approximate value of σF itself is obtained (to good approximation) by dividing the plotted "SD factor", the square root of the RHS of Eq. (31), by the square root of the total number of foreground counts ($c^{1/2}$). Considered as a function of the true fraction ($0\le f_0\le 1$), the numerator the variance factor is the square root of a downward-opening parabola whose value at $f_0=0$ is $x_A(1+x_B)$ and at $f_0=1$ is $x_B(1+x_A)$. Thus the variance is always positive over the interval ($0\le f_0\le 1$). The difference between the two ordinates corresponding to $f_0=0$ and $f_0=1$ is Δx (which could be either positive or negative). The maximum ordinate is attained at $f_0=(1+\Delta x)/2$ (which is not necessarily within the valid interval $0\le f_0\le 1$), where its value is $(1+\Delta x)^2/4+x_A(1+x_B)$.

We now can revisit our statistical significance "n-sigma" discussion (Section 4.3), with instrument background effects included. Although $\eta^2=f^2/\sigma^2F$, it is algebraically more convenient to first consider its inverse.

$$\eta^{-2}=\sigma^2F/f^2=(1/f-1)g_1(C)=[Cg_1(C)]\,[(1+x)/(f_0+x_A)-1]/(c+k)$$

Multiplying the equation through by the foreground counts (c) and re-arranging terms as we just did in obtaining (f) and (σ^2F), we get

$$c\eta^{-2}/[Cg_1(C)]=(1-f_0+x_B)/[(f_0+x_A)(1+x)] \tag{32}$$

Again, the RHS depends only upon the background/(total foreground) ratios. Then the square root of the RHS of the inverse we call the n-sigma factor. Since $[Cg_1(C)]^{1/2} \approx 1$ is usually a good approximation, we take the RHS as an approximation for $\eta/c^{1/2}$. The RHS is plotted *vs.* f_0 in Figure 5 (right panel).

$$\eta/c^{1/2} \approx (f_0+x_A)^{1/2}/[(1-f_0+x_B)(1+x)]^{1/2} \qquad (33)$$

The n-sigma value itself (η) is obtained by multiplying the ordinate by the square root of the total foreground counts ($c^{1/2}$).

All of the foregoing background analysis was conducted in terms of "true" ensemble parameters, but how can we estimate (f_0) and (η) using the observed counts (M*,N*)? It may not be worth the effort for very low counts, because unless S*>>k (total *observed* counts exceeding the total *background* counts), the Poisson *fluctuations* in the background contamination ($\sigma^2k_A=k_A$ and $\sigma^2k_B=k_B$) will likely be severe enough to preclude obtaining a useful estimate (f_0*) of the ensemble value for the foreground abundance fraction (f_0) from the observed counts (M*) and S*=M*+N*.

8. Numerical simulations of the Group Abundance Fraction

It may be helpful to the reader to relate all of the foregoing calculations and relationships to some numerical simulations of the Group Abundance Fraction F*=M*/S*. Figure 6 illustrates clearly the nature of the Poisson fluctuations of the group abundance fraction when we suppress occurrences of S*=0. It presents the result of 100 Poisson (random-number-generated) simulations of individual observations of M^*_k and N^*_k, with $S^*_k=M^*_k+N^*_k$ ($1 \le k \le 100$). This was a *conditional* simulation in which we *excluded* trials with M*=N*=S*=0. The results then should be consistent with our conditional bivariate Poisson probability (S≥1) distribution given by Eq. (14). The ensemble parameters for these 100 trials were always set at C=5 counts (for the combined A and B processes) and f=0.2 (which implies that the ensemble value fC=1 count for A-process). In the upper panel, the results of the 100 trials are presented *vs* the trial index (k) as the group abundance fraction. Its estimated uncertainty shown by error bars) $F^*_k \pm \sigma F^*_k$ calculated from our algebraic estimators F*=M*/(M*+N*) and $\sigma F^*=[F^*(1-F^*)/(S^*-1)]^{1/2}$. The lower panel plots the simulated counts M^*_k (lower points) and S^*_k (upper points), and it is evident that about 1/3 of the M*-counts equal zero. It is also clear that the S* (total counts) fluctuate significantly about ensemble value (C=5), justifying our use of the full bivariate Poisson distribution, as opposed to assuming that S*=C (exactly).

Returning to the upper panel of Figure 6, the ensemble value of the group ensemble fraction (f=0.2) is indicated by a horizontal dashed line, as are the ensemble expectation values of f±σF with $(\sigma F)^2=\sigma^2F=f(1-f)g_1(C)=0.04112$ for C=5 giving the ensemble expectation value for the SD of σF=0.2028. The average values for the 100-trial run agrees with the expectation values to within a fraction of 1%. A few dozen trials have fractions (F*) above the one-sigma level, but the majority of their lower error bars reach down to the line f=0.2 (demonstrating some consistency with the estimated uncertainties). On the other extreme, almost 1/3 of the trials had F*=0 because M*=0 (compare with the lower panel). It is precisely because of these F*=0 trials that there also had to be a set of F*-values lying above F*= f+σf (and thus compensating for the many F*=0 values), since our estimator demands that the over-all average expectation value <F>=f.

Thus Fig. 6 illustrates the relationships among our three classes of numbers: firstly, the ensemble parameters (f, C), secondly the expectation values <F>=f and $\sigma^2F=f(1-f)g_1(C)$, which are functions of the ensemble parameters (f,C), and thirdly, the estimators F* and F*(1-F*)/(S*-1) for <F> and σ^2F, respectively. It is remarkable that even with the large fluctuations (δS^*_k) in the total counts (S^*_k), wherein δS*~±C=±5.0 counts, the *biased* estimator (σ^2F^*) for the variance of the group abundance fraction (f) from the *conditional* bivariate probability distribution has turned out to be i(σ^2F^*) dentical to the *unbiased* estimator ($\sigma^2F^*_S$) for the *binomial* probability distribution with wherein C estimated by S* itself. See Eqs. (8) and (20) and the discussions thereof.

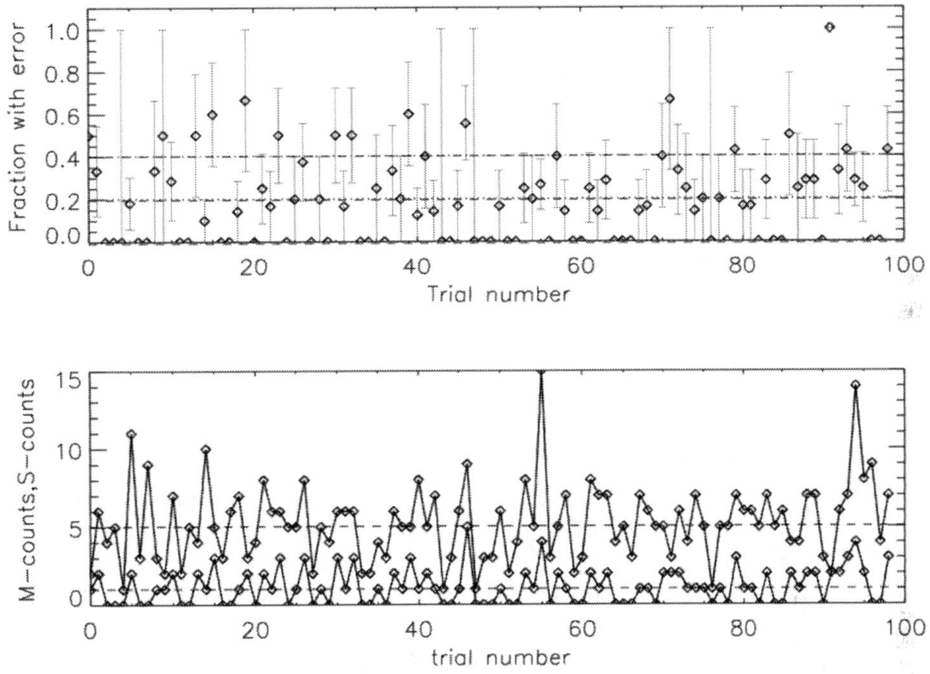

Figure 6. Results of 100 Poisson conditional random-number "trials" (excluding those with S*=0). *Upper panel*. The observed group abundance fraction F*=M*/S*=M*/(M*+N*) (diamonds) with two-sided error bars $F^*\pm\sigma^2F^*$ estimated by $\sigma^2F^*=F^*(1-F^*)/(S^*-1)$. The horizontal lines are the ensemble values $f\pm\sigma f$ for $\sigma^2 f=f(1-f)g_1(C)$ with f=0.2 and C=5. Note the numerous points with F*=0 (corresponding to M*=0) along the bottom axis. These zero counts are a consequence of the Poisson fluctuations about M*~fC=1 in this low-count regime. *Lower panel*. The randomly-generated values of M* and S* for the ensemble parameters: M-counts=(0.2)(5)=1.0 and total counts C=5.

8.1. Cumulative distribution function for F: Poisson realizations specified by (f,C)

We use Figure 7 make a final point. One might imagine that there is a well-behaved probability distribution function for the group abundance fraction. Indeed, one does exist, but only in the strict limit $C\rightarrow\infty$. However, it may not be generally realized that for small values of C, the probability distribution is a rather peculiar and ill-behaved mathematical object. We can demonstrate this with numerical simulations of the *cumulative* unconditional distribution function Z(F;f,C). This is the probability that the observed fraction (F*) does not exceed the value F, *i.e.*, that $F^*\leq F$. As is well-known, Z(F;f,C) may be simulated by using a Poisson random number generator to produce a large number (K) of pairs (M^*_k,N^*_k)

using the same ensemble parameters C_A=fC and C_B=(1-f)C for 1≤k≤K. We ran K-30,000 trials. One then sorts the corresponding fractions (F^*_k) in ascending order from zero to one and plots them vs. their ordering index (divided by K). Such a plot is shown in the left-hand panel of Figure 7 for C=5 counts and an ensemble fraction f=0.2. Since the abscissa gives the fraction of all of the realizations where F* was less than or equal to the ordinate (F), the cumulative distribution function (CDF), the probability that F*≤F as a function of F, is immediately obtained by interchanging the ordinate and the abscissa, as displayed in the right-hand panel of Figure 7.

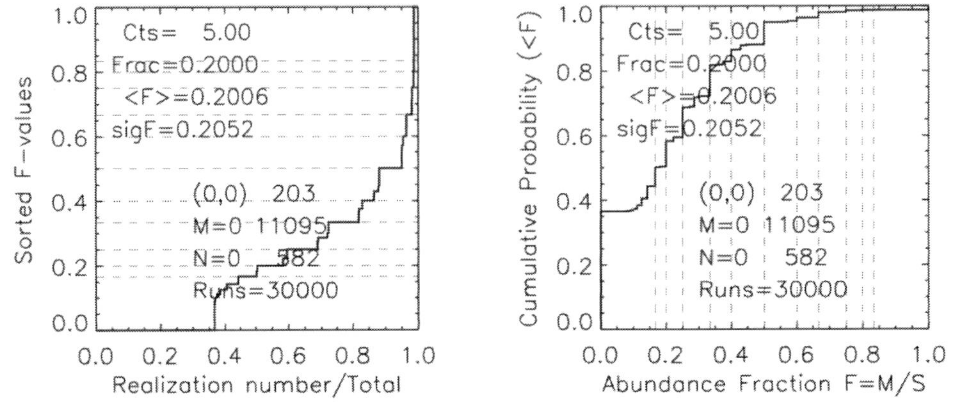

Figure 7. Poisson simulation (constructed from 30,000 trials of M* and N*) of the cumulative probability distribution function Z(F|f,C) for the Group Abundance Fraction F=M/(M+N) for ensemble parameters f=0.2 and C=5 total counts. *Left panel:* Sort of the K=30,000 values of F* plotted by the normalized realization number (k/30,000). *Right panel:* Cumulative probability distribution Z(F|f,C), the fraction of occurences with F*<F. The discontinuities in the function (marked by dashed lines) are produced by different pairs of counts (M,S) "near" M=1 and S=5 having the same ratio F=M/S. Consequently, for low numbers of total counts (S), the ordinary probability density (normally given by dZ/dF) will contain "spikes" at the discontinuities of Z.

What immediately strikes the eye is the plethora of discontinuous jumps in the CDF. These discontinuities are not an artifact of the simulation, because it was performed with K=30,000 Poisson realizations. The jumps are an intrinsic property of the CDF, and their cause is revealed by the vertical dashed lines (which correspond to the horizontal dashed lines in the original plot in the left-hand panel). They have been drawn where F=M/S takes the values of a ratio of small integers: 0.16667=1/6, 0.20=1/5, 0.25=1/4, 0.33333=1/3, 0.40=2/5, 0.5=1/2, *etc*. The vertical jumps occur at particular values of F in the right-hand plot, because there several realizations that give the same value of F for different pairs (M,N) in the left-hand plot. Thus pairs with the *same* (F) may have a *different* number total counts S=M+N. Such multiplicities *cannot* appear in the *binomial approximation* wherein S is set to the specific observed value (S*). See Eq. (8). The jumps in F=M/S due multiplicity in S can only appear in the mathematical form of the CDF when the exact distribution function $P_S(C)$ is retained in the conditional distribution function U(M|S) given by Eq. (14).

For the case simulated, we chose C=5 and f=0.2, so the most likely value of M* is fC=1. This means that the realizations F*=M*/S* are quite likely to take values with M* near 1 and S* near 5, like 1/6, 1/5 (or 2/10), 1/4 (or 2/8), 1/3 (or 2/6 or 3/9), 2/5, 1/2 (or 2/4 or 3/6 or 4/8). The higher likelihood for multiplicities of F=1/3 or F=1/2 yields larger discontinuities than at F=2/5. Note that this effect was not produced by our choosing integer values for the ensemble parameters C=5 and fC=1. The discontinuities would appear just as strongly if we had chosen C=5.35 and fC=0.8 or 1.2. The strong discontinuities always appear at values of F=M/N where M and N are small numbers (as long as C itself is a small number). Of course, the discontinuities become undetectably small for C>>1 because then there are many rational fractions (no longer restricted to the ratio of small integers) in the vicinity of any value of F.

When one has a CDF for a continuous process (like a Gaussian), then one can obtain the usual probability "density" (or probability distribution function, PDF) by differentiating the CDF with respect to its argument. However, in our situation with C a small integer (as revealed by the RH plot of Fig. 7), such a differentiation will produce a series of delta-function-like "spikes" in the PDF at each of the jumps. In other words, one cannot develop a "smooth" approximation to the CDF by any fitting of smooth (finite-order) polynomials in Z(F), because it is effectively not differentiable.

Despite the intrinsic "roughness" of the CDF, our expressions for the ensemble expectation values $\langle F\rangle=f$ and $\sigma^2F=f(1-f)g_1(C)$ are *exact*, because no approximations were made in the summation calculations over the joint Poisson probability distribution. The numbers presented within the RH panel of Fig 7 are the *computed* values from the 30,000 realizations of $\langle F\rangle=0.2006$ and its SD is $\sigma F=0.2052$. Our ensemble expectation values are $f=0.2000$ and $\sigma^2F=f(1-f)g_1(C)=(0.2)(0.8)(0.287)=0.04112$ (taken from the exact formulas), giving the ensemble value for the SD of $\sigma F=0.2028$. The agreement of both is within 0.1%.

9. Summary and Conclusion

As we anticipated from the outset in Section 2, the results of this paper are quite simply summarized by the *algebraic* expression, from Eq. (24), for the estimator (in terms of the measured counts M* and N* with total S* from two independent Poisson processes) for the ensemble expectation value (f) of the group abundance fraction F=M/(M+N) and its one-sigma uncertainty (σF)

$$f\pm\sigma F \rightarrow M^*/S^* \pm [F^*(1-F^*)/(S^*-1)]^{1/2}$$

The estimator for (f) is statistically *unbiased*, while that for the standard deviation (σF) is *conservatively biased*, in that (on average over the ensemble), it always *overestimates* the "true" value (σF) for very low total counts, but approaches it asymptotically beyond S*>5. See Fig. 3. These results were obtained by introducing an exactly valid *conditional* Poisson joint probability distribution function U(M,S) satisfying the requirement of *non-zero* total measured counts (S*≥1). See Eq. (14). The interesting relationship to earlier results [1] based on an approximately valid *binomial* distribution is traced in Section 3 and Appendices A and B. The final three sections (6, 7, and 8) of the paper are devoted to topics not (to our knowledge) addressed similarly in the literature: the relationship (due to instrument response functions) between the measured group abundance fraction for the counts and that for the incident intensities; the effect of *backgrounds* in the counting detectors on the estimate of the *foreground* group abundance fraction; and the peculiar mathematical properties of the *cumulative probability distribution function* for the group abundance fraction.

Acknowledgements This research was partially supported by the Voyager Interstellar Mission under NASA Contract NNX07AB02G, by the Cassini Mission under Task Order 003 of NASA Contract NAS5-97271, and by the ACE Mission under NASA Grant NNX13AR20G (all to the Johns Hopkins University). I thank Dr. Tycho Von Rosenvinge of NASA/GSFC and Dr. Alan Cummings of the Space Radiation Laboratory/CalTech for directing my attention to the confidence limit analyses of Niel Gehrels [2] and De Angelis and Iori [3], respectively. I am particularly indebted to the Editor, Dr. Gary P. Zank, for allotting me sufficient pages for a full development and explanation of the ideas in this paper.

References

[1] Lindgren, B W 1976 *Statistical Theory* (3rd edition), Macmillan (New York)
[2] Gehrels, N 1986 Confidence limits for small number of events in astrophysical data, *Astrophys. J.*, **303**, 336-346
[3] De Angelis, A and Iori, M 1987 Errors on ratios of small numbers, *Nucl. Inst. Methods Physics. Res.*

A, **260**, 451-454
[4] Roelof, E C 2012 Statistical analysis of sparse data: Space plasma measurements, *AIP Conf. Proc.*, **1436**, 265-272

Appendix A. Confidence limits approach

The joint probability distribution W(M,S-M) given in Eq. (6) was the starting point for a thorough confidence limit analysis of the group abundance fraction by Gehrels [2]. However, he then avoided using the Poisson distribution of the total counts $P_S(C)$ by setting the population parameter (C) precisely equal to the observed number of total counts (S*). He noted that the remaining *binomial distribution* for (M) was a function solely of the ensemble abundance fraction (f). He then tabulated single-sided uncertainties (f_U and f_L) for a specific confidence limit (CL) using the following equations for a *binomial distribution* (re-expressed in our current notation)

$$\sum_{M=0,M^*-1} B(M|S^*)\, f_L{}^M(1-f_L)^{S^*-M} = CL$$

$$\sum_{M=0,M^*} B(M|S^*)\, f_U{}^M(1-f_U)^{S^*-M} = 1-CL \qquad (A.1)$$

where M* and S* are the *observed* counts. The latter equation is equivalent to

$$\sum_{M=M^*+1,S^*} B(M|S^*)\, f_U{}^M(1-f_U)^{S^*-M} = CL \qquad \text{because } \sum_{M=0,S^*} B(M|S^*)\, f^M(1-f)^{S^*-M} = 1$$

for any value of the ensemble fraction parameter (f). These equations needed to be solved numerically for the parameters f_L and f_U as functions of M* and S*. He also took F*=M*/(M*+N*) as the estimator for the ensemble fraction (f), so his confidence limits were $f_L<F^*<f_U$. He states that when both M* and N* "are both small, these equations will give somewhat conservative upper limits", as a result of fixing the ensemble value for the total counts (C) equal to the observed number of counts (S*). Extensive tables were presented of the computations. In a later paper [3], other authors took a similar approach to the binomial distribution, but introduced Bayesian estimation in order to also incorporate the Poisson distribution $P_S(C)$. Again, tables of computed quantities were presented.

We saw (in Section 5) that our new results, which take quite a different approach that *exactly* incorporates the Poisson distribution $P_S(C)$, are nonetheless consistent with Gehrels' [2] confidence limits to quite good accuracy. However, our simple algebraic formula of Eq. (24) is much simpler to apply than the computed tables.

Appendix B. Moments of the binomial distribution

We begin with a basic iterative identity for the binomial coefficient B(M|S)

$$M\, B(M|S) = M\, S!/M!(S-M)! = S!/(M-1)!(S-M)! = S\, B(M'|S')$$

However, we must first ensure that M≥1 so that the factorial (M-1)! in the denominator has a non-negative argument. Then, since we already require S≥1 for the conditional probabilities, both M'=M-1>0 and S'=S-1>0 in the identity. A similar useful identity is

$$M(M-1)B(M|S) = M(M-1)\, S!/M!(S-M)! = S!/(M-2)!(S-M)! = S(S-1)\, B(M''|S'')$$

in which M"=M-2 and S"=S-2 requires that M≥2 and S≥2. We now evaluate several moments containing M^k, with k=0,1,2 over the binomial distribution with (S) total counts, as indicated by the notation $\langle M^k\rangle_S$ introduced with Eq. (8).

Journal of Physics: Conference Series **577** (2015) 012026 doi:10.1088/1742-6596/577/1/012026

$$<M^0>_S = \sum_{M=0,S} B(M|S)\ f^M(1-f)^{S-M} = [f+(1-f)]^S = 1 \qquad (B.1)$$

For the first moment (M itself), the first term (M=0) in the summation vanishes, so the original range $0 \le M \le S$ becomes $1 \le M \le S$ which (because $S \ge 1$) can be shifted to $0 \le M' \le S'$.

$$<M^1>_S = \sum_{M=1,S} MB(M|S)\ f^M(1-f)^{S-M} = \sum_{M=1,S} S\ B(M'|S')\ f^M(1-f)^{S-M}$$

$$= \sum_{M'=0,S'} SB(M'|S')\ f^{M'+1}(1-f)^{S'-M'} = Sf \sum_{M'=0,S'} B(M'|S')\ f^{M'}(1-f)^{S'-M'}$$

$$= Sf\ [f+(1-f)]^{S'} = Sf \qquad 1 \le M \le S \text{ and } S \ge 1 \qquad (B.2)$$

Similarly, for the compound moment $M(M-1)=M^2-M$, the first two terms in the summation vanish, so the original range $0 \le M \le S$ becomes $2 \le M \le S$ which (because $S \ge 1$) can be shifted to $0 \le M'' \le S''$ because $2 \le M \le S$ requires that that $S \ge 2$.

$$<M(M-1)>_S = \sum_{M=2,S} M(M-1)B(M|S)\ f^M(1-f)^{S-M} = \sum_{M=2,S} S(S-1)\ B(M''|S'')\ f^M(1-f)^{S-M}$$

$$= S(S-1) \sum_{M''=0,S''} B(M'|S')\ f^{M''+2}(1-f)^{S''-M''} = S(S-1)f^2 \qquad 2 \le M \le S \text{ and } S \ge 2$$

Therefore

$$<M^2>_S = <M(M-1)>_S + <M>_S = S(S-1)f^2 + Sf \qquad 2 \le M \le S \text{ and } S \ge 2 \qquad (B.3)$$

and we also obtain the compound moment $M(S-M)=SM-M^2$.

$$<M(S-M)>_S = S<M>_S - <M^2>_S = S(fS) - (fS)^2-(Sf)(1-f) = f(1-f)S(S-1) \qquad (B.4)$$

in which the first term in the expanded factors is restricted by $1 \le M \le S$ and $S \ge 1$, while the second and third are restricted by $2 \le M \le S$ and $S \ge 2$.

Appendix C. Properties of the functions $I_k(C)$ and $g_k(C)$

Consider the infinite sum (absolutely convergent for k non-negative)

$$I_k(x) = \sum_{r=1,\infty} x^r/r!r^k \qquad \text{where} \quad I_0(x) = \exp(x)-1$$

The normalized summations then define our set of probability functions

$$g_k(x) = I_k(x)/I_0(x) \qquad (C.1)$$

For the range $x<<1$, we can make use of the generating function for the Bernoulli number (B_n)

$$x/[\exp(x)-1] = x/I_0(x) = \sum_{n=0,\infty} B_n x^n/n! \qquad B_n=1,-1/2,1/6,\ldots \text{for } n=0,1,2\ldots$$

with Abel's summation formula for the product of two convergent power series.

$$g_k(x) = I_k(x)/I_0(x) = (1/x)I_k(x) \sum_{n=0,\infty} B_n x^n/n! = \sum_{r=0,\infty} x^r/(r+1)!(r+1)^k \sum_{n=0,\infty} B_n x^n/n!$$

$$= \sum_{r=0,\infty} x^{(r+n)}/(r+1)!(r+1)^k \sum_{n=0,\infty} B_n x^n/n!$$

$$= \sum_{s=0,\infty} x^s \sum_{n=0,s} B_n/n!(s-n+1)!(s-n+1)^k \qquad s=r+n$$

$$= B_0 + (B_0/2!2^k + B_1)x + (B_0/3!3^k + B_1/2!2^k + B_2/2!)x^2 + O(x^3)$$

Since the first few Bernoulli numbers are $B_0=1$, $B_1=-1/2$, $B_2=1/6$, we have the behavior near zero

$$g_k(x) = 1 - (x/2)(1-1/2^k) + (x^2/12)(1 - 3/2^k + 2/3^k) + O(x^3)$$

$$g_0(x) = 1 \qquad \text{identically}$$

$$g_1(x) = 1-(x/2)(1-1/2)+(x^2/12)(1-3/2+2/3) +\ldots = 1-x/4+x^2/72+\ldots$$

$$g_2(x) = 1-(x/2)(1-1/4)+(x^2/12)(1-3/4+2/9) +\ldots = 1-3x/8-17x^2/432+\ldots$$

$$g_1(x)-g_2(x) = x/8+23x^2/432+\ldots$$

$$1- g_2(x)/g_2(x) = x/8+73x^2/864+\ldots \qquad (C.2)$$

At the other extreme of the range $x>>1$, we can make use of the identity (obvious from the definition)

$$dI_k(x)/dx = (1/x)I_{k-1}(x) \qquad (C.3)$$

combined with l'Hospital's rule (because the limits of the derivatives of the numerator and denominator exist).

$$lim(x\to\infty)\ x^k g_k(x) = lim(x\to\infty)\ x^k I_k(x)/I_0(x)$$

$$= lim(x\to\infty)\ [kx^{k-1}I_k + x^k(1/x)I_k(x)]/dI_0(x)/dx$$

$$= lim(x\to\infty)\ [(k/x)(x^k g_k) + x^{k-1}I_{k-1}]/(d\,ln I_0/dx)$$

The first limit on the RHS is $O(k/x)$ of the limit on the LHS, and hence for $x>>k$ it can be neglected. As for the denominator, since $ln I_0(x) = ln[\exp(x)] + ln[1-\exp(-x)]$, we have

$$d\,ln I_0/dx = 1 + 1/[\exp(x)-1] = 1 + O(\exp(-x))$$

and we are left with the recursive relation

$$lim(x\to\infty)\ x^k g_k(x) = lim(x\to\infty)\ x^{k-1} g_{k-1}(x)$$

which can be solved by induction, since for $k=1$, we have

$$x^{k-1}g_{k-1}(x)=g_0(x)=I_0(x)/I_0(x)=1$$

for all x (by its definition). Thus the general limiting relation (for all $k\geq 0$) is

$$lim(x\to\infty)\ x^k g_k(x) = 1 \qquad (C.4)$$

A similar analysis of the function $x(x^k g_k-1)$ reveals that

$$lim(x\to\infty)x(x^k g_k-1) = lim(x\to\infty)[x(x^k I_k-I_0)/I_0]$$

$$= [lim(x\to\infty)\ d(x^{k+1}I_k)/dx-d/dx(xI_0)]/\ [lim(x\to\infty)dI_0/dx]$$

$$= [lim(x\to\infty)\ (k+1)x^k I_k+x^k I_{k-1}-I_0-x(I_0-1)]/\ [lim(x\to\infty)(I_0-1)]$$

$$= lim(x\to\infty)\ [(k+1)x^k g_k+x^k g_{k-1}-1-x] = lim(x\to\infty)\ [(k+1)x^k g_k +x(x^{k-1}g_{k-1}-1)]$$

Since $lim(x\to\infty)\ x^k g_k=1$ for all $k\geq 1$ and $x>>k$, we obtain the recursion relation

$$lim(x\to\infty)x(x^k g_k-1) = k+1 + lim(x\to\infty)x(x^{k-1}g_{k-1}-1)$$

Again we proceed by induction. Starting with k=1 and noting that $x^0 g_0=1$ identically for all x, we have

$$lim(x\to\infty)x(xg_1-1) = 2$$

because $x^0 g_0(x)=1$ identically for all values of x. Then for arbitrary k

$$lim(x\to\infty)x^k(xg_k-1) = \textstyle\sum_{k'=1,k}(k'+1) = k(k+1)/2 + k = k(k+3)/2 \qquad (C.5)$$

Therefore the behavior of $g_k(x)$ for $x>>1$ is given by what appears to be an asymptotic expansion

$$g_k(x) \sim 1/x^k + k(k+3)/2x^{k+1} + \ldots \qquad (C.6)$$

Consequently, the functions $g_k(x)$ behave quite simply, decreasing monotonically away from $g_k(0)=1$ for $x<<1$ then decreasing monotonically towards zero as $1/x^k$ for $x>>1$. In particular, for the function of immediate interest, $g_1(C)\approx 1/C$ is quite a good approximation over the open interval $C>1$. While $g_1(1)=0.76684$ and $g_1(1.5)=0.99887$, $g_1(C)$ never exceeds $1/C$ by more than~32% for $2\leq C\leq 5$. The ratio has decreased to only 10% higher by C=12 and then asymptotically approaches $1/C+2/C^2$ for higher values (see Fig. 1).

AUTHOR INDEX